SETON HALL UNIVERSITY
3 3073 10282169 9

T5-DID-655

ICONS OF EVOLUTION

Recent Titles in
Greenwood Icons

Icons of Horror and the Supernatural: An Encyclopedia of Our Worst Nightmares
Edited by S.T. Joshi

Icons of Business: An Encyclopedia of Mavericks, Movers, and Shakers
Edited by Kateri Drexler

Icons of Hip Hop: An Encyclopedia of the Movement, Music, and Culture
Edited by Mickey Hess

ICONS OF EVOLUTION

An Encyclopedia of People, Evidence, and Controversies

VOLUME 2

Edited by Brian Regal

Greenwood Icons

GREENWOOD PRESS
Westport, Connecticut · London

Libray of Congress Cataloging-in-Publication Data

Icons of evolution: an encyclopedia of people, evidence, and controversies / edited by Brian Regal.
p. cm.—(Greenwood icons)
Includes bibliographical references and index.
ISBN 978–0–313–33911–0 (set: alk. paper)—ISBN 978–0–313–33912–7 (v. 1: alk. paper)—ISBN 978–0–313–33913–4 (v. 2: alk. paper)
1. Evolution (Biology)—Encyclopedias. I. Regal, Brian.
QH360.2.I33 2008
576.803—dc22 2007033049

British Library Cataloguing in Publication Data is available.

Library of Congress Catalog Card Number: 2007033049
ISBN-13: 978-0-313-33911-0 (set)
978-0-313-33912-7 (vol. 1)
978-0-313-33913-4 (vol. 2)

First published in 2008

Greenwood Press, 88 Post Road West, Westport, CT 06881
An imprint of Greenwood Publishing Group, Inc.
www.greenwood.com

Printed in the United States of America

The paper used in this book complies with the Permanent Paper Standard issued by the National Information Standards Organization (Z39.48-1984).

10 9 8 7 6 5 4 3 2 1

Table of Contents

Preface and Acknowledgments

The purpose of this two-volume compilation is to give the reader a set of clear, concise, readable, understandable, and fascinating explanations for the evidence for evolution and why that evidence is so important for understanding how life on earth came to be and how it spread and diversified. It shows the reality behind some of the most famous symbols of the evolutionary process and evolutionary studies. The book is geared toward readers new to the subject and to those who have some knowledge of it but want to expand their understanding of the topic with some more technical explanations. Readers may be surprised that some of these icons are not quite what they thought they were. There are many misconceptions about these icons, claims made for and against them. For example, was Charles Darwin an atheist? Was *Archaeopteryx* the first bird? What *is* a missing link? What is the significance of all those horse fossils? What does "survival of the fittest" really mean? What ever happened to Peking Man? And others.

As this book is a compilation of topics, a sort of "best of," readers are encouraged to argue over which topics should have or should not have been included. This particular collection—by no means a definitive one—tends to lean heavily toward geology and anthropology with little discussion of the important role of plants in evolution and, besides peppered moths, little mention of insects or fish. The arrangement of the chapters will also probably not satisfy everyone. They were shuffled and reshuffled a number of times before reaching this line up. I tried to follow a roughly chronological order so that the individual chapters would have a certain logical progression with a balance between individuals, ideas, and artifacts. There is also overlap between different chapters. This is only natural as these icons do not exist in a vacuum but are part of a much wider continuum. The reader will also no doubt notice that not all scientists and historians always agree on things. This should not be seen as a problem but as evidence of the robust and dynamic nature of the scientific enterprise and historical analysis. As new discoveries are made, ideas change. This is how scholars operate. As with everything in science and history nothing is ever final; there are always new tomorrows which bring the possibility of new knowledge. To do otherwise would be to stagnate and never learn anything new. The book includes a chronology of events in evolution history, an extensive bibliography so that readers can go to original sources and see where the proofs are, and short biographies of the authors.

The topics covered in these volumes are not the only icons involved here. In their research into evolutionary mechanics and their writings on the history and social implications of evolution many of the authors in this collection are considered icons themselves, and others will be thought of that way in the future. When choosing authors I wanted a mix of well-known researchers, writers, and teachers and up-and-coming new scholars. They have written their chapters each from a different prospective and in different styles. They alternate between historical narratives, scientific dispositions, and philosophical discourses. While each chapter is designed to stand as a separate unit, they can be taken together as an entire story.

Volume 1 begins with Olof Ljungström's discussion of why evolution is the cornerstone of biology and acts as an overview of the history of evolutionary thought. This helps give some context for the rest of the chapters. This is followed by chapters on the early giants: Bowdoin Van Riper on Charles Darwin, Lisa Nocks on T.H. Huxley, and Dawn Digrius on Gregor Mendel. Peter Bowler then explains what "survival of the fittest" means and why it is so misunderstood. Then come some of the most famous icons of human evolution: Marianne Sommer on Neanderthal Man, Pat Shipman on Java Man, and Brian Regal on the disappearance of Peking Man. David Rudge and Christine Janis follow with discussions of evolution in action with the peppered moth and the horse series. Finally comes Anne Katrine Gjerløff on the Taung Child of Africa with the volume rounded out by a discussion of the history of monkey trials and the ongoing assault against the teaching of evolution from some quarters.

Volume 2 shows that evolution is not just rocks and fossils but involves interpretation and analysis of those fossils as well as a way of studying the descent of living organisms by techniques only dreamed of in the nineteenth century. After David Fastovsky takes a look at our old friends the dinosaurs and Bent Lindow gives a history of flight and the fossil *Archaeopteryx*, Olof Ljungström explains the fossil record and how it is organized and what it tells about the history of life. Bowdoin Van Riper explains another cornerstone of evolutionary studies, radiometric dating. He explains how we know how old the fossils are. The infamous hoax

Piltdown Man is discussed and a crack is taken at figuring out who the culprit is. Then Jeffrey Schwartz and Adam Wilkins explain how modern biologists figure out how to arrange living things into an orderly fashion and how the modern notion of evolution and genetics mix to form the current view of how evolution works. Rounding out the list of icons are the life of Louis Leakey by Anne Katrine Gjerløff and the most famous human fossil since the Neanderthal Man, Lucy, by Holly Dunsworth, and Christopher Stringer explains how humans got out of Africa. Finally, comes intelligent design and the newest assault on evolution, which, it turns out, is not so new after all.

I would like to thank Kevin Downing who first brought this project to me and allowed me to transform it from an enormous one-author job into an enormous multi-author job. That decision allowed me to tap some of the finest authors and thinkers—scientists, historians, and philosophers—on evolution in the world. It was a great honor and privilege to work with them. They did an outstanding job and put up with me constantly harassing them to get things done, despite the fact that they all had many other commitments and projects they were working on. Also thanks to everyone at Greenwood Publishing for all their hard work.

It should be known that there is another book called *Icons of Evolution*; however, it is a work geared specifically to refute evolution and to show that scientists are wrong in their conclusions. Many of the topics that book attacks are supported in this one with copious amounts of evidence. That leaves the reader with an interesting project on their hands: Compare the two and see where they stand.

—Brian Regal
New York, June 2007

The Dinosaurs

David Fastovsky

INTRODUCTION

Dinosaurs are famous icons of evolution, but ironically, their notoriety derives from the fact that they embody evolutionary failure rather than success. The notion of their failure is linked to their ultimate extinction; why did they go extinct, after all, if they were evolutionarily successful? The extinction, however, is widely perceived as the inexorable result of yet another failing of dinosaurs: their irredeemable stupidity. So the image of dinosaurs is one of immanent stupidity leading to extinction, which itself is rightly perceived as the result of biological failure.

The iconography of dinosaurs is far richer than that, however. Its cultural component is discussed comprehensively in W.J.T. Mitchell's *The Last Dinosaur Book*. Yet, the scientific ideas—including those dealing with evolution—are not absolutely divorced from the cultural ones (scientists, after all, exist within a culture), and aspects of the cultural viewpoint necessarily bleed into the scientific ideas. Given the reciprocal interdependence of cultural and scientific ideas, many of the images presented here will be cultural, presumably based upon the prevailing science. In the case of dinosaurs, however, the cultural images appear to have lasted long after the science upon which they were based became obsolete.

Here, I restrict myself to the iconography of dinosaurs as exemplars of evolution. In that context, the very antiquity of dinosaurs suggests that they are primitive, and there is a popular feeling that they deserved to be supplanted by "better" creatures, e.g., mammals. The very size of dinosaurs, rather than being viewed as a hallmark of evolutionary success, is commonly regarded as emblematic of inadequacy. Indeed, the Oxford

Dictionary and Thesaurus defines the word "dinosaur" as a "large, unwieldy system or organization" (Abate, 1997, p. 208).

And yet for all that, at least one dinosaur, *Archaeopteryx*, is commonly portrayed as a "missing link"—proof positive that evolution has occurred. Since its discovery, the fossil has appeared, to many observers, to bridge the gap between "reptiles" and birds, in exactly the transitional way predicted by Darwin in 1859. So in this case, while dinosaurs are generally regarded as icons of failed evolution, here is one that is perceived as the ultimate validation of the theory.

These, then, are the images with which I shall contend in this chapter. I say "contend," because in virtually every instance they arise from a misunderstanding of dinosaurs, or evolution, or both. I start with the idea of extinction-as-failure, following it with stupidity and size as both a cause of extinction and an evolutionary Achilles's heel. I then summarize some of the features that characterize selected dinosaurs—features that highlight the fact that dinosaurs are among the most highly evolved creatures that have ever existed. Finally, I will address the question of *Archaeopteryx* as missing link and as the ultimate validation of Darwinian evolution—a different kind of icon, but an icon nonetheless. Readers will note that the iconography of *Archaeopteryx* is also addressed in Chapter 14 of this book.

DINOSAURS, EXTINCTION, AND FAILURE

Our best data suggest that non-avian dinosaurs, that is, all dinosaurs excluding birds, went extinct about 65.5 million years ago (Mya) at the Cretaceous-Tertiary (K/T) boundary (e.g., the moment in time between the Mesozoic and Cenozoic Eras). While the ultimate cause of the extinction remains conjectural, several facts constrain the otherwise astronomically large range of possible explanations for the extinction. These include:

1. The K/T extinctions took the lives of many, many organisms—both plant and animal. Most significant perhaps were the marine extinctions which, insofar as they are understood, apparently reduced the world's oceans to 10 percent of their previous biological activity. In North America, where diversity across the K/T boundary has been carefully reconstructed, plants apparently underwent an over eighty percent extinction. Likewise, insects appear to have undergone significant extinctions. Protists, insects, and plants constitute a fundamental component of the total biomass, by comparison to which the extinction of a few dinosaurs is largely irrelevant. Likewise, from the vantage point of global organic productivity, the specter of nearly lifeless oceans reduces the dinosaur extinction to effective insignificance.

2. At a scale with a precision of tens of thousands of years, all of the extinctions were geologically synchronous as well as instantaneous. While time intervals measured on tens of thousands of years are admittedly far longer than the time scales upon which ecological events occur, they nevertheless allow us to rule out whole classes of processes, such as sea-level regression, which operate on million-year timescales.
3. Within the past twenty-five years, the evidence has become overwhelming that birds are dinosaurs. Not just descended from dinosaurs (although they were); rather, birds are living dinosaurs, bearing all of the shared, derived morphological characters that characterize Dinosauria. That being the case, it is now generally agreed that Dinosauria was not eradicated from Earth. *Avian* dinosaurs (birds) survived the end of the Mesozoic, and only *non-avian* dinosaurs fell prey to the K/T extinction.
4. Among terrestrial vertebrates, dinosaurs weren't the only victims of a catastrophe. Strikingly, mammals also underwent a startling diminution in diversity (and, presumably number), with marsupials being particularly hard hit. Placental mammals, though surviving the K/T boundary, nonetheless suffered noteworthy extinctions. Although the avian record is very poor, they, too, appear to have undergone significant extinctions. Perhaps reflecting this, a taxonomic inflexion occurs between the known birds of the latest Cretaceous, and those of the early part of the Cenozoic.
5. Researchers generally agree that among terrestrial vertebrates, aquatic habitat and, to a lesser extent, small size, are best correlated with survival.
6. Dinosaurs and mammals first appeared at approximately the same time on earth. For the next 160 million years, while dinosaurs radiated into an extraordinarily diverse variety of forms, mammals retained a comparatively subdued diversity and never exceeded house-cat size. The mammalian radiation took place *after* the extinction of the dinosaurs, and our best evidence suggests that it was the dinosaur extinction that allowed mammals to radiate into ecospace previously occupied by dinosaurs.

The global synchronicity and instantaneity of the extinctions, occurring on land as well as within the world's oceans, require a global cause. Moreover, that cause was by no possible stretch of the imagination, dinosaur-specific. It had to be a globally pervasive, geologically instantaneous catastrophe, and for this reason, the Chicxulub asteroid impact is commonly implicated in the K/T extinctions. It is also for this reason that paleontologists, writing about causes of the extinction, describe extinction patterns as "concordant" with the asteroid impact-as-cause. A vocal minority of geoscientists still invokes end-Cretaceous volcanism in the extinction; regardless, were there stronger evidence for pervasive pyroclastic volcanism at the end of the Cretaceous, the pattern of the extinctions would also be widely viewed as concordant with this putative cause.

So where do these basic patterns of survivorship, extinction rate, and magnitude leave us? Clearly the "fault" for the extinction can hardly be laid at the dinosaurs' feet. With wholesale extinction affecting so many other organisms, and with an external—likely, extraterrestrial—abiotic cause for the extinctions, the cause should not be sought and cannot to be found in some property or susceptibility immanent to Dinosauria.

Or should it? All non-avian dinosaurs went extinct, and that, by definition, is biological failure. To use paleontologist D.M. Raup's causal dichotomy, was that failure the result of "bad luck, or bad genes?" Paleontologist P.M. Sheehan and I argued in the journal *Geology* that survivorship was dependent upon detritus feeding—and non-avian dinosaurs were clearly dependent upon primary production. Geologist D.W. Robertson and colleagues suggested in the Geological Society of America *Bulletin* that an ability to find refugia was the key to survival. Because of their size, some non-avian dinosaurs, at least, were potentially harder pressed than birds and mammals to find refugia. In either causal scenario (detritus-feeding or the need to find refugia), it would seem that aspects of non-avian dinosaurs, such as size and a dependence upon primary production, may have made them more susceptible than either birds or mammals to extinction.

But as we have seen, both birds and mammals also underwent significant extinction. They were not so well buffered against the cause of the extinction to pass through unscathed—as crocodiles and several others types of vertebrates apparently did. So, returning to the question of extinction and implicit failure, it is fair to state that bad luck must have played a very significant role in the ultimate fate of non-avian dinosaurs. But with avian dinosaurs still living, and non-avian dinosaurs extinct as a result of extraordinary causes, the stereotype of non-avian dinosaurs as archetypical icons of failure can't be right.

It should perhaps be thought of this way: Both dinosaurs and mammals suffered significant extinctions at the K/T boundary, and both groups produced exemplars that managed to get through. It's just that the dinosaurs were so much more impressive and diverse to begin with, that the loss of the non-avian forms magnifies the tragedy of their extinction.

This point can be reinforced from yet a different perspective. The earliest non-avian dinosaurs are known date from 228 Mya; the most recent date from 65.5 Mya. This gives the group approximately a 163 million-year tenure on Earth. With birds as dinosaurs, dinosaurs have been on earth at least 228 million years. This makes dinosaurs among the most long-lived groups of terrestrial vertebrates and, most important, *not yet extinct*! By comparison homonoids have been around for an estimated 26 million years; homonids for between 6 and 7 million years, *Homo* for 2–3 million years, and *Homo sapiens* for 250,000 years. Moreover, modern birds are a far more diverse group than are modern mammals, and thus yet again, in spite of the K/T extinction, dinosaurs have managed to radiate and dominate—by number and diversity, at least—the vertebrate world. The image of dinosaurs as icons of evolutionary failure as a result of some of their number having gone extinct is as utterly bankrupt as it is possible for an image to be.

PRIMITIVENESS, SIZE, AND STUPIDITY

The notion of primitive-as-inferior comes originated with the idea of progress. And progress, at least in terms of the human condition, is an idea associated with the Enlightenment. In an evolutionary context, the science of which was rooted in the Enlightenment, a common, but very flawed, reading of Darwin has it that organisms proceed from the simple to the complex, making progress as they do so. That idea, combined with a misunderstanding of evolution by natural selection, has led to the mistaken view that nature is continually honing and refining organisms such that those around today are far better adapted, and more sophisticated than—in short, superior to—their forebears.

This misunderstanding provoked, in a sense, an entire book from S.J. Gould: *Wonderful Life*. The early pages of this work present image after image of *scala naturae*—primitive and crude forms evolving linearly to more advanced, superior, and familiar forms. That the most modern forms are more advanced is effectively tautological—especially when the term "advanced" is used in its biological context to reflect more, rather than less, evolved. That modern forms are generally more familiar is undoubted. But that they are superior is a whole other question.

If natural selection was continually honing and refining organisms in a static earth, it would be conceivable that organisms today would be superior—e.g., more fit—than their forebears. But, geoscientists spent much of the latter half of the twentieth century demonstrating that the earth and its environments are anything but static, and so, as Leigh Van Valen articulated in the "Red Queen Hypothesis," organisms must "run in place" (e.g., evolve) just to maintain the level of fitness achieved by preceding generations; that is, to retain their position on what Sewell Wright termed the "adaptive landscape." With variable rates of both organic evolution and environmental evolution, it is conceivable—even likely—that descendents could be *less* fit than their ancestors. While it may be comforting to suppose that modern organisms are superior in some way to generations of antecedents from the geological past, there is in fact nothing whatsoever to suggest that this is the case.

Wonderful Life also contrasted "contingency"—the notion that evolution, if it were replayed, would lead to a different end point—with determination, in which evolution is always directed toward a particular type of end point. Two commonly cited examples of deterministic evolution are the tendency toward increasing complexity through time, and the tendency toward increasing intelligence. Both of these examples have relevance in dinosaurs.

But before I address the particular case of dinosaurs, I want to consider these two examples of determinism. Life viewed at a distance appears to support the notion that evolution means increasing complexity. Indeed, the step from prokaryote to eukaryote—even *if* it occurred through prokaryotic endosymbiosis—was something of a quantum leap in complexity. When that step occurred, sometime around 2 billion years ago, life had already had 1.8 billion years of evolution behind it. Multicellularity appears to have occurred sometime around 0.7 billion years ago, and it can be said that shortly after 543 Mya, life in its full complexity had evolved. After all, is there really such a big difference between a trilobite and a mammal? The fact is, the really tough problems—complex locomotory, sensory response, and volitional systems—were solved at the time of the appearance of the first trilobite over 530 Mya. By far, the preponderance of evolutionary time passed with very little increase in complexity, after the eukaryotic step.

Several other observations are troublesome for the idea that evolution is a march toward increasing complexity. We know now that the complexity of the genome bears little relationship to the apparent complexity of the organism. Yet, as noted by B. Bryson in *A Short History of Nearly Everything*, some ferns have more than 600 chromosomes while humans have 46. One would be hard-pressed to argue that a fern is thirteen times as complex as a human! Interestingly enough, however, given the antiquity of the group, it would be much easier to argue that a fern is thirteen times more evolved than a human—and that's the point: Evolution is *not* about increasing complexity.

In fact, the argument could be made that the entire history of vertebrate evolution is one of *decreasing* complexity. This is certainly the case when it comes to comparative osteology, in which the dominant and repeated trends are for the *loss* of bony elements. Digits reduce constantly and independently, from initially eight in very primitive tetrapods, to five, and then to a variety of different combinations: most distinctively, two in artiodactyls and in some dinosaur hands, and one in horses. These represent highly evolved—advanced—conditions, yet, they do not appear sequentially: tyrannosaurids (including *T. rex*) reduced the digits on the manus to two 65 million years before humans appeared on earth, sporting the rather conservative (e.g., primitive) condition of five digits on the foot and the hand! Likewise, the volume (and sometimes number) of bony elements decreases across the board in vertebrates: in fish, particularly in the agnathans (compare an osteostracan with a modern lamprey, a Devonian dipnoan with a modern one, or a shark with its presumed antecedents); in amphibians (compare a Permian temnospondyl with a modern lissamphibian); and in reptiles (compare a Pennsylvanian amniote with a modern squamate). Bone reduction—a trend toward simplicity—characterizes much of vertebrate evolution.

Intelligence is a particularly human quality, and is thus prized particularly by humans. Yet I am unable to reconcile observations in the world around me with a deterministic view of the desirability of ever-increasing intelligence. Clearly, intelligence is a meaningless quality for the vast preponderance of the biosphere: plants, protists, bacteria, fungi, and many, many macro-invertebrates. These organisms, who, as opposed to vertebrates, constitute the real weight and magnitude of the biosphere, have not developed vertebrate intelligence levels despite 3.8 billion years of evolution. They have not done so even despite rubbing shoulders with vertebrates, as it were, for 500 million of those 3.8 billion years. It would be absurd to argue that Cambrian bacteria were less intelligent than modern ones. Even in the case of insects, in which the value of intelligence in the lifestyle can easily be imagined, it would be hard to argue that a Devonian cockroach was less intelligent than a modern one. Moreover, cockroaches have not shown the slightest inclination toward increasing intelligence over the past 350 million years, despite the fact that a human reading of their ethology suggests that more intelligence could potentially render them more fit. Their intelligence seems to be quite enough for what they do. I address the question of dinosaur intelligence below.

On the basis of these and a myriad of other examples, my view is that the notion of determinism as a pervasive quality in evolution is bankrupt. Arguments for it run counter to the historical biological record. In the absence of evidence for the linearity implied by determinism, e.g., the *scala naturae*, it is specious to imply that dinosaurs, by virtue of having appeared in the distant past, must somehow be more primitive or inferior than animals currently alive. Indeed, dinosaurs—avian or no—were stunningly advanced, that is, highly evolved, creatures by virtually every conceivable measure. In the following sections, I take a cursory stroll through the glories of dinosaur paleobiology, highlighting their evolutionary advancement and sophistication.

LEAVING "REPTILES" BEHIND

Non-avian dinosaurs were characterized by a high degree of specialization, which is itself the reflection of a significant amount of evolution away from the generalized amniotic tetrapodan (e.g., reptilian) condition.

That condition, or "grade," specifies a variety of attributes, including obligate quadrupedality, five digits on the hand and foot, scales, an ectothermic metabolism, a fully roofed skull, small—medium size, a carnivorous and/or insectivorous diet, the presence of an amnion, and a rudimentary repertoire of intraspecific behaviors. By contrast, even the most primitive, as well as earliest, non-avian dinosaurs, had moved away from many aspects of the primitive reptilian grade. While these forms—such as *Herrerasaurus*, *Eoraptor*, *Saturnalia*, and *Saurikosaurus*—all were of small to medium body size, these animals had opened up their skull roof for a more powerful bite, adopted a bipedal stance, and developed the fully erect posture that characterizes living birds and mammals. That last fact alone caused paleontologist R.T. Bakker to speculate that all dinosaurs were warm-blooded, including even the earliest members of the group. The earliest, most primitive dinosaurs, though technically reptiles, were very different from what we commonly think of when we imagine living reptiles. Their appearance is in fact more akin to birds than to living reptiles, a fact that is reflected in their phylogenetic relationships: Dinosaurs are more closely related to birds than to any living reptile. This point is explored in L. Dingus and T. Rowe's popular book *The Mistaken Extinction*, and it is one to which I will return later in this chapter.

What's in a Name

The names usually given dinosaurs often come from where they are found or because they resemble modern animals (*Albertosaurus* from Alberta, Canada, or *Psittacosaurus*, the parrot, for example). One of the first dinosaurs recognized as such was *Iguanodon*. It was named because its teeth resembled those of the modern *Iguana*. When artist Waterhouse Hawkins drew the first pictures of the incomplete skeleton of *Iguanodon* he placed what looked like a horn on the creature's nose. When more complete skeletons were found it was discovered that the "horn" was actually a thumb spike. It was also realized that *Iguanodon* was not simply a giant version of the modern *Iguana*. A group of dinosaurs similar in appearance to the *Iguanadons* are hadrosaurs: so-called because they were first found at Haddonfield, New Jersey, USA. Because of the distinctive shape of their skulls and broad, flat faces these creatures are often referred to popularly as duck-billed dinosaurs. The remarkable resemblance of the faces of these creatures to modern ducks caused many early depictions of hadrosaurs and their kin to show them flopping around in lakes like giant waterfowl. It is now thought that duck-bills spent relatively little time in or near bodies of water other than to drink. So remember that just because it looks like a duck and has a duck-like name it does not mean it is a duck. A good lesson to learn about many things associated with evolution studies.

Much of the popular misunderstanding—and the paralysis of thinking about dinosaurs that characterized the last quarter of the nineteenth century and the first three-quarters of the twentieth century—stems from basic misconceptions about the term "reptile." The word was first coined by Linnaeus in the non-evolutionary intellectual climate of the eighteenth century. It designated a group of living organisms—snakes, lizards, turtles, crocodiles, and the tuatara—whose unification came to be based upon the shared characteristic of laying eggs with an amniotic membrane around the embryo. The fact that both birds and mammals also have amniotic membranes around their embryos should have alerted scientists that "reptiles" not only includes snakes and lizards, and so forth, but also by definition birds and mammals. That expanded definition of Reptilia would have then allowed paleontologists to imagine a far greater range of physiological and behavioral options for dinosaurs. Instead, however, for over 100 years all interpretations of dinosaurs were tied to the living Reptilia of Linnaeus, and the thought that dinosaurs might actually be something very different was really not entertained until the 1960s. By then, the iconic dinosaur, a beast deserving of extinction by virtue of its fundamental primitiveness, stupidity, and non-adaptability, was well entrenched in the popular mind.

Before we look at the remarkable suites of specializations that characterize the major groups of dinosaurs, we need to address two iconic qualities of dinosaurian evolution: intelligence and cold-bloodedness.

INTELLIGENCE

Dinosaur intelligence—perhaps the most pervasive embodiment of dinosaur evolutionary inadequacy—is largely misunderstood and underappreciated. While it is true that sauropods, the magnificent giants of the Mesozoic, had remarkably low brain-to-weight ratios, allometry as well as the empirical evidence of their sustained, successful stay on earth (Middle Jurassic—latest Cretaceous) suggest that their brains were more than enough to carry them not only through the basic functions, but, apparently through more advanced behaviors such as herding and group nest–building. Allometry is the idea that as an organism changes in size, its constituent parts will not necessarily change in size in a 1:1 ratio with the overall size change. In the case of brains, as noted by paleontologist J.A. Hopson, "vertebrate brain size varies in proportion to the 2/3 power of body size" (1980, p. 287), which means that dinosaur brains will appear proportionally smaller and smaller as the animals become larger and larger. Sauropod brains, for example, look ridiculously small—but only if you're human-sized.

Non-avian dinosaurs are a remarkably diverse group of animals and, as might be anticipated, had a large range of brain sizes, even when allometry is considered. A different—and perhaps better—measure of intelligence than brain-to-body-weight ratios is encephalization quotient (EQ). EQ is the measure of the deviation from the predicted brain size, given its body size, which an organism possesses using the brain size of a living archosaur (crocodile) as a baseline. Hopson (1980) calculated EQs for various groups of non-avian dinosaurs and found that, with one exception, all fell comfortably within the range of living reptiles. The one exception was maniraptorans—the small, carnivorous dinosaurs epitomized by *Velociraptor* and *Deinonychus*. These dinosaurs all had EQs within the lower part of the living bird and mammal range.

In short, then, the reputation of dinosaurs for relentless stupidity—and for evolutionary failure because of it—is utterly undeserved. The phyletic longevity of non-avian dinosaur suggests that intellectual deficiencies did not hamper the group, and our best quantitative estimates of their intelligence put them in the same range as, or higher than, living reptiles. Strikingly, the mammals that coexisted in the Mesozoic with non-avian dinosaurs do not appear to have had EQs that significantly set them apart from contemporary dinosaurs. Lack of intelligence was not likely a factor in the evolutionary success, or the extinction of the non-avian dinosaurs.

IN COLD BLOOD

"Warm-bloodedness" is a quality that characterizes living so-called "higher" vertebrates: birds and mammals. In fact, the blood of warm-blooded animals is not actually warmer than "cold-blooded" animals when they reach their ideal operating temperatures; "warm-blooded" animals differ from their cold-blooded brethren in part because they utilize metabolic pathways that that allow them to maintain a temperature gradient against external temperatures. For this reason the term that is generally applied to warm-blooded animals is "endotherms;" their cold-blooded counterparts are called "ectotherms," in reference to the fact that their temperature modifies with ambient temperatures. As the primitive condition in vertebrates—reflected, for example, in living reptiles—is ectothermy, an endothermic metabolism is considered evolutionarily advanced.

The idea that dinosaurs might be endotherms is hardly new. In fact, it originated in 1842 with the introduction of the term "Dinosauria" by Richard Owen. Nonetheless, the reptilian ("saurian") ancestry of dinosaurs came to dominate thinking, such that within thirty or so years after Owen named them, and for the next 100 years, Dinosauria came to embody retrograde, non-dynamic, overspecialized evolution; the ultimate icon of an evolutionary dead end.

The change in thinking about dinosaur metabolism(s) has been well documented and has been the inspiration for a spate of popular books, some of which are listed at the end of this chapter. Bakker observed that all living endotherms have a fully erect stance. He proposed that this relationship ought to be extended such that all organisms, dead or alive, with a fully erect stance might be endothermic. Less successful (but nonetheless brilliant) was his use of inferred predator/prey ratios to infer dinosaur endothermy. Many scientists have noted the well-developed chewing specializations of dinosaurs (see below); likewise, the striking skeletal designs of many dinosaurs appear more bird-like than "reptile"-like, and suggest bird levels of metabolic output. Oxygen isotopes have been used to measure the temperature fluctuations undergone by body tissues; bone histology has been used to obtain insights into rates of growth and development, which also have ramifications for elevated metabolic rates.

From today's vantage point, the evidence for dinosaur endothermy is mixed or inconclusive. Some scientists argue that the very mixed nature of the evidence suggests that dinosaurs maintained some kind of physiology intermediate between ecotherm and endotherm. Others suggest that different dinosaurs maintained different physiologies, noting, for example, that immense sauropods can hardly be expected to have maintained the same metabolic levels as small, evidently active creatures like dromaeosaurs. Large dinosaurs may have been ectothermic; yet, because of their large size, could have effectively maintained relatively constant temperatures. Smaller dinosaurs may have been endotherms in the bird and mammal sense, developing plumage as insulation to maintain their body temperatures (see below). One very interesting possibility is that some dinosaurs began their lives as ectotherms and, upon reaching adulthood and large size, became endotherms.

In short, while there isn't complete agreement about which metabolism(s) dinosaurs actually possessed, there is a general acknowledgement that the "reptilian" model for dinosaur metabolism—ectothermy as exemplified in living crocodiles—is unlikely. One thing is certain: The stereotypical iconic image of the cold-blooded saurian, although surviving in the popular press and psyche, is as extinct as non-avian dinosaurs.

NOT YOUR FATHER'S DINOSAURS

Dinosaurs radiated into remarkable variety of forms during their 160 million-year tenure on earth. In the following section, I discuss each of the major groups of dinosaurs separately, emphasizing their departures from the "reptilian" grade of evolution described above. The point, of course, will be to evaluate (and ultimately reject) dinosaurs as icons of primitiveness or evolutionary failure. We now know that even the most primitive forms known were not your parents' "reptiles."

Dinosaurs are conventionally divided into two great groups: Ornithsichia and Saurischia, first by the English paleontologist and anatomist, Harry Seeley, in 1887. Remarkably, throughout all the phylogenetic fluxes of the intervening years, Seeley's great division stuck, but as new discoveries continue to be made, dinosaurs continue to astound with the range of their diversity and specializations. In recent years, these two great groups—and the more familiar names that characterize dinosaurs—have been augmented by phylogenetic systematic analysis to include some less familiar groups. Because some of the less familiar groups are identified by striking specializations, I use those groups as well as the more familiar ones.

GENASAURIA

Genasauria ("gena" = cheek) is the name given to a group that includes familiar dinosaurs such as duckbills, iguanodonts, stegosaurs, ankylosaurs, ceratopsians, and pachycephalosaurs. The dominant quality of this group of dinosaurs is the presence of cheeks, recognized in fossil skulls by the tooth row being inset toward the midline of the skull (to allow space for the existence of cheeks). Cheeks keep food from falling out of the mouth during chewing, and for this reason, the presence of cheeks can only suggest one thing: that all these dinosaurs chewed their food.

Chewing, as it turns out, is a remarkable invention—most vertebrates don't do it (think, for example, of birds)—that was invented only one other time in the history of vertebrates: by mammals. So while chewing may seem like a prosaic quality to us (we chew, after all), it is in fact a very foreign way to do business, and involves a lot of evolution and a whole suite of adaptations, in order to make it work.

But genasaurs have these in spades. Like their mammalian counterparts, the heavy lifting was done by a tightly packed block of cheek teeth (we call them molars in mammals) with tight occlusion that ground away on plant matter—tough, gymnospermous plant matter in the case of dinosaurs—so that material was pulverized. But unlike mammals, who only have one set of adult teeth wear down, genasaurs had constant tooth replacement throughout their lives, so their teeth were never the limiting factor on nutrient intake.

Genasaurs also developed an adaptation that parallels the incisors of mammals: a strong, sharp beak, which was used to crop vegetation, before it was sent to the back of the mouth to the block of highly specialized cheek teeth for grinding. Like their mammalian herbivorous counterparts, genasaurs also had a small toothless gap between the upper and lower beak at the front of the mouth. In mammals, this region is used by the tongue for the manipulation of food; there is no reason to suppose that it was used differently in genasaurs. Finally, chewing required big muscles, and places for those muscles to attach. In both mammals and genasaurs, the lower jaw developed a large protuberance called the coronoid process, which acted as an attachment site and lever arm for powerfully shutting the lower jaw.

Chewing, then, is a noteworthy adaptation, not only for the striking parallelism displayed between mammals and genasaurs, but also for its un-reptile-ness. "Reptiles" and birds don't chew; they bite food and send it backward, generally storing it temporarily in a sac-like crop, before masticating it in a muscular offshoot of the stomach called the gizzard. In genasaurs, the function of the gizzard has been supplanted, to a greater or lesser degree, by the mouth.

Within Genasauria, there are a variety of other indicators of exceptional evolutionary advancement. Particularly noteworthy is what is known of parental care in the group. The archetypical genasaur exhibiting parental care is the "duck-billed" dinosaur *Maiasaura*. In this case, apparent rookeries have been found, also implying significant social behavior, and some kind of extended parental care along the lines of birds is indicated. Nests are known for the small ornithopod *Orodromeus*; however, the extent of parental care is unclear in this genus. In the case of the Mongolian ceratopsian dinosaur *Protoceratops*, eggs, embryos, and nests are known, as well as compelling evidence to suggest that these animals raised their young at the nest. The evidence consists of the discovery of two different nests, each containing multiple juveniles. Within each nest, the siblings' level of advancement is approximately similar, but between the nests, the siblings appear to be significantly older in one than the siblings appear in the other nest. This discovery implies growth at the nest, which was likely mediated by parental care. As we shall see, some non-genasaur dinosaurs exhibited apparent hallmarks of parental care as well.

Other genasaurs had other sophisticated specializations. Ankylosaurs evolved a complex pavement of armor more than 40 million years before mammalian glypodonts and armadillos attempted the same trick, and more than 100 million years before humans developed chain mail. Ankylosaurs also developed massive protective armor via thickening the skull roofing bones, closing all the openings in the skull roof, and then cleverly lightened the skull by developing large vacuities within. Iguanodonts anticipated human hands by developing an opposable fifth digit on the hand. Some stegosaurs are now widely suspected of using their plates not for protection or defense, but for a kind of radiation-based thermoregulation.

Ceratopsians and ornithopods give every indication of having had very active social lives. Large bone beds, complex trackways, and rookeries all attest to herding in both of these groups. A variety of cranial adornments—including horns and crests—suggest, such as is seen in modern mammals, intraspecific competition. In some cases—notably in duckbills and in *Protoceratops*—sexual dimorphism can be recognized and, strikingly, the appearance of adult size correlates with the fullest development of some of these features. This implies that the cranial features are linked with the onset of sexual maturity, and that in turn means that it's a good bet that these features were involved in complex intraspecific behaviors. Along with the visual cues that horns, frills, and crests may have provided, evidence suggests that the hollow crests, at least, may have acted as amplifiers for various dinosaurian vocalizations.

OUTSIDE OF GENASAURIA

The evolutionary story outside of Genasauria is different but no less impressive. The iconic dullards of the dinosaur world are of course sauropods, the long-necked, long-tailed herbivorous quadrupeds epitomized by brontosaurus. In the popular view, their presumed lack of intelligence coupled with their extraordinary size yielded a primitive animal of legendary inactivity and stupidity; giant brontosaurs who supported their massive bulk in the steamy swamps of the Mesozoic, idling their thought-free hours away contentedly munching soft water-logged vegetation.

It would be hard to more completely misunderstand any group of animals! Sauropods were clearly fully terrestrial animals, and are known from every continent except Antarctica—where it is reasonable to predict that they may yet be found. They apparently inhabited a variety of terrestrial settings—from rivers to deserts—and did so for 140 or so million years, suggesting that whatever they were doing must have been done very well. Their large size (the largest known fossils suggest that lengths of 120 feet weren't all that rare) and great weight must have presented extraordinary biomechanical problems. The little that we know of them suggests some remarkable solutions to those problems.

Humans often marvel over the small size of sauropod heads and brains. But considered from a design standpoint, what fool would put a large head at the end of an approximately thirty-foot neck? Small heads mean small brains, but corrections for allometry, as we have seen, suggest that sauropod brain functions were within the range of living reptiles. The presence of well-developed trackways suggests that sauropods were gregarious animals, indicating that the "tiny" brain was capable of managing social behavior.

Because the sauropod mouth was clearly not set up for chewing, we (chewing animals) have myopically dismissed it as having "weak" teeth suitable only to marsh vegetation. The teeth tend to be more toward the front of the mouth—in camarasaurs and brachiosaurs the teeth are robust and spatulate. In diplodocids they are restricted to the very front of the mouth, and are pencil-like. The teeth of no sauropod show much in the way of grinding surfaces or occlusion. Indeed, chewing was likely not much of an option. But if we toss out the chewing model, the mouth was obviously designed for cropping of chunks of vegetation and sending them posterior for mastication in a stony gizzard: the very thing that birds do today! These mouths (and teeth) fed, for 140 million years, the largest beasts ever to walk the face of the earth; empirically, they were extremely functional.

Behind the skull are a series of astoundingly complex cervical vertebrae. These are design masterpieces: each with a complicated series of projections unparalleled in the rest of the animal kingdom, designed to allow what must have been a rather sophisticated musculature to support the long neck.

Strikingly, the vertebrate and upper parts of the ribs in sauropods are generally hollow, a feature interpreted by most paleontologists to be a weight-saving device. These hollowed bones contrast with the bones of the pelvis, legs, and feet, which are filled with dense, spongy bone. Here, then, the design of the bones nicely matches the needs of the animals; where great strength to support the body was required, the bones are robust and solid. On the other hand, where lightness is desirable, the bones were hollow.

Sauropods are obvious candidates for the most majestic terrestrial animals of all time. Many of their unique solutions to the physiological and biomechanical problems of life elude us. Even their size makes them difficult to manage: collecting, preparing, and mounting a sauropod is a daunting proposition both in terms of time and money. It is paradoxical that because the sauropods are ultimately so unfamiliar to us, we tend to dismiss them as icons of primitiveness, when they ought to be reckoned as icons of the most sophisticated and advanced evolution.

The popular conception of theropods—the bipedal, carnivorous clan of dinosaurs—does the group more justice than it does sauropods. Somehow the very carnivory of theropods conveys a more engaging popular image than that of herbivorous dinosaurs. Moreover, almost forty years of literature and popular media linking theropods with birds has called attention to spectacular adaptations, high levels of activity, and intelligence possessed by many members of the group, particularly the small- and mid-sized forms. Yet, in many respects the popular image still badly underestimates them.

Although the iconic dinosaur is hardly a runner, theropods were all cursorial animals, and possibly the fleetest bipeds the world has ever seen. The design of the rear legs, whether one is inspecting the six-foot *Coelophysis*, or the forty-five-foot *Tyrannosaurus*, is all about efficient bipedal locomotion. Here, the entire leg mechanism moves in a plane parallel to the vertical midline (saggital) plane, focusing the full muscular energy of the leg movement directly into propulsion.

Tyrannosaurids (*Tyrannosaurus* and its near relatives) were certainly among the largest carnivores ever to have walked the earth, and in the popular view, these dinosaurs and similar forms have become icons for Ultimate Carnivory. In fact, however, the image of *T.rex*-as-killing-machine has been under close scrutiny since the early part of the twentieth century. The extreme size of the dinosaur required the evolution of a number of biomechanical adaptations, not the least of which are its (by theropodan standards, at least) comparatively stout and short legs. Estimates of its running ability, based upon stride length as well as the inferred required propulsive musculature, suggest that it could not have reached the high speeds achieved by contemporary lighter-bodied dinosaurs, such as the ultra-gracile ostrich-like ornithomimids. Thus, while the animal was fast enough for whatever it actually did, its top speed may not have been quite snappy enough for chasing prey. Moreover, its teeth are strikingly bulbous, and don't seem as amenable to the shearing steak-knife model that likely characterized the undoubted aggressive predators among theropods. Based upon this, and upon other aspects of the morphology, therefore, it has been suggested that perhaps *T. rex* was a scavenger. While this question is still unsettled, it is worth noting that the ultimate icon of killing, *T. rex*, may not even have done its own killing.

Despite the notoriety of tyrannosaurids, the small- to medium-bodied theropods were the unambiguous hunter-predators of the dinosaur world. These animals were all lightly built with strikingly gracile skeletons. The hands and feet were equipped with well-developed claws; in some cases (deinonychosaurs), the toes had one or two particularly large claws capable of being rotated along an arc slightly short of 180 degrees. In such animals, the foot thus performed the dual function of support and evisceration. In the rest, the sharp-clawed foot with its three forward-pointing toes (digits II, III, and IV) and semi-reversed toe (digit I) was clearly as grasping organ, and likely used for prey manipulation as well as running.

The hands were equally adept at prey manipulation. The striking adaptation in the hand—with the exception of its very size, which was approximately equal to the length of the forearm—was a long, semi-opposable thumb. Thus the large hands, tipped with intimidating, curved claws, had a highly evolved grasping function. Even the number of fingers in the hand indicates significant evolution: Unlike humans, which show the primitive condition of five fingers in the hand, these advanced theropods had the more advanced condition of three fingers on each hand.

A most unusual adaptation of deinonychosaurs occurs in the tail. There, each vertebra has elongate, rod-like processes that extended across many vertebrae. The sum of all these overlapping processes is to lock the tail into a straight, inflexible rod, the only remaining movement of which occurs at its base. With this adaptation utterly unknown in the modern world, we can only speculate about its function; yet, a look at those long arms terminating in large, elegant, grasping hands suggests that the rigid tail must have served as a dynamic counterbalance to the powerful thrusts and swipes of the arms. Behaving in this manner would be inconceivable for the iconic reptilian dinosaur predator.

In virtually every detail, advanced theropods—tenanuran coleurosaurs to specialists—reveal remarkable adaptations. Along with the features described above, highly specialized, enigmatic forms, such as oviraptorsaurids, and *Mononykus*, have a variety of distinctive, evolved features whose function, although not well understood, clearly must have had important roles in their behavior. Such features would include the fused hand in *Mononykus*, and the toothless, pneumaticized skull of *Oviraptor*. And, while we do not have a complete story for all theropods, in the case of *Oviraptor*, at least, the dinosaur evidently incubated its eggs, in fine bird style. These are hardly the hallmarks of a primitive, non-adaptive group of animals.

This takes us to the evolution of birds, which are icons of evolutionary advancement in all the ways that dinosaurs are not. Birds, of course, are often portrayed as creatures whose every adaptation is keenly honed by natural selection for flight. Nobody would characterize a bird as a "dinosaur" in the pejorative sense of the word. Yet, a modern evolutionary viewpoint suggests that birds not only evolved from dinosaurs, but in fact *are* specialized theropods. Indeed, many of the supposed flying adaptations of birds are in fact inherited from their very earthbound theropodan ancestors (see below). Among the features that appear to be exclusively for flight, but are in fact inherited from non-flying non-avian theropods are feathers, hollow bones, and the furcula (wishbone). Dinosaurs as iconically primitive is obviously antithetical to the fact that *avian* dinosaurs are widely considered among the most highly evolved, specialized creatures on earth.

ARCHAEOPTERYX, MISSING LINKS AND THE EVOLUTION OF BIRDS

The key player in the story of how birds are theropods is the feathered theropod *Archaeopteryx lithographica*, an animal that appears to mix so-called "reptilian" features with bird features, and has thus been portrayed as the ultimate icon of Darwinian evolution: a "missing link" between reptiles and birds. In the following section, I examine neither

Skeleton of huge carnivorous dinosaur, Allosaurus. American Museum of Natural History. (Courtesy, United States Geological Survey.)

the unambiguous dinosaurian ancestry of *Archaeopteryx*, nor the last twenty years of remarkable discoveries of feathered dinosaurs that have reshaped our view of the evolution of birds. Instead, I address a more specialized aspect of the origin of birds and their relationships to non-avian theropods: the characterization of *Archaeopteryx* as a "missing link" between reptiles and dinosaurs. For it is in this regard that the fossil has been truly misunderstood.

Charles Darwin first proposed his theory of evolution by natural selection in 1858 (and published *On the Origin of Species* in 1859). The theory, like all meaningful scientific hypotheses, made explicit, testable predictions, that is, predictions that could be falsified via observations. Among the many testable predictions built into the theory was the notion that, as evolution progressed, a smooth gradation of morphologies would accrue in the historical record. Darwin didn't have at his disposal the insights of genetics and population ecology, fields that came to supply a mechanistic basis for this theory, and so his image was not one of overlapping allele frequencies in successive ancestor-descendent populations, such as is now how generally regarded as a true reflection of the mechanistic details of evolutionary sequences.

His theory postulated intermediate morphologies, and in the chapter titled "Difficulties on Theory," he asked this question, "Why, if species have descended from other species by insensibly fine gradations, so we not see everywhere innumerable transitional forms?" (Darwin, 1964, p. 171). His answer was in part in the imperfection of the fossil record and in the inadequacy of paleontological collections. Imagine what a boost his theory received, then, when in 1861 the fossil *Archaeopteryx* was discovered and proclaimed an example of the very transition fossils he sought! The dinosaur-bird connection was almost immediately made by Darwin's friend and advocate, T.H. Huxley: "Surely there is nothing very wild or illegitimate in the hypothesis that the phylum of the Class of Aves has its foot in the Dinosaurian Reptiles" (quoted in Feduccia, 1980, p. 31). The argument held sway until the South African paleontologist Robert Broom pointed out in the early 1900s that dinosaurs were too "specialized," and that a primitive archosaur, *Euparkeria*, could also serve as the ancestral to birds (as well as dinosaurs). The "reptilian" (e.g., primitive) nature of *Euparkeria* seemed to better match the reptilian nature of dinosaurs. And so the "pseudosuchian thecodont" (primitive archosaur) hypothesis of bird ancestry was proposed, and held sway until the late 1960s, when Yale paleontologist J.H. Ostrom first described the small, carnivorous theropod *Deinonychus*.

Ostrom was struck by the number of morphological parallels between *Deinonychus* and *Archaeopteryx*, similarities that he developed in detail in a large monograph on *Archaeopteryx* (1975). The bird-dinosaur relationship was back with a vengeance, given a modern, cladistic treatment by Yale paleontologist J.A. Gauthier and innumerable subsequent technical articles, and enshrined in popular book form by paleontologists L. Dingus and T. Rowe and anthropologist P. Shipman. Now, most—but not all—systematists agree that *Archaeopteryx* is rightfully called a bird, and that its ancestry is to be found somewhere within the group of dinosaurs that includes *Deinonychus*. But how are we to understand that in terms of the more iconic ancestors and missing links?

The answer, as has been the case throughout this chapter, is that it is the expectations embodied in the icon that are flawed, and not the science. The real problem begins with the aggressive search for—and need to find—ancestors. Darwin made the point—as paleontologists and evolutionary biologists have done since—that the chances of fossilizing the actual *ancestor* of any group—even a species—are really quite small (see Chapter 15). This is because:

- As Darwin argued, evolution proceeds via the isolation of a population and its subsequent adaptation to new settings. Under selective pressures engendered by a new setting, the small, isolated population morphologically diverges from the larger, ancestral population until a new species had evolved within the new population. In the "new synthesis" of genetics, paleontology, and population ecology of the 1920s and 1930s, this became enshrined as the allopatric model of speciation. Allopatric speciation, therefore, suggests that the proportion of the total fauna occupied by transition forms is really very small.

- Fossilization is an uncommon process, requiring a happy coincidence of factors that impede bacterial degradation. Common means by which this occurs are fossilization in arid climates, that is, settings in which there already is low biotic productivity, under conditions of rapid sedimentation (burial); fossilization in hypersaline environments in which biotic productivity is depressed; and fossilization by rapid burial followed by intensive mineral replacement and/or precipitation. Most organisms that die, however, are bacterially degraded, recycled, and don't get fossilized.
- The episodic nature of sedimentation is such that very little of the time that actually passed is represented by deposition. Most of the time that has passed is instead contained within gaps within the record (unconformities). In the terrestrial realm, this is particularly aggravated, because deposition occurs most commonly during storm events, but most of the *living* that organisms do occurs during the hiatuses between storms. What organism, after all, thrives during storms?

Unfortunately, many of the expectations that ancestors will be found in the fossil record came from the paleontologists, themselves. Too often, particularly in the past, paleontologists have spoken of antecedent fossils as "ancestors," particularly with regard to high-visibility groups such as hominids. As the fossil record richened through collection and study, time and time again, putative "ancestors" have been recognized as actually offshoots from the so-called "main line" of evolution. This is largely because the concept of a "main line" in evolution is itself likely flawed as well. To understand this, we must delve into the difference between *orthogenesis* and *cladogenesis.*

Orthogenesis is straight-line evolution; the idea that a lineage can be traced directly from ancestor to descendent in a straight line. A classic example of orthogenesis is the now rather dated sequence first proposed in 1879 by paleontologist O.C. Marsh to explain the appearance of modern horses: *Orohippus* (Eocene) --> *Mesohippus* (Miocene) --> *Miohippus* (Miocene) --> *Hipparion* (Pliocene) --> *Pliohippus* (Pliocene) --> *Equus* (Recent).

Time and study eventually disqualified this viewpoint (see "Neo-Darwinism"), as each of the "ancestors" was ruled out as being directly ancestral because of one or another specialization it possessed. As noted by paleontologist B. MacFadden in his extended treatment of fossil horses, most paleontologists and evolutionary biologists now believe that speciation events are likely better represented by cladogenesis, that is, that lineages evolve in a branching form. Horse evolution, therefore, would appear in a more bush-like form, with the branching endpoints being the various known fossil genera, rather than the linear sequence itemized above (see "The Horse Series," vol. 1).

In contrast to orthogenetic evolution, cladogenetic evolution suggests that the fossils that are found are not likely to be ancestral, but rather only to embody some of the ancestral characters. This corresponds nicely with the empirical fossil record, in which, predictably, rarely known indeed are ancestors to particular organisms. Given allopatric speciation, the rarity of fossil preservation, the incompleteness of the sedimentary record, and cladogenesis as the dominant mode of evolution, how could it be otherwise?

As it turned out, this viewpoint was happily concordant with the adoption of cladograms as the best means of reconstructing evolutionary relationships (see "Cladistics"). Cladograms, it will be recalled, are testable, branching diagrams depicting hierarchies of shared, diagnostic characters in nature. The possession of shared diagnostic characters is inferred to indicative of relationship, and thus phylogenetic relationships can be constructed based upon the hierarchies of shared, diagnostic (or derived) characters. Non-diagnostic characters are characterized as "primitive," which makes no judgment about the sophistication or fitness of a character, but rather only refers to the fact that the character represents the more general, or ancestral, condition. Differing arrangements of the characters produce different cladograms, and thus in a phylogenetic sense, these can be considered as competing phylogenetic hypotheses. The preferred hypothesis (and thus, the preferred phylogenetic reconstruction) is the one that is the most parsimonious, that is, that requires the least steps.

All of this, then, makes the very *idea* of a "missing link" something of a red herring. If evolution proceeds by cladogenesis, then true "missing links" can't really exist, because only orthogenetic evolution possesses the linear, chain-like attributes that allow for the image of missing links to have meaning. In essence, then, the term "missing link," if it is ever used by professionals, has come to mean something other than an ancestor in an orthogenetic evolutionary chain. Modern evolutionary biologists and paleontologists, when they speak of "intermediate" taxa, are actually referring to mixed advanced and primitive suites of characters rather than missing links. Occasionally, the term "grade" is used to denote the amount that a particular form has evolved from that which is inferred to be the ancestral condition. But the word "grade" is not phylogenetic; and the fact that a particular organism has reached say, a intermediate grade of evolution only means that it possesses a mixed—or *intermediate*—suite of primitive and derived characters.

In short, the term "missing link" as a professional term of art is essentially nineteenth and early twentieth century in outlook. Unfortunately, like a bad meal, in the popular press the term "missing link" continues to repeat and repeat long after its fundamental validity has been rejected. The term resides in the very iconography of the *scala naturae* that Gould excoriated in 1989, and, unfortunately, has become entrenched in high school textbooks. Creationist author J. Wells, for example, quotes two recent (1998, 1999) high school textbook examples that use missing link imagery, and indeed his own rejection of *Archaeopteryx* as the key fossil in understanding the origins of birds derives not from the irrelevance of *Archaeopteryx*, but rather from Wells's misunderstanding of cladogenesis and the extreme unlikelihood of finding "missing links" in the first place.

How, then, should one view *Archaeopteryx*? A cladistic treatment of bird evolution such as was carried out by J.A. Gauthier, or reported in L. Dingus and T. Rowe's *The Mistaken Extinction* (1988), shows that many of the characteristics that appear to uniquely characterize living birds (e.g., the diagnostic characters of birds) are in fact distributed across the cladogram among a variety of dinosaur taxa. Page 357 shows the distribution of characters between the closest group to *Archaeopteryx* (dromaeosaurs), *Archaeopteryx*, and living birds. It highlights the fact that *Archaeopteryx* possesses a mixed suite of primitive and derived characters such that it can be said to represent—as was first observed almost 150 years ago by T.H. Huxley—an intermediate or transitional grade between non-avian dinosaurs and birds.

CONCLUSION

Dinosaurs are popularly viewed icons of failed evolution. This misconception is fueled by a suite of misunderstandings including:

1. Their small brains consigned them to oblivion;
2. Their large size consigned them to oblivion;
3. They were reptiles, which means that they were primitive, cold-blooded animals that ultimately couldn't compete with the more superior, warm-blooded, intelligent mammals;
4. Their extinction was the obvious and predictable result of their inadequacy; and
5. They left no descendents.

As I have attempted to show in this chapter, each one of these ideas is simply not correct. Indeed, the longevity and tenacity of these images is remarkable, given that they run utterly counter to current scientific thinking. Their persistence is all more remarkable because, unlike fifty or so years ago, there is a media glut of information about dinosaurs—television shows; IMAX movies; feature-length movies; innumerable lavishly illustrated, glossy coffee-table books, magazine articles, and popular books by well-known paleontologists—all of it highlighting modern interpretations of dinosaur paleobiology. Indeed, it would be hard not to be aware of the waves of change that have passed over ideas about dinosaurs. Why, then, are the antiquated viewpoints so well entrenched?

Dromaeosaurs	***Archaeopteryx***	**Living Birds**
Teeth (+)	Teeth (+)	Teeth (–)
Braincase slightly enlarged	Braincase slightly enlarged	Enlarged braincase
Tail long, well-developed; stiffened	Tail long, well-developed; stiffened	Pygostyle (+)
Large, asymmetrical hand: three-fingered; digits I, II, & III; semi-opposable digits I; claws	Large, asymmetrical hand: three-fingered; digits I, II, & III; semi-opposable digit I; claws	Large hand; three fused digits (I, II, III) (carpometacarpus); ontogenetically derived from an asymmetrical, three-fingered, clawed hand
Semi-lunate carpels	Semi-lunate carpels	Highly modified wrist structure
Fully erect stance (as in Dinosauria)	Fully erect stance (as in Dinosauria)	Fully erect stance (as in Dinosauria)
Bipedal	Bipedal	Bipedal
Foot with tridactylous weight-bearing portion and fourth toe reduced and to back (or side); unguals clawed	Foot with tridactylous weight-bearing portion and fourth toe reduced and to back (or side); unguals clawed	Foot with tridactylous weight-bearing portion and fourth toe reduced and to back; unguals clawed
Three unfused, closely appressed metatarsals (digits II, III, and IV); digit V (–) or vestigial	Three unfused, closely appressed metatarsals (digits II, III, and IV); digit V (–) or vestigial	Three closely appressed metatarsals (digits II, III, and IV) fused during ontogeny; digit V (–) or vestigial
Ascending process on the astragalus	Ascending process on the astragalus	Ascending process on the astragalus visible ontogenetically; lost in adult
Attenuate fibula	Attenuate fibula	Attenuate fibula
Hollow bones; pneumatic in some forms	Hollow bones; likely pneumatic	Pneumatic bones
Furcula (wishbone)	Furcula (wishbone)	Furcula (wishbone)
Sternum small; flat	Sternum small; flat	Carinate sternum
Pelvis unfused	Pelvis unfused	Synsacrum
Pubis pointing down; anterior portion of footplate absent; ischium reduced in length	Pubis pointing down; anterior portion of footplate absent; ischium reduced in length	Pubis pointing back; anterior portion of footplate absent; rotates from downward orientation during ontogeny
All vertebrae (+)	All vertebrae (+)	Some vertebrae (–)
(–) of avian skeleto-muscular flight specializations of shoulder region; animal non-flying	(–) of avian skeleto-muscular flight specializations of shoulder region; flight nonetheless possible	Avian skeleto-muscular flight specializations of shoulder region
Feathers (+) Known in several members of the group	Feathers (+)	Feathers (+)

It is easy to construct a facile explanation about the lack of flexibility of the educational system. Teachers who teach in the old way, with no particular incentive to "get it right," simply convey what they, themselves learned. Dinosaurs are popularly caricatured in advertisements, or used as attractants for products ("cereal-box" dinosaurs), where the goal is not to get the science correct, but to sell the product. And it is likely the case that the product is better sold using imagery that is familiar to the buyers—many of whom grew up with a now-antiquated understanding of what dinosaurs are. So the images are continually rejuvenated, when they should have long ago been forgotten and relegated to the scrap heap of abandoned ideas.

I sometimes wonder, however, if the antiquated perceptions of dinosaurs exert a kind of psychological hold on us that transcends either the media or science. Indeed, they may serve as a kind of counterpoint to us as mammals and if so, they might be said to represent a kind of alter ego. In this dichotomized view they are the Other; paralleling mammals as "dominant" terrestrial vertebrates, and yet opposite in every discernible aspect of their biology and behavior.

The whole non-avian dinosaur story as it is popularly portrayed can be thought of as a morality parable. In this fiction, the spectacular rise of dinosaurs is embodied in their very large size and diversity; indeed, in this view they inherited an Eden-like earth, put there exclusively for them to live and thrive. But that kind of size and limitless domination implies a certain kind of arrogance, and so their fall can be seen as somehow deserved—indeed, immanent to who they are and what they did. Identifying them as the Other, we find satisfaction in the fact that we and our way survived and that the Other—with its parallel, foreign world—did not. As morality plays continue to give satisfaction long after the stories they tell are well known to their audiences, so it is that this story continues to resonate long after the images and theories upon which it is based have been consigned to scientific oblivion.

My view is that looking for morality within, or anthropomorphizing, the natural world is foolish. Dinosaurs are best viewed, I believe, as organisms in natural systems devoid of higher-order meaning. Yet they—and their world—provide important lessons for how the earth and its biota respond to change. The iconography of dinosaurs, therefore, ought to embody the magnificent successions of former worlds, engendering, as we look across 228 million years of time, a deeper appreciation of our own.

BIBLIOGRAPHY

Abate, F. (ed.). *The Oxford Desk Dictionary and Thesaurus* (American Edition) (New York: Oxford University Press, 1997).

Darwin, C. *On the Origin of Species* (facsimile of the First Edition) (Cambridge, MA: Harvard University Press, 1964).
Feduccia, A. *The Age of Birds* (Cambridge, MA: Harvard University Press, 1980).
Hopson, J.A. "Relative Brain Size in Dinosaurs, Implications for Dinosaurian Endothermy," in Thomas, R.D.K., and E.D. Olson (eds.), *A Cold Look at the Warm-blooded Dinosaurs: AAAS Selected Symposium* no. 28(1980): pp. 287–310.
Moore, R.C., C.G. Lalicker, and A.G. Fischer. *Invertebrate Fossils* (New York: McGraw-Hill, 1952).

FURTHER READING

Farlow, J. O., and Brett-Surman, M. K. (eds). *The Complete Dinosaur* (Bloomington: Indiana University Press, 1997).
Fastovsky, D. E., and Weishampel, D. B. *The Evolution and Extinction of the Dinosaurs* (Cambridge, MA: Cambridge University Press, 2005).
Weishampel, D. B., Dodson, P., and Osmólska, H. (eds.). *The Dinosauria* (Berkeley: University of California Press, 2004).

Archaeopteryx

Bent E.K. Lindow

Some 150 million years ago the dead body of a small, feathered animal sank to the bottom of a warm, tropical lagoon. The plumage of the carcass was soaked with water, and the heavy weight dragged it downward, until it settled gently into the fine-grained mud of the bottom. The animal was a stranger to the area, and its kin normally lived on the nearby islands, but this one had ended up in the lagoon for reasons unknown. Perhaps the animal had been blown into the area by a storm and drowned in the water; perhaps the already-dead corpse had fallen into a river inland, and drifted out into the lagoon by accident.

In nature, scavengers will normally quickly set upon such a cadaver and consume the soft parts, such as muscles, intestines, skin, and feathers, and scatter and destroy the bones until nothing remains. But there were no scavengers living on the muddy bottom of this particular lagoon. The environment was extremely hostile; there was almost no oxygen present and the water had an extremely high content of salt, which prevented almost anything from living there. This preserved the carcass from destruction. More fine-grained mud, brought in by a storm, settled on top of the carcass and buried it. While bacteria and decay finally consumed the soft parts of the body and the feathers, the delicate bones survived and an imprint of the feathers had been left in the mud. Over time, the lagoon slowly filled up, and more mud was deposited on top of the layer which contained the carcass.

Millions of years passed, the mud turned into limestone, and the bones likewise fossilized. The animal which ended up at the bottom of the lagoon had become a fossil. Finally, geologic movements raised the deposits of the former lagoon back above sea level in the area around what would one day become the town of Solnhofen in the German state of Bavaria. In the nineteenth century, humans quarried the limestone for use in the printing industry as lithographic slate. During their work, fossils would occasionally turn up on the surface of the limestone and the quarrymen would sell these accidental discoveries as curiosities to visitors, or, in the case of this specific fossil, to a local doctor. Some fossils went on to museums around the world to be studied and described by scientists and exhibited to the public. Others went into the collections of private citizens, to be marveled at by their owners and proudly displayed to visitors. But this particular fossil, along with later discoveries of animals of its kind, would represent especially powerful proof of a new scientific theory.

Scientists named the fossil *Archaeopteryx*, meaning "ancient wing," and it would turn out to be the earliest known bird, a representative of one of the most successful groups of vertebrates—the backboned animals. As the fossil was also a near-perfect transitional form between two major groups of animals, reptiles and birds, the timing of its discovery and scientific announcement in 1861, was especially fortuitous. It was just two years after the publication of Charles Darwin's *On the Origin of Species*; the one publication which established evolution as a solid theory and evolutionary biology as a science. *Archaeopteryx* became one of the strongest proofs of the evolutionary process, despite the fact that it came under attack from anti-evolutionists almost from the day of its discovery. Today, it remains the earliest known bird and as a result it has been at the center of discussions of macroevolution for almost 150 years. Further discoveries of new specimens of *Archaeopteryx* have both refuted dubious claims of fossil forgery by anti-evolutionists and have helped spur research into important biological questions, such as the origin of bird flight.

THE FOSSIL MATERIAL

> It is an icon—a holy relic of the past that has become a powerful symbol of the evolutionary process itself.
>
> –Shipman, 1998

So what is *Archaeopteryx*? It is the earliest known bird, about the size of a crow or magpie, and lived 150 million years ago. This was during the middle part of the age of dinosaurs. Today, ten skeletons of *Archaeopteryx* are known. All known specimens derive from the same geological deposits in Bavaria in southern Germany. Most of the specimens are kept in collections or on display in museums around Europe and North America, while a few are privately owned and one is unfortunately lost. Each individual specimen has been informally named after the town where they are kept and numbered after the order in which they were recognized as being an *Archaeopteryx*. Thus paleontologists talk about the "London" or "First" specimen; the "Berlin" or "Second" specimen; the "Maxberg" or "Third" specimen, and so forth. This naming helps researchers to know exactly *which* specimen is being referred to in scientific papers or discussions, since there are differences in size, anatomy, and the number of bones preserved in each.

Gerhard Heilmanns reconstruction of a male *Archaeopteryx* courting a female from his 1926 edition of *The Origin of Birds*. The original watercolor painting is today in the collections of the Geological Museum in Copenhagen. (© Bent Lindow. Used by permission of the Natural History Museum of Denmark.)

The most complete specimen is kept at the Humboldt Museum für Naturkunde in Berlin, and provides the best view of the overall anatomy of the skeleton of *Archaeopteryx*.

The Berlin specimen, discovered in 1877, has been widely commented on by many writers as probably being one of, if not the most, beautiful fossil in the world.

The fossil is lying on the flat surface of a pale yellowish-gray limestone slab, which measures 15.5 inches by 19 inches. At the top of the slab, its forelimbs lie spread out toward the left and right. The upside-down skull is set on a sharply backward-bent neck. A long bony tail is visible on the lower left part of the slab. To the immediate right of the tail, the two long, clawed legs extend from the hip. Most remarkable are the clear and distinct impressions of feathers extending from the arms, and along each side of the tail. Vague impressions of feathers are also present along the thighbones. Close inspection of the feather impressions on the arms reveal that they look exactly like the feathers on the wings of modern birds. Another very bird-like feature of *Archaeopteryx* is the presence of a wishbone. This U-shaped bone is part of the shoulder girdle and is found at the top of the ribcage, spanning the shoulder joints. Today, a wishbone is only found in birds, and no other living group of vertebrates possesses one. Although the wishbone is not visible in the Berlin specimen, it is present in other specimens, such as the London one. However, the rest of the skeleton is not very bird-like. The skull of *Archaeopteryx* is very reptilian, specifically like that of a small meat-eating dinosaur. *Archaeopteryx* did not have a bill, but had tiny, pointed teeth. There are between twelve and thirteen teeth present in each upper jaw and eleven to twelve in each lower jaw of the fossils where the skull is preserved. The arms of *Archaeopteryx* are long and slender, and have three free fingers with long, curved claws on the end of each. This is quite unlike modern birds, where the bones of the fingers are much shorter, are fused and grown together and most have lost the claws. The long, bony tail is also different from that found in living birds. Their long tails are actually mostly made up of feathers; the actual bony part of the tail is a short stub at the rump, called a "pygostyle." The pygostyle is a lump of tailbones, which grow and fuse together while the bird is still within the egg. This lump functions as an anchor and attachment for the muscles, which move the tail feathers. Compared to the heavy, bony tail of *Archaeopteryx*, the tail in modern birds is much lighter, thus making flight easier.

Like all the other specimens of *Archaeopteryx*, the Berlin one is still embedded in its limestone slab. Paleontologists have not tried to remove the fragile bones and mount them in a freestanding display. Part of the explanation for this is that the bones might break if removed, but more importantly because it would destroy the limestone and the all-important imprints of feathers. However, some of the limestone around some of the bones has been carefully removed, allowing the paleontologists to study details, which could reveal more about the anatomy and relationships of *Archaeopteryx*. Care has been taken to do it in areas of the fossils where it would not damage the feather imprints. In effect, the skeleton of each specimen of *Archaeopteryx* is lying in the same position as when the carcass ended up on the bottom of the lagoon some 150 million years ago.

To have ten specimens of the same fossil animal means that it is actually quite well known. Most fossil animals are only known from a single or a few skeletons, where much is missing. Many are known only from isolated bones. This is due to the extremely rare circumstances in which fossil preservation happens; it is estimated that out of 100,000 animals living today, only one has even a remote change of becoming a fossil one day.

THE DISCOVERY OF *ARCHAEOPTERYX*

> The fossil Bird with the long tail & fingers to it wings . . . is by far the greatest fossil of recent times.
>
> –Darwin, 1863

In 1859 the first edition of Charles Darwin's *On the Origin of Species by Means of Natural Selection* was published (see "Charles Darwin," vol. 1). The theory of evolution described in the book established evolutionary biology as a science and caused dramatic changes in the general view of the living world and especially the place of humans within nature. Darwin's new theory immediately found itself under attack from conservative supporters of the old, anti-evolutionistic view of the world. These attacks were often centered on the apparent lack of intermediate forms, living or fossil, between different groups of animals. Darwin himself had already commented on this problem in Chapter 6 of *Origin of Species*, titled "Difficulties on Theory." However, he noted that the "absence or rarity of transitional varieties" could basically be explained by two circumstances: First of all, intermediate forms would, according to the theory, be quickly competed out of existence by their better-adapted descendants. This would account for the lack of living transitional forms. Second, the lack of intermediate forms among fossils was explained by the "imperfections in the geological record," to which Darwin devoted the entire Chapter 9 of *Origin of Species*. He quite correctly noted that there are vast gaps in the geological record, where we do not have any suitable deposits with fossils. This is either because no deposits were simply laid down at the time, or because later geologic events and erosion has destroyed them. Finally, only a few areas have the correct environment which allows for the preservation of fossils. A sandy beach, for example, is not a good place for the preservation of animals or plants as fossils. The waves continually move, remove, and shift the sand thereby destroying any remains within. All things considered, we can only hope to discover a fraction of the animals or plants which once lived as fossils. If the transitional form between two groups lived in area, where no suitable geologic layers where deposited during its time, we will never know about it.

Nonetheless, as stated above, opponents of the theory continued to point to the lack of intermediate forms. One example was professor and geologist Louis Agassiz of Harvard University, a staunch anti-evolutionist. In a critical review of *Origin of Species* published in 1860, Agassiz used the "the definiteness of the characters of the class of Birds" in his argumentation against Darwin's new theory. Briefly put, he noted that birds were too different from any other group of animals, living or fossil. There were no known "intermediate forms," which had both features of a bird and another animal. Yet one year later, the first specimen of *Archaeopteryx* was discovered, an animal which looked more like a reptile than a bird, yet was clearly an intermediate form between the two groups, and shattered Agassiz's argument completely.

A harbinger of what was to come appeared already in 1860, when a worker in a quarry near the town of Solnhofen in southern Germany

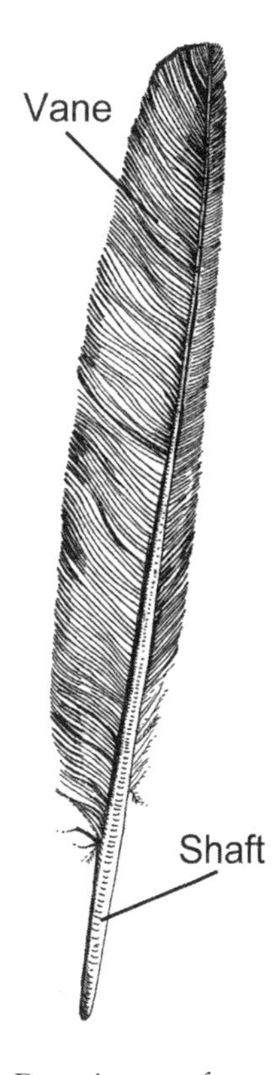

Drawing of a flight feather of a modern bird. Note how the vanes, the main surfaces of the feather, on each side of the central shaft are asymmetric, i.e. one is broader than the other. This is an adaptation for aerodynamic functions; asymmetric feathers are found only in birds which are able to fly or glide. (© Anne Haastrup Hansen. Used by permission.)

discovered a fossil feather in the limestone. The limestone in the area was mined and used in a printing process known as lithographic printing. "Lithography" literally means "stone-writing," and pictures are painted or drawn in ink onto the surface of the stone. A sheet of paper is then placed on top of the ink-covered stone and the two are pressed together. In this way, the picture on the stone is transferred to the paper. This process was extensively used in the nineteenth century to print pictures, and is still used by some artists today. The Solnhofen limestone is renowned for its very smooth surface, which allows extremely fine lines to drawn and printed. To avoid damaging the surface of the limestone slabs, they must be mined by hand and delicately split using hammers. Occasionally fossils of animals such as fish, shells of extinct squid-like animals, or flying reptiles would turn up on the surface of the slabs. Slabs with fossils were not suitable for printing, but were sold to interested visitors, and increasingly from the end of the eighteenth century, to museums and private collectors all over Europe.

The fossil feather was described scientifically by German paleontologist Hermann von Meyer in 1861. Most astonishing, although millions of years old, the fossil feather was completely modern-looking and matched the flight feather of a modern bird perfectly. It had imprints of a central stiff shaft, and the vanes of the feather were asymmetric, meaning one was wider than the other. The latter feature indicated that the feather belonged to an animal capable of flying or gliding. All together, this single fossil feather indicated the presence of birds in the geologic prehistory long before anyone had expected it. And just one month later, von Meyer reported on a new discovery from the limestone: "At the same time I am hearing from the Chief Judge, Mr. Witte, that a nearly complete skeleton of an animal covered with feathers was found in the lithographic slate . . . *Archaeopteryx lithographica* is a name that I deem appropriate for the designation of the animal." In accordance with the international laws on the naming of animals and plants von Meyer had given the animal its official scientific Latin name meaning "Ancient wing of lithographic stone."

The actual fossil specimen had, as usual, been discovered by local quarrymen. They had turned it over to a local doctor, Carl Friedrich Häberlein, as payment for services rendered. As the quarrymen were poor, they could only "pay" in fossils, which Häberlein could then hope to sell on. Häberlein was a widower with eight children, and was in need of money to support his family, and, among others, to pay a dowry for his daughter's upcoming wedding. Through his services to the quarrymen, he had acquired a large collection of fossils and in 1862 he put the entire collection of fossils, including the

Archaeopteryx, up for sale. Häberlein's fossils were bought by the British Museum in London, much to the chagrin of several German museums. Häberlein received the sum of 700 British Pounds, of which 450 pounds alone were for the *Archaeopteryx* fossil. This was a vast sum; at the time, 700 British Pounds would equal ten to twenty times the yearly wage of a skilled worker.

The fossil arrived at the British Museum on 1 October 1862 and the task of scientifically studying and describing it went to the famous anatomist and paleontologist Richard Owen. Owen published countless papers on living and fossil animals from all over the world, such as recently discovered lungfish and the extinct giant flightless Moa birds of New Zealand. Owen is also the man who invented the term "dinosaur" as a joint description for the group of large fossil reptiles, which had just started to be known at the time. He wasted no time in studying the specimen and presenting his discoveries in a lecture to the Royal Society on 20 November the same year. However, Owen was strongly opposed to Darwin's ideas about evolution. Thus, while describing *Archaeopteryx* as a bird, he did not in any way think of it as a transitional form between birds and reptiles. This was noted by British paleontologist Hugh Falconer, who attended the lecture and commented on Owens description as a "slip-shod and hasty account" in a letter to Darwin in January 1863. Unlike Owen, Falconer immediately recognized the fossil as an intermediate form between birds and reptiles and continued in his letter to Darwin: "Had the Solenhofen Quarries been commissioned—by august command—to turn out a strange being à la Darwin—it could not have executed the behest more handsomely—than with the *Archaeopteryx*."

Darwin was delighted by the news of the fossil and quickly wrote back to Falconer asking for more. While *Archaeopteryx* strongly supported Darwin's theory on evolution, he did not actually mention it much in the later editions of *Origin of Species*. Yet privately he was delighted by the fossil, as witnessed by the quote at the top of the chapter, which derives from a letter he wrote in 1863 to Professor James Dana.

Interestingly, using his theory of evolution as a base, Darwin had actually predicted an important characteristic of the wing of *Archaeopteryx* two years before it was discovered. In a letter to the English geologist Sir Charles Lyell in 1859, he described his considerations about the origins of the "bastard wing" in modern birds. This is a tiny, but important part of the wing. It is also called an "alula" or "thumb wing" and consists of three small feathers, which are attached to the tiny thumb bone. The alula is very important for the flight ability of birds, as the bird can extend and retract the feathers and thus control the flow of air over the wing during flight. Darwin predicted that the alula was a much-reduced version of what had once been a much more well-developed part of the wing in prehistoric birds. He noted that if an older fossil bird should one day be discovered, it should display several large and well-developed feathers on the thumb finger. And indeed, *Archaeopteryx* fulfilled this prophecy completely.

Richard Owen published a longer and more detailed description of *Archaeopteryx* in 1863. However, the renowned English zoologist Thomas Henry Huxley criticized his work heavily in a publication five years later. He pointing out several grave anatomical mistakes on Owen's part; for example he noted that Owen had mistaken the left leg for the right and vice versa. Huxley was one of the chief champions of Darwin's new evolutionary theory and was both scientifically and personally opposed to Owen (see "T. H. Huxley," vol. 1). The mistakes in Owen's description probably stemmed from his hastiness in describing the animal. During his career, he published many, many scientific papers on a wide variety of animals. This came at the price of often being a bit too superficial in his studies and descriptions, where he should have been more thorough. *Archaeopteryx* was one of those cases. To Owen's defense it must be said, that the London specimen of *Archaeopteryx* is not in the same state of preservation as the magnificent Berlin specimen described above. The London fossil represents a carcass that rotted and floated around on the surface of the lagoon for a long time. Many of its bones were lost as they dropped from the carcass while it drifted or are lying in unnatural positions. But impressions of feathers are still visible, and enough of the bones are present to deduce the nature of the animal.

Huxley used *Archaeopteryx* as a prime example in promoting of Darwin's theory, pointing out that its anatomy was intermediate between reptiles and birds. He compared the skeleton of *Archaeopteryx* to the dinosaurs known at the time, but also to ostriches and found many similarities. Huxley especially compared *Archaeopteryx* to a chicken-sized meat-eating dinosaur called *Compsognathus*, which had also been discovered in the Solnhofen limestone. Based on his comparisons, he made a further prediction about the skull of *Archaeopteryx*. The skull of the London specimen is missing and the shape of the jaws were not known. In his description, Owen had predicted that the animal must have had a toothless bill. This prediction was clearly based on Owen's view of *Archaeopteryx* as a true bird, not an intermediate form. His argument was that *Archaeopteryx* needed to clean and preen its feathers, which, according to Owen, could only be done with a bill. However, on the slab of limestone containing the London *Archaeopteryx*, another scientist, Sir John Evans, had later discovered and described another piece of bone; a small part of a jaw with four teeth. It was debated whether this piece belonged to *Archaeopteryx* or if it was a part of another animal. Slabs of fossils sometimes contain the remains of more than one animal, and it was not unlikely that parts of another carcass could have ended up on the lagoon bottom together with the *Archaeopteryx*. However, Huxley interpreted *Archaeopteryx* as intermediate between reptiles and birds. He suggested that if a better-preserved fossil of *Archaeopteryx* was discovered, it would turn out to have teeth, a reptilian feature, and not a bill. Huxley continued expounding his views on the similarities between dinosaurs and birds in more scientific paper from 1870 titled "Further Evidence of the Affinity between the Dinosaurian Reptiles and Birds." The paper is especially interesting, as Huxley does not only make comparisons with small dinosaurs like *Compsognathus,* but also the large (twenty-six feet) meat-eating dinosaur *Megalosaurus.* Despite the size difference, Huxley noted there were many anatomical similarities between birds and dinosaurs, especially in the legs and the hip.

The next specimen of *Archaeopteryx* was discovered in 1877. This was acquired by Dr. Häberlein's son, Ernst. He was tax consultant, and probably obtained fossil through his contacts with quarry owners. Knowing that the second specimen of an already-famous fossil which was in the midst of several heated scientific discussions would fetch a good price, Häberlein announced the fossil publicly and put it up for sale. He also actively contacted various museums around the world. At one point he tried to sell the new *Archaeopteryx* and the rest of his collection of Solnhofen fossils to the Yale Peabody Museum in the United States for 10,000 U.S. dollars. This was a huge sum at the time, and would equate to several million dollars today. The director of the Yale Peabody Museum at the time was the famous American paleontologist Othniel Charles Marsh. Marsh is chiefly known for mounting many expeditions to the American West, and describing numerous dinosaurs and extinct mammals in the infamous "Bone Wars" with his North American rival Edward Drinker Cope. However, Marsh was rather tight-fisted, and apparently did not respond to Häberlein's initial offer. Instead he offered Häberlein the much lower price of 1,000 German Marks through a middle man, for just the *Archaeopteryx*. This offer was turned down by Häberlein, and Marsh thus missed a singular scientific opportunity to purchase what would become one of the world's most famous fossils.

Meanwhile, German paleontologists were anxious to avoid having such an important fossil leave the country. Lacking the funds themselves, they contacted industrial magnate Werner Siemens who bought it for 20,000 German Marks, and donated it to the Humboldt Museum in Berlin. Thus the fossil stayed in Germany, and has since been known as the "Berlin *Archaeopteryx*." As mentioned above, it is one of the all-time greatest icons of evolution and probably the most widely published of all fossils. The skull of the Berlin *Archaeopteryx* is complete and revealed that it had indeed jaws with teeth, just a Huxley had suggested years before. The scientific description of the Berlin specimen was not published until 1897, by paleontologist Wilhelm Dames. Dames found that some anatomical features of the Berlin specimen were different from the London one. He thus described it as completely different genus and species, called *Archaeornis siemensii*, Latin for "Siemens's Ancient Bird." With the new scientific name, he honored Werner Siemens who paid for and donated the specimen.

Since then, further specimens of *Archaeopteryx* have been discovered, all of which stem from the same area in Bavaria. The third specimen was discovered in 1955, and the latest was published in 1995. All-in-all, ten different specimens are officially known to exist, although others may be hidden away in private fossil collections. One of the specimens was actually discovered in 1855, but was not recognized as an *Archaeopteryx* by a paleontologist until 1970—the strange story is told in the sidebar.

Mistaken Identity

Actually, the first specimen of *Archaeopteryx* was discovered in the Solnhofen limestone already in 1855 and described in a scientific paper in 1857—four years before the "London" specimen was even discovered! This specimen was later sold to the Teyler Museum in the town of Haarlem in the Netherlands. However, in the original paper, this "Haarlem specimen" was described as a pterosaur, an extinct flying reptile unrelated to birds. The paleontologist who described it, Hermann von Meyer, was in fact the same man who would announce and name *Archaeopteryx lithographica* in 1861! How could this kind of mistake happen? In retrospect, it was very understandable. First of all, the Haarlem specimen only consists of fragmentary wing and leg bones, and its feather impressions are extremely faint. Second, pterosaurs were not uncommon fossils in the Solnhofen deposits, and thus it was easy to assume that the new fossil represented a new kind of pterosaur. Third, *Origin of Species* had not been published, and thus the very idea of a transitional form between reptiles and birds was simply not part of the mindset of any scientist.

The Haarlem specimen did not come to the world's attention until 1970, when American paleontologist John H. Ostrom visited the Teyler Museum to study pterosaurs. Instead, he found an *Archaeopteryx*—only the fourth specimen known at the time. Ostrom's discovery spurred his further research into *Archaeopteryx* and the origin of birds and flight.

But this would not be the only time an *Archaeopteryx* would be mistaken for something else. The fifth "Eichstätt" and sixth "Solnhofen" specimens are both almost complete and well preserved, but with extremely faint feather impressions. Nonetheless, both were initially identified as young animals of the small meat-eating dinosaur *Compsognathus*, which is also known from the Solnhofen limestone. Only later did paleontologists recognize them as being specimens of *Archaeopteryx*. These two misidentifications demonstrate an important point: Without the characteristic long feathers of the wings, the skeleton of *Archaeopteryx* looks exactly like that of a small dinosaur, because while *Archaeopteryx* is the earliest known bird, it is also a small feathered dinosaur—perfectly intermediate between birds and dinosaurs.

Archaeopteryx also quickly became well known outside of scientific circles. For example, it appeared as a character in a French stage play in 1897. Since its discovery, *Archaeopteryx* has also featured in many popular books and textbooks on evolution as a prime example of an intermediate form between two animal groups. *Archaeopteryx* is considered vital to our understanding of the evolution of birds as a group, and the origin of bird flight. As a result of this, there is also a certain amount of prestige associated with publishing papers describing new discoveries or theories on *Archaeopteryx* among paleontologists and zoologists. This is witnessed by fact that the announcement of the latest, tenth specimen of *Archaeopteryx*, appeared in the 6 December 2005 issue of the prestigious scientific journal *Science*.

THE LIFE AND DEATH OF *ARCHAEOPTERYX*

> Let us imagine ourselves standing on one of the large islands in the Jurassic sea . . . a feathered creature launches itself from the top of a tree-fern.
>
> –Heilmann, 1926

A prime goal of paleontologists is to understand how extinct animals functioned, lived, interacted with each other, and eventually died. No other fossil animal has been so extensively studied in these regards as *Archaeopteryx*, except perhaps *Tyrannosaurus rex*, and some of our own hominid ancestors.

First of all: How did *Archaeopteryx* look when it was alive? Usually all that is left of a fossil are the hard parts, the bones, but in the case of *Archaeopteryx* we have more clues from its exceptional preservation. It must have looked very much like a modern bird; its body, wings, and upper part of the legs were covered in feathers, and just like in modern birds, the lower shinbone and foot was covered in scales. Interestingly, there are no traces of feathers around the skull of any of the fossils of *Archaeopteryx*. This could be because they were lost as the carcass floated around the lagoon. Another possibility is that the head of

Archaeopteryx was naked; we can imagine something either looking like a modern vulture or covered with scales like a reptile. The colors of *Archaeopteryx* are impossible to say, as it is not preserved in fossils. However, given the wide variety of colors in modern birds, we are spoiled for choice.

A closer look at the wings reveals, apart from the very un-bird-like fingers with long claws, that the long flight feathers on the wings are asymmetric—a condition which shows that *Archaeopteryx* was able to fly actively. All modern flying birds have asymmetric flight feathers; the flight feathers of birds who cannot fly and feathers which not used for flying (such those covering the body, for example) are symmetrical with vanes are of equal width. An important question immediately follows: How good a flyer was *Archaeopteryx*? Studies of the skeleton, especially the bones of the wing and shoulder girdle, where the necessary muscles for flight are situated, show that *Archaeopteryx* was capable of flapping flight. However, the size and shapes of the bones indicate it had relatively weak flight muscles. We can therefore assume that its method of flight was primarily gliding, with some flapping. Computations of its flight speed have shown that *Archaeopteryx* was relatively quick when flying, at some twenty-six feet per second. However, it was not very maneuverable. Research has also shown that *Archaeopteryx* was not capable of taking off directly from the ground, unlike modern birds. Its wing muscles were not strong enough to give it the required initial speed for take off. However, the situation was different if *Archaeopteryx* first climbed a tree and then took off by jumping from a branch. During the initial "controlled fall" of the jump, it would be able to get enough speed to become airborne and then begin flying. The skeleton and anatomy of *Archaeopteryx* support this conclusion. It has small size, which, combined with a hand which was good at grasping and possessed pointed claws, shows that *Archaeopteryx* was a good, swift climber.

The picture that has emerged of *Archaeopteryx* is one of an animal which primarily ran around on the ground and searched for food. If threatened by predators, it would swiftly run to a nearby tree or other vegetation and climb it. From here, it could take off by jumping from a branch and fly away to safety.

Another question posed by paleontologists and zoologists since its discovery is: How many kinds or species of *Archaeopteryx* are there? There appears to be a wide range of sizes within the fossils, ranging from the small Eichstätt specimen which is only half the size of the large London specimen. Does this means that they represent different species or is it just because one animal is young and the other is adult? To reveal the answer, researchers have studied minute details of the skeletons, such as the differences in proportions of the limbs or the shape of the teeth. The question is not merely academic. If there is just one species, it could indicate that *Archaeopteryx* was a relatively rare and possibly newly evolved form 150 million years ago. However, if there was different species of the same animal living together in the same small geographic area, it would indicate that birds had been around for some time, and had evolved and diversified into separate ecological niches at the time. The teeth of most of the fossil specimens indicate that *Archaeopteryx* ate insects with relatively soft bodies. However, the teeth of the Munich specimen have slightly more pointed tips, indicating that it might have eaten insects with tougher carapaces. This had led to it being considered a different species called *Archaeopteryx bavarica*. The Solnhofen specimen, which is very large, may have been capable of catching and eating small vertebrates and it has been proposed to be a completely different genus, called *Wellnhoferia*. However, researchers do not completely agree on the number of species, and today it is considered that there are between two and four different species of *Archaeopteryx*.

THE RELATIONSHIPS OF *ARCHAEOPTERYX*

> Were it not for those remarkable feather imprints, today both specimens would be identified unquestionably as coelurosaurian theropods [meat-eating dinosaurs].
>
> –Ostrom, 1976

Archaeopteryx represents the earliest known bird and thus the origin of bird flight. It is therefore crucial to answering two important questions: How did bird flight evolve? And why did it evolve? To answer these questions satisfactorily, one must realize that *Archaeopteryx* actually represents just one step, albeit a crucial one, in the evolution of bird flight. To unravel the entire history of this remarkable adaptation, it is therefore also necessary to know what came before *Archaeopteryx*. Which group of animals did it evolve from? Which adaptations, which could later be used for flying, were present in these ancestors? And why? Placing the origin of *Archaeopteryx* and birds within different groups will lead to different theories for the evolution of bird flight. To discover the correct sequence of events leading to the evolution of bird flight also necessitates that we correctly pin down which group of extinct animals *Archaeopteryx* evolved from.

This was clear to the researchers who studied *Archaeopteryx* and supported the evolutionary view of the world immediately after its discovery. It was clear that it was an intermediate form between birds and reptiles, but the question remained: Which reptiles? Researchers began studying and comparing the bones and anatomy of extinct and living groups of reptiles for clues to the origins of *Archaeopteryx*.

As mentioned above, Thomas Huxley presented the first substantial, well-researched inputs in the debate in 1868 and 1870. He not only compared *Archaeopteryx* to dinosaurs and various extinct reptiles, but also to living birds. A small, chicken-sized, meat-eating dinosaur called *Compsognathus*, which had also been discovered in the Solnhofen limestone, especially caught Huxley's attention. He pointed out the many bird-like characters of *Compsognathus* and other dinosaurs known at the time, and the dinosaur-like qualities of *Archaeopteryx* and birds in general. Huxley's suggestion that birds might have evolved from dinosaurs was widely discussed in the following years. Not all researchers agreed with Huxley. Some argued that the advanced characters of *Archaeopteryx* must have taken much longer to evolve, and thus indicated that its origins lay with groups older than dinosaurs; groups which could have been ancestors of both dinosaurs and birds.

One of the problems at the time was that relatively few fossil vertebrates were known at the time, and almost nothing from outside Europe. It was not until from the 1870s onward that huge numbers of dinosaurs and other extinct reptiles were discovered in North America. Most of these were large and impressive species, which easily caught the public and scientific attention. There was a dearth of fossils of very small forms, which could include specimens that might shed further light on the relations of birds. Overall, this meant that theories and hypotheses of the origin of birds had to be built on scant material.

Another problem was that fossils documenting the evolution of birds after *Archaeopteryx* were also very few. In fact, only two other fossil birds from the age of the dinosaurs were well known at the time. Both were from 70 to 90 million-year-old deposits in North America and thus much younger than *Archaeopteryx*. One of them, *Ichtyornis*, was the size of a gull, and its skeleton was relatively modern looking; for example, the wings had lost their claws and the finger bones of the wing were fused together. It also had a short pygostyle tail. The other one, *Hesperornis*, was very different: It looked superficially like a one-meter-long wingless penguin, and was a very specialized diving bird which used its powerful hind limbs to swim with. While the overall anatomy of the skeletons of *Ichtyornis* and *Hesperornis* were quite evolved, both were primitive in one regard: Just like *Archaeopteryx*, they still had teeth in their jaws. Unfortunately between these two fossil forms and *Archaeopteryx*, there was a gap of some 60 to 80 million years where nothing was known about bird evolution. This meant that discussions on the evolution of birds and bird flight centered on *Archaeopteryx* and resulted in a lot of hypothetical theorizing about intermediate forms.

Another group of fossils discovered in late 1800s and earliest 1900s became contenders for the title of bird ancestors. They were called the pseudosuchians ("false crocodiles") and are a mixed group of reptiles some of which are the ancestors of modern crocodiles. They were widespread in the early part of the Triassic period, 230 to 250 million years ago, just before the rise of the dinosaurs. The pseudosuchians were not actually very bird-like, but their skeletons are generally very primitive and could therefore easily be constructed as evolving into something looking like *Archaeopteryx*.

The person who effectively shut the debate on bird origins down for almost fifty years was the Danish artist Gerhard Heilmann. Heilmann was an extremely talented freelance artist who, among others, illustrated several books, and the series of Danish banknotes in use between 1913 and 1945. He was also an interested amateur bird-watcher, and illustrated several books on birds. Heilmann also became interested in the question of the origin of birds, but when he found that there was no agreement between the professional researchers, he started to conduct his own research into the problem to solve it. This resulted in a series of popular articles in Danish in the *Journal of the Danish Ornithological Society* titled "Our Current Knowledge on the Origin of Birds" from 1913 to 1916. Heilmann used his artistic skills to the full, and lavishly illustrated the articles with beautiful figures. In the papers, he delved into every aspect of anatomy of birds and various living and extinct reptiles; he not only compared skeletons and bones, but also the evolution of embryos and various organs, and the structure of feathers and scales.

When comparing *Archaeopteryx* and birds to various living and extinct reptiles, Heilmann's studies initially led him to the same conclusions as Thomas H. Huxley had some forty years before: *Archaeopteryx* most closely resembled dinosaurs, specifically a group of small two-legged meat-eating dinosaurs called coelurosaurs. He noted similarities in the skulls, the legs, hips, the proportions of the arm to the leg, and even favorably compared footprints of birds with fossil footprints of dinosaurs. In fact, Heilmann piled similarity upon similarity and fact upon fact, which could support a close relationship between dinosaurs and birds. However, he then went on to reject the theory of the dinosaurian ancestry of birds completely by invoking "Dollo's Law." At the time Heilmann was writing, some paleontologists and evolutionary biologists tried to formulate a number of "laws." These were intended to be incontestable statements and rules, which could be used to govern the research within their field, in clear emulation of the "laws" of physics (Newton's laws, etc.). One of these laws was named after the Belgian paleontologist Louis Dollo. It basically stated that once a group of animals in the course of evolution had "lost" an organ or other anatomical structure (for example, a tail or a specific bone) then it could not re-evolve that organ later. An organ or structure could of course get a new function during evolution and as a result develop a new shape, but the original organ or structure could not reappear or revert to its original function. This "law" could be used to test theories of evolution, which stated that one group of animals had developed from another. If all the organs and structures in the descendants were present, although in a primitive shape, in the proposed ancestors, then the theory might be correct. However if a structure or an organ was present in a group of animals, but not in their proposed ancestors, and the structure could not be shown to have evolved from one already present in the proposed ancestors, the theory was wrong.

For Heilmann the structure that was missing in dinosaurs were the clavicles. Clavicles are a pair of bones in the shoulder girdle, which in birds have fused into a unique structure: the wishbone. Among other functions, the wishbone supports the flight muscles of the wing while the bird is flying; *Archaeopteryx* has a wishbone. However, at the time of Heilmann's writing, no one had described clavicles or a wishbone in a dinosaur. Instead, using "Dollo's Law," Heilmann concluded that because of the apparent lack of this singular feature, dinosaurs could not be the ancestors of birds. All the similar features and structures, which birds and dinosaurs shared, must instead be the results of convergent evolution. Convergent evolution is the process whereby two otherwise unrelated groups of animals have developed superficially similar features and structures, because their mode of life is similar.

Instead Heilmann supported the pseudosuchians as bird ancestors. Not because they actually were more bird-like than the dinosaurs, but because they did not lack any key features, as dinosaurs apparently did. Pseudosuchians were known to have square, block-like clavicles, which although not very wishbone-looking at all, had the potential to evolve into a wishbone. The same was true for the relatively unspecialized skull of the pseudosuchians, which could gradually evolve into that of *Archaeopteryx*. Heilmann constructed a hypothetical intermediate between a pseudosuchian and *Archaeopteryx*, which he dubbed "the pro-avian" or "before-bird." He also speculated about the lifestyle of this "pro-avian," which basically looked like four-legged reptile with long, fringed scales on the arms, legs, and tail. The scales would enable it to glide between trees and would later evolve into feathers. Heilmann was not the first researcher who constructed a hypothetical intermediate "pro-avian" between reptiles and birds. However, his specific reconstruction would influence most of the later ones of this purely theoretical animal, which has never been discovered as a fossil.

As mentioned above, the series of papers in the *Journal of the Danish Ornithological Society* were written in Danish and thus had a fairly limited audience and Heilmann's studies might have had little further impact. However, in the course of these, Heilmann had corresponded with a number of leading international paleontologists and zoologists around the world. They in turn encouraged him to publish his studies in

Reconstruction of the small four winged meat-eating dinosaur Microraptor, which was about the size of a blackbird. (© Anne Haastrup Hansen. Used by permission.)

English also, and Heilmann set out to revise his material. He also traveled to Berlin to study the specimen there, and made some new anatomical discoveries. The result was a 208-page book, titled *The Origin of Birds*, which was published in 1926. It was a re-edited and improved version of the series of popular papers but with the same overall conclusion. Heilmann's well illustrated and apparently very thoroughly researched book convinced everybody, and it appeared that the final word in the debate had been said. By modern standards Heilmann's book contains some mistakes, and is somewhat superficial, but it still contains a number of interesting insights and wonderful illustrations and is well worth a read, which is witnessed by the fact that it was reprinted as late as 1972.

In the following years research into dinosaurs and other fossil animals went into decline due to the economic difficulties of the Great Depression in the 1930s and later World War II. Concurrently, the public image of dinosaurs gradually changed from lively, active animals to cold-blooded, sluggish evolutionary failures, which were doomed to extinction. Research in vertebrate paleontology focused on mammals and their origins instead.

The debate on the origin of birds was not reopened until the early 1970s. This was occasioned by the discovery of a new and very bird-like kind of meat-eating dinosaur in the late 1960s by the American paleontologist John H. Ostrom of Yale University. These new dinosaurs were called dromaeosaurs. They were relatively small, but had a skull with a relatively large brain, long arms, and a long, stiff tail. Finally, they possessed a giant sickle-shaped claw on each foot, which prompted Ostrom to give the first dromaeosaur he described the Latin name *Deinonychus*, which means "Terrible claw." To Ostrom, the whole anatomy of the animal's skeleton produced a picture of a very active, aggressive hunter, which used the large claws on its feet to kick deep wounds in its prey, while using its stiff tail like a balancing rod. Ostrom went further with this new information, and started critically reviewing the accepted assumptions about the biology of dinosaurs. For more than forty years, dinosaurs had been as having a reptile-like ectothermic or "cold-blooded" physiology. Simply stated, reptiles are unable to produce their own body heat, but need an external source of heat, such as the sun, to warm their body before they can become physically active. In contrast, endothermic or "warm-blooded" animals such as mammals and birds are able to produce their own body heat. This means that endothermic animals can be fully active during the night or in cold conditions, which gives them a distinct evolutionary advantage over ectothermic animals. The drawback to being an endothermic animal is that they need approximately ten times more food than ectothermic ones. Based on his new studies, Ostrom suggested that the dinosaurs had also been endothermic or "warm-blooded." His suggestions immediately raised a huge debate among paleontologists.

In 1970 Ostrom discovered a "new" specimen of *Archaeopteryx*, during his visit to a museum in the Dutch town of Haarlem (see sidebar). During his studies and description of this new specimen, he started noticing many anatomical similarities between *Archaeopteryx* and the dromaeosaurs, the new meat-eating dinosaurs he had just discovered. There were minute details, such as the almost exact similarities in the proportions of the arms and the shape of the bones of the wrists, shoulders, hip, and foot. To Ostrom, the dinosaurs and especially the dromaeosaurs began looking more and more bird-like, and he began to suspect that Heilmann and previous researchers had been wrong in dismissing the dinosaurs as ancestors of birds. But there was still the question of the absence of clavicles in dinosaurs. Or was there? In fact, paleontologists *had* described at least three dinosaurs with clavicles. The first had been described in 1924, another in 1936, and finally one in 1972 in the meat-eater *Velociraptor*. *Velociraptor* would later turn out to be a dromaeosaur. Based on this evidence, Ostrom stated in a scientific paper in 1976, that there was no longer any evidence against dinosaurs as ancestors of *Archaeopteryx* and thus all modern birds; in fact the evidence for the dinosaur-bird link was much better than that supporting the "pseudosuchian" hypothesis.

Ostrom's suggestions set off a new heated debate about the origin of birds. Basically, researchers were split into three camps: One group promoted dinosaurs as ancestors, another defended the traditional "pseudosuchian" theory, and finally one group suggested that the origins of birds should be found among the so-called crocodylomorphs.

The "crocodylomorphs" were a group of crocodile-like reptiles, which include the ancestors of living crocodiles. As little new fossil bird material had been discovered since Heilmann wrote his book, *Archaeopteryx* once again became the natural focus of theories. The debate raged and, most important, spurred much research into previously ignored or neglected areas concerning the anatomy of birds, dinosaurs, and reptiles and the mechanics of flying in birds. In turn this meant that much new information was gathered, and many previous false assumptions were corrected. The debate culminated temporarily in a scientific meeting in 1984 in the town of Eichstätt in Germany. The meeting was housed at Jura Museum, where one of the *Archaeopteryx* specimens was kept. It lasted five days while paleontologists, biologists, zoologists, ornithologist, geologists, and other researchers discussed every aspect of *Archaeopteryx* and the origin of birds: Did *Archaeopteryx* fly or glide? And if so, how well could it fly? What kind of environment did it live in? How was the fossil preserved? What are the closest relatives of *Archaeopteryx* and modern birds? How did flight evolve in birds and other vertebrates? The event was very remarkable, as until then no single fossil had had an entire five-day-long conference dedicated to it. It also resulted in a 380-page book with scientific papers about *Archaeopteryx* and the issues discussed at the meeting. The conference failed to produce a general agreement on the ancestry of *Archaeopteryx* and birds. However, it did see the demise of "crocodylomorph theory." Research showed that details in the anatomy of the skull prevented a close relationship between *Archaeopteryx* and crocodylomorphs.

In the years after the conference the evidence for a dinosaurian ancestry of birds steadily mounted. Many dinosaurs were discovered to have hollow bones with air sacs. Air sacs are cavities connected to the lungs of birds, which making their oxygen intake much more efficient than that of reptiles and mammals. Air sacs also intrude into the bones and lighten them. They were previously thought to be unique to birds, but turned out to be extremely widespread among dinosaurs. Not only meat-eating dinosaurs possessed them; the distantly related gigantic, four-legged, long-necked, and long-tailed sauropods turned out to have bones riddled with air sacs. In the 1990s another stunning discovery was made; several kinds of meat-eating dinosaurs do in fact have wishbones! Not just clavicles, but actual wishbones. This evidence completely obliterated Heilmann's only argument against a dinosaur origin of birds. Of course, today there are many more kinds of dinosaurs known than in Heilmann's time, especially smaller forms. One cannot help wonder, that if Heilmann had had today's information available to him, he would probably have come to a different conclusion.

Furthermore, from the 1980s onward, modern computer-based analyses of relationships, called cladistics (see "Cladistics") repeatedly revealed that the closest relatives to *Archaeopteryx* and birds could be found among the theropods; the meat-eating dinosaurs. Specifically one group: the dromaeosaurs. The characters uniting dromaeosaurs with birds are many. Dromaeosaurs have relatively long arms, around fifty to eighty percent of the length of their legs. This is longer than most other dinosaurs and almost as long as the arms of early birds such as *Archaeopteryx*. They have very long hands; the second finger is as long as upper arm. And as was recently discovered in the tenth "Thermopolis" specimen of *Archaeopteryx*, both it and the dromaeosaurs have a toe with a sickle claw on their foot, although the claw of *Archaeopteryx* is not as big as in dromaeosaurs.

The final piece of evidence, which definitively proved the theory of a dinosaur origin of birds arrived in mid-1990s. In the Liaoning province in Northeast China, spectacular new fossils were discovered in 125-million-year-old geological deposits, approximately 25 million years younger than the ones at Solnhofen which yielded *Archaeopteryx*. Like the geological deposits at Solnhofen in Germany, extraordinary and poisonous circumstances had allowed the preservation of fossils. Unlike Solnhofen, this was not an ancient lagoon, but a lake. Around 125 million years ago in the present Liaoning area nearby volcanoes would erupt from time to time and spew a deadly cocktail of poisonous gases and ash onto the lake and its surroundings. The animals which lived around and above the lake were killed off by the ash and gases, and their dead bodies ended up in the mud at the bottom of the lake: birds, dinosaurs, mammals, and pterosaurs. Again the lifeless environment at the bottom of the lake would preserve the animals as exquisite fossils. Thousand of birds with their feathers and mammals with their fur and whiskers preserved as black imprints and pterosaurs showing remains of soft tissue of the wings. But most important: small meat-eating dinosaurs with clear black imprints of fossil feathers along the body and on their arms! At first, opponents of the bird-dinosaur link were quick to suggest that the impressions were faked or something completely different. However, detailed comparison with the feather impressions on the fossil birds (which no one doubted were feathers, since they were found on what were clearly extinct birds), showed that the impressions around the dinosaurs were clearly the fossil remains of feathers. This revelation was probably the best evidence for the theory that birds evolved from dinosaurs. It also meant that the popular image of at least the smaller dinosaurs had to be changed. Instead of the two-legged scaly reptiles, they could now be envisioned as two-legged feathered bird-like creatures.

In contrast to the gathering evidence for a dinosaurian origin of birds, the proponents of the "pseudosuchian" origin of birds failed to produce any convincing evidence in favor of their theory. Although they continued to make attacks on the "dinosaurian" theory, no fossils turned up which convincingly supported the "pseudosuchian" theory. The proponents did not produce any cladistic analyses, which showed that *Archaeopteryx* and birds was more closely related to the "pseudosuchians." This clearly showed the major weakness of the "pseudosuchian" theory; it was completely unable stand up to rigorous modern, computer-based testing. In fact later cladistic analyses of the "pseudosuchians" have clearly revealed that they are not a natural group, but an assembly of unrelated animals. The anatomical characters used as "evidence" for the pseudosuchian origins of birds come from various animals that did not have anything to do with each other; a further nail in the coffin of this particular theory.

Nor has a four-winged pseudosuchian "proavian" with elongated scales on legs and arms turned up as a fossil in the intervening years. Ironically a dinosaur, which fits the characteristics of the "proavian" has. In 2003 Chinese paleontologists described a new tiny dromaeosaur, called *Microraptor. Microraptor* also derives from the Liaoning deposits and is the size of a blackbird. What is especially interesting about it, is the fact that has long feathers attached not only to its arms but also to its legs—it has four wings! And the flight feathers have asymmetric vanes, showing that they belong to animal which could glide and possibly fly. Just what would be expected of the hypothetical "proavian." However, cladistic analyses of relationship indicate that *Microraptor* or its kind did not develop into birds. Although closely related, it represents an evolutionary lineage, which went its own way in the development of gliding or flying and later became extinct without leaving any descendants.

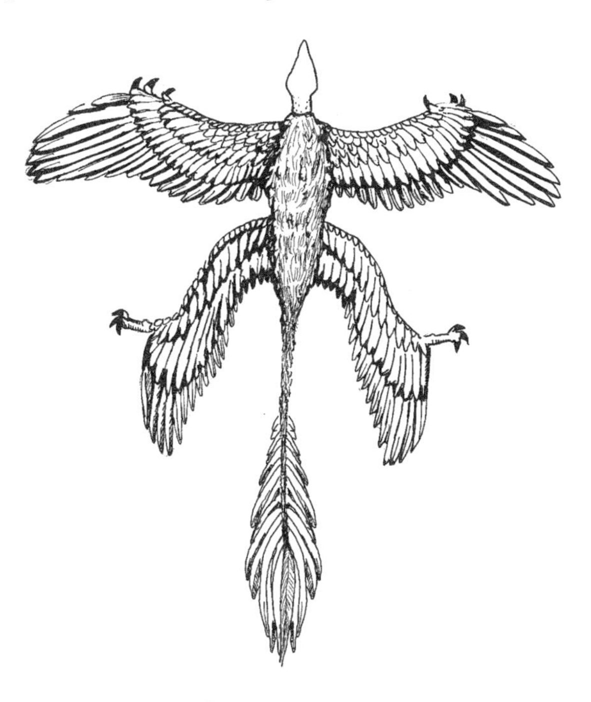
Reconstruction of Microraptor as seen from above. (© Anne Haastrup Hansen. Used by permission.)

Since John Ostrom reopened the debate on bird origins in the 1969, research into new fossils and re-interpretation of old ones have conclusively shown that dinosaurs are the ancestors of birds. Interestingly, this conclusion is exactly the same which Thomas Huxley reached in 1868–1870, less than nine years after the first specimen of *Archaeopteryx* was discovered. However, as described above, for a long period thereafter paleontologists and zoologists subscribed to different and erroneous theories of bird origins due to lack of well-preserved fossils and incorrect scientific approaches.

The fossil evidence which has been gathered since 1969 has revealed that none of the anatomical features usually thought unique to birds, such hollow bones with air sacs, a wishbone, and finally feathers, are all found among dinosaurs. Another typical "bird feature," the toothless bill, was not present in the earliest known bird, *Archaeopteryx*. Instead it appears that this feature has evolved a number of times in later birds, the earliest known example being the 125-million-year-old Chinese form *Confuciusornis*. While no dromaeosaurs have yet been discovered with a toothless bill, a number of related groups of dinosaurs, called oviraptorids and struthiomimids, did evolve toothless bills.

The rediscovery of the fact that birds are the direct descendents of the dinosaurs has several implications. First of all, we should not consider birds only as "birds," but also as dinosaurs, albeit highly evolved ones. In turn this means that the dinosaurs did not become extinct some 65 million years ago. On the contrary, they are alive and kicking and are one of the most successful groups of vertebrates with an estimated 9,600 species living today. Furthermore, this realization that birds *are* dinosaurs means that paleontologists and zoologists can now drastically expand our knowledge of the extinct dinosaurs. They are no longer restricted to basing theories of behavior, physiology, and zoology just on studying fossil bones. Instead, it is now possible to make direct comparisons with living representatives of the group, the birds. For example, fossil "nesting colonies" have been excavated, where several dinosaurs of the same species have been discovered buried in and around a group of nests with eggs. This behavior becomes easier to understand, because we can see the exact same kind of behavior in living birds, such as gulls. The ability to make well-founded interpretations of dinosaur behavior and biology by making direct comparisons to their living relatives is perhaps the most of important result of the rediscovery of the dinosaur-bird link.

ARCHAEOPTERYX AND THE EVOLUTION OF BIRD FLIGHT

> No other fossils have had more impact on the progress of biological thought than those of *Archaeopteryx*
>
> –Elzanowski, 2002

As far as is known, during the long evolutionary history of vertebrates, only three groups have attained "true flight": birds, bats, and the extinct pterosaurs. "True" flight is also called "active" or "flapping" flight because the animal actively beats is wings when it moves through the air. The beating of wings creates a physical force known as lift, which allows the animal to stay in the air and move forward at the same time. This is opposed to simple gliding, where the animal does not beat its wings and, although able to control direction to a certain degree while gliding in the horizontal direction, is not able to gain height.

In pterosaurs and bats the wings are made up of a single part, a skin membrane which extends between the fingers and the body. In birds the wing is made up of many individual parts: feathers. Thus the study of the evolution of flight in birds is inextricably linked to the evolution of feathers. However, it is important to realize that feathers were not originally intended for the use of flight. Unfortunately, many researchers have tried to explain the evolution of feathers by only considering them as adaptations for flying. This stems from our common, but mistaken, assumption that evolution tends to be directed toward some "end result." This is nonsense, as evolution does not "plan ahead" and partially evolve a structure for "later use." Instead, evolution proceeds through many steps in random directions, during which various biological structures are improved, maintained, or reduced. Structures can have several functions, and during each step each and every structure must function as part of an integrated whole. During each step, a structure or combination of structures, which gives the individual animal a slight advantage over its kin, will have a greater likelihood of being passed on to its descendants. Over time within a group of animals, this results in more optimal configurations of structures suppressing the lesser ones. However, a structure which originally had one function can begin to be used for other functions—this process has been called "exaption." The exaption of one organ or group of organs for a new purpose often allowed animals to expand into new environments and marked the appearance of radically new groups of animals. For example, 375 million years ago a group of fish gradually began entering shallow water swamps, which were clogged by branches and other obstacles in the form of fallen vegetation. In turn, their fins gradually evolved into stronger, limb-like organs, which allowed the fish to move more efficiently among the obstacles. Coupled with the evolution of lungs for true air-breathing, their new limbs were then "exapted" to act as true legs for walking on land. This allowed them to expand further into the vast, unoccupied environment of dry land and marked the origin of the tetrapods—the four-limbed vertebrates. In the case of birds, feathers appear to originally have developed for insulation against warm and cold. They were later successively exapted for nesting coverage, then display, then for braking and steering while jumping and gliding between trees and branches, and finally, for true flight.

Archaeopteryx has figured heavily in the debate on the origins of bird flight ever since its discovery and continuing up to today. For example, in the relative recent book *Taking Wing* from 1998, Pat Shipman reviewed the origin and mechanics of flight through *Archaeopteryx* and the history of its discovery and the scientific debates it spurred. However, the main problem with *Archaeopteryx* is that, on its own, it does not actually provide many clues to the history of bird flight. To understand the complete evolutionary history of birds, we also need to know what came before and after *Archaeopteryx*. As described above, until the recent discoveries in China, the lack of fossils which could describe these stages hampered our understanding.

Today, if we combine our knowledge on the flight ability of *Archaeopteryx* with studies of other fossil dinosaurs and birds, we get the following approximate sequence for the evolution of feathers and of bird flight:

The first stage was the appearance of primitive "proto-feathers" in small meat-eating dinosaurs. Fossil evidence for this stage derives from the magnificently preserved Chinese dinosaurs from Liaoning. Although geologically younger than *Archaeopteryx*, modern cladistic analyses of relationships have shown that they represent a more primitive group of meat-eating dinosaurs, which originated before *Archaeopteryx*. Information on the plumage of these fossils can thus give information on the earlier, primitive stage of feathers, long before the evolution of the specialized flight feathers seen in *Archaeopteryx*. Studies of the shape of the "proto-feathers" in the Chinese dinosaurs have shown them to be hair-like, cylindrical structures. This cylindrical shape is reflected by the early growth of feathers in living birds. When a feather first develops in the skin of the bird, it is in the shape of a cylinder which is curled upon itself and encased in a cylindrical sheath. Only later does the feather fold out into a flat structure when it breaks out the sheath and is free of the skin. This early cylindrical stage harks back millions of years, when feathers first appeared among their ancestors. The evolutionary origin of the proto-feather structure was a fortuitous mutation of the scales in a group of small meat-eating dinosaurs, which resulted in the appearance of longer, fur-like scales on their bodies. These were better at keeping the animal insulated than ordinary scales and gave the small, early dinosaur an evolutionary advantage.

The next stage is marked by the appearance of longer feathers along the arm and tail in fossils of more advanced meat-eating dinosaurs. These structures were a further development of the proto-feathers into long, branching ones more akin to the feathers seen in modern birds. The exact reason for this development is not known, but a reasonable hypothesis has been made recently by Thomas Hopp and Mark Orsen: The long feathers on the arms were used for nesting coverage when brooding. Fossils of dinosaurs have been found in Mongolia literally sitting on their nests, where they were buried alive during a sandstorm while trying to protect their eggs. The size of the nest is such that it is impossible for the dinosaur to protect it all its eggs against the effects of sun and rain. However, if the arm is covered in long feathers, the animal is able to cover all its eggs and nestlings, just as nesting and brooding birds do today. This would also immediately result in a selective pressure toward longer feathers on the arms, because animals with long ones would get more surviving offspring than animals with shorter feathers. At the same time, the longer feathers on arms and tail could also be used for display purposes—the use of one feature for several tasks is quite common in the animal world. For example male dinosaurs can easily be imagined to make displays with their feathers during the mating season to attract mates, or the feathers could be used to try to look bigger and scare off predators, just as birds also do today.

Having attained longer feathers on the arms, the next stage of exaption could have happened while the dinosaur was jumping between branches in trees. Many small meat-eating dinosaurs have all the characteristics of good tree climbers: lightweight bodies and relatively long arms with grasping hands and sharp claws. While jumping between or down from branches and tree trunks, the long feathers on the tail and arms would increase their surface, and thus function as useful airbrakes or steering mechanisms. Again this would immediately result in selective pressure to evolve more complex feathers, with shafts and small hooks within the barbs of the feather, which would help to keep a stiff surface.

From jumping between branches using the feathers on arms and tail for balance and steering, there is not far to go to the next stage: simple gliding. This is the process where an animal jumps from a tall structure and uses some parts of its body as a surface or wing to fly with in the horizontal direction. During the glide, the animal has some control over the direction of the glide, but is unable to gain height. Through time, a number of vertebrates have reached this stage, for example flying squirrels, which use a fold of skin suspended between their forearms and legs as a wing. *Microraptor* certainly represents this stage in the evolution of flight, although it was not directly on the evolutionary line leading to birds. Rather it is "branch" of the dinosaurian tree, which would eventually prove a dead end.

The next stage after simple gliding is called "active" or "flapping" flight because the animal beats is wings, when it moves through the air. The beating of wings creates the physical force known as lift, allows the animal to stay in the air, or gain height, and move forward at the same time. As mentioned above, within the vertebrates, only birds, bats, and pterosaurs have reached this stage. *Archaeopteryx* represents an interesting sub-stage in this development, as its wings and skeleton clearly show it was capable of active flight, but not of taking off from the ground under its own power. This next crucial sub-stage was achieved no later than 25 million years after *Archaeopteryx*, as is evidenced by a number of birds from the above-mentioned Chinese deposits. Studies of their skeletons clearly show anatomical adaptations in their shoulder girdle, which would allow them to take off from the ground under their own power.

Once birds had achieved this stage, they were quickly able to diversify into a variety of forms, although they had to share the sky with the pterosaurs, until these became extinct at the end of the Cretaceous period, 65 million years ago. Since then, birds have effectively ruled the air with insignificant competition from their mammalian counterparts, the bats, which evolved as recently as 54 million years ago.

ARCHAEOPTERYX AS AN ADVOCATE FOR THE THEORY OF EVOLUTION

> In conclusion, I must add a few words to ward off Darwinian misinterpretation of our new Saurian. At first glance of the *Griphosaurus* we might certainly form a notion that we had before us an intermediate creature, engaged in the transition from the Saurian to the bird. Darwin and his adherents will probably employ the new discovery as an exceedingly welcome occurrence for the justification of their strange views upon the transformation of animals. But in this they will be wrong.
>
> –Wagner, 1861

Archaeopteryx has been at the forefront of the struggle between science and anti-evolutionism since immediately after its discovery. The above statement was made by a German geologist, Professor Johann Andreas Wagner, immediately after von Meyer had published his initial notification of the new fossil, but before the *Archaeopteryx* had been even properly described in a scientific paper. Wagner had not even seen the fossil and furthermore committed a scientific *faux pas* by giving the fossil a new name, *Griphosaurus problematicus* (meaning "problematic riddle lizard"), despite the fact that von Meyer had already named it according to established scientific conventions. The clearly transitional nature of the animal of course flew in the face of anti-evolutionists—here was a near-perfect example of a fossil which is intermediate between two animal groups. As mentioned above, anti-evolutionist Sir Richard Owen, who had the honor of publishing the first proper scientific description of *Archaeopteryx*, tried to describe it as a modern bird, and not comment on its clear intermediate reptilian characteristics. In the end he just left himself open to severe corrections by his peers, such as Thomas H. Huxley.

The most recurring type of anti-evolutionist attack on *Archaeopteryx* is the accusation that the feather imprints seen on the fossil are forgeries. These accusations began shortly after the fossil appeared, partly as a result of the fact that Carl Häberlein did not allow anyone to photograph, nor make a detailed drawing of the specimen before he had sold it. The most recent attack of this kind began when a group of physicists led by Fred Hoyle claimed that *Archaeopteryx* was no bird. Instead it was a small dinosaur skeleton, which had been fitted out with feathers. Their conspiracy theory went that forgers in the nineteenth century, possibly the Häberlein family, had "improved" the fossil to make more money. The forgers had made impressions of modern feathers into a soft material around skeleton of small dinosaur and sold it off as a fossil bird. At the time of the intensive discussion on Darwin's brand new theory of evolution, any intermediate-looking fossil between birds and reptiles would of course dramatically increase its price. The attack was followed closely by fundamentalist creationists, who would like to see one of the greatest proofs of the theory of evolution turn out to be a simple forgery. However, it quickly turned out that several of the claims made by the Hoyle group were extremely badly researched. For example, they stated that feathers were only present on the London and Berlin specimens. However, as was pointed out by paleontologist Siegfried Rietschel, there were also feather impressions on the Teyler, Maxberg, and Eichstätt specimens of *Archaeopteryx*, the latter two of which had been discovered after the Häberleins were long dead and gone (thus they could not have forged these two). Also, repeated attempts at recreating the feather impressions using materials available to putative forgers during the nineteenth and twentieth centuries consistently failed to produce the intricate, microscopic details seen in the feathers of the fossil. Geologists studied the composition of the limestone in and around the fossil feathers and compared it both to the limestone on the rest of the fossil and in other limestone slabs from the Solnhofen area. They found no differences, neither in composition nor in the structure, again indicating that the impressions are not later additions by forgers.

The accusations by the Hoyle group are frequently still touted by anti-evolutionists as indications that *Archaeopteryx* is "dodgy evidence" of evolution. They have, not surprisingly, failed to mention the heap of scientific evidence and publications which later have definitely proved that *Archaeopteryx* is not a fake. And of course, more *Archaeopteryx* fossils with feather impressions have been excavated since then. These include the magnificent "Thermopolis" specimen first described in 2005, which has very well-preserved feathers. Again, these cannot be the work of the original forgers. Furthermore, the recent discoveries of dinosaurs and birds with imprints of fossil feathers from China, which are completely unconnected to the German ones, add further support to the dinosaur-bird link and place any accusations by the anti-evolutionists well within the large box containing very untenable conspiracy theories.

CONCLUSION

The importance of *Archaeopteryx* as an icon of evolution cannot be understated. From a paleontological and zoological viewpoint it marks the origin of a very successful, major vertebrate group: birds. Also, despite nearly 150 years of continued fossil discoveries, no fossil bird has been found that is older and more primitive than *Archaeopteryx*. Thus it remains the geologically oldest known bird. The question of the true relationships of *Archaeopteryx* has spurred countless studies of areas related to the origin of birds, bird flight, and feathers, which in turn have given us a much deeper insight into subjects such as bird flight, feather evolution, and the macro-evolutionary processes which lead to the emergence of radically new animal groups. It is now also possible for paleontologists to gain a much clearer understanding of the biology and physiology of the dinosaurs, through direct studies of their living descendants, the birds.

From a historical point of view the discovery of *Archaeopteryx* just two years after the publication of the first edition of *Origin of Species*, could not have been more fortuitous. Its discovery came at the time when the theory of evolution was beginning to dramatically change the common view of the world. *Archaeopteryx*, more than anything else, helped vindicate Darwin's theory, despite the fact that anti-evolutionists tried to disprove its nature almost exactly from the moment it appeared on the scientific stage. Despite these attacks, it has stood the test of time and is one of the best examples, if not *the* best example, of a Darwinian "transitory form."

Finally, from a purely aesthetic point of view, some of the specimens of *Archaeopteryx*, such as the Berlin one, have the important iconic quality of being simply beautiful to look at.

FURTHER READING

Currie, P. J., Koppelhus, E. B., Shugar, M. A. and Wright, J. L. *Feathered Dragons. Studies on the Transition from Dinosaurs to Birds* (Bloomington: Indiana University Press, 2004).

Heilmann, G. *The Origin of Birds* (reprint) (Mineola, NY: Dover, 1972).

Proctor, N. S., and Lynch, P. J. *Manual of Ornithology* (New Haven, CT: Yale University Press, 1993).

Shipman, P. *Taking Wing:* Archaeopteryx *and the Evolution of Bird Flight* (New York: Simon and Schuster, 1998).

The Fossil Record

Olof Ljungström

There always was a limestone mountain just outside Paris. Historically it used to be an execution ground, where bodies were dumped in one of the numerous lime pits. This included Christian martyrs, including St. Denis, the patron saint of Paris. Consequently the mountain became known as "Martyr Mountain," Montmarte. This was long before the modern metropolis caught up and engulfed it in the nineteenth century, when it also became Paris's main venue for risqué entertainment.

But there were other bodies to be found on Martyrs' Mountain, beside those of criminals and saints (past and present). The limestone is choked with the fossils of long-dead organisms. This was realized by the quarry-men working there, whose interest was whetted by the possibility of selling fine specimens to noblemen with an interest in natural history. The Revolution of 1789 put a sudden end to this interest in fossils, but it was not long until, in 1807, two men equipped with a new scientific agenda, again started haunting the chalk-pits of Montmartre. This time they bore middle-class names like Cuvier and Brongniart, just like the men who had made the Revolution, and they worked in a new public institution of science, the Paris Museum of Natural History, erected by the revolutionaries for the benefit of the people through the state.

The objective of this chapter is to provide a brief history of the rise of paleontology in the early nineteenth century, and the fossils that form its main area of study. It also includes as general discussion of what fossils are and the uses to which they have subsequently been put. It concludes with a brief discussion of a part of the debate over evolution and the fossil record that in recent years has been conducted between scientists working on macro- and micro-evolution. On the one hand, a number of paleontologists strive to make the theory of evolution work better for their macro-evolutionary concerns, and on the other a group of evolutionists more concerned with experimentalism and micro-evolution. The former have formulated additional mechanisms for evolution to complement natural selection, while the latter uphold natural selection as the primary mechanism of evolution through slow adaptation.

In the first decade of the nineteenth century fossils became a main material of study for the professor of comparative anatomy of the Paris Museum for Natural History, Georges Frédéric Cuvier (1769–1832), and his colleague the professor of mineralogy, Alexandre Théophile Brongniart (1770–1847). Their interest was part of a new departure in the research at the museum with Cuvier's activities as its center of gravity. He had first decided to prove the "reality of extinction," in Martin Rudwick's words, of all sorts of curious prehistoric species of animals (Rudwick, 1992). A survey of the state of zoological knowledge made him decide to take the elephants as his point of departure. They were too big to be easily overlooked, so future discoveries of hitherto unknown species of them were unlikely. At the same time curious remains of elephants were known from both North America and Siberia, where they certainly did not live at present. Surveying this material as completely as possible Cuvier established not only that the Indian and African elephants were distinct species, with considerable differences in, for instance, the structure of the teeth, but there had once existed a number of species of elephant distinct from either, most importantly the American mastodon and the mammoth, common to the Old and New World alike. The elephant problem had been one of his reasons to start delving into the Montmartre limestone in the first place.

Having established the reality of extinction in the case of elephants, Cuvier's research had progressed to an interest in those other fossils being unearthed. He had brought his mineralogist colleague Brongniart along as a specialist on the stratigraphic chronology of rocks necessary for their relative dating. Together they collected an impressive set of prehistoric fossils of now-extinct mammals. We know them as examples of the mammalian mega-fauna which replaced the dinosaurs upon their sudden extinction. Cuvier reconstructed these alien animals, various kinds of *paleotherium* (ancient beasts), as he named them, articulating their fossilized bones on paper, filling in muscle and hide. He thus initiated the science of (vertebrate) paleontology as an extension of his chosen field of comparative vertebrate anatomy.

At the same time Alexandre Brongniart struck upon a still-valid principle of paleontology, that of "index fossils." What Brongniart realized was that not only could the strata of rocks be ordered in a relative chronology, older at the bottom, younger on top, but as some of these strata were fossil-bearing, the fossils they contained could also be ordered in a relative chronology. The next step was to identify certain key fossils that would only appear in one set of strata. All other fossils appearing next to them could be assumed to be contemporary. By tying all other fossils to these "indexes" it would be possible to make a relative chronology of prehistoric life. Index fossils are themselves very common and very unassuming, typically a bivalve shell of some kind. Their special interest is derived from their usefulness.

The findings of the two men were jointly published in 1812 as *Recherches sur les ossemens fossils* (Researches on Fossil Bones). To be fair, the notion of "index fossils" was independently cropping up elsewhere at the same time. In England the geological surveyor William Smith drew the same conclusions as Brongniart, and published them in 1816. The Italian natural historian Giovanni Battista Brocchi in 1814 had published a treatise on how the study of marine fossils allowed the reconstruction of the prehistoric world.

Cuvier's reconstructed *paleotherium* and Brongniart's index fossils would both count as breakthrough achievements in the subsequent history of paleontology and the study of the fossil record in order to reconstruct the prehistoric fauna and flora of the world. One might leave it at that, thinking that this was simply the effect of individual brilliance, or happy coincidence. Modern history of science generally considers such an approach too facile. The objective is rather to work out what scientists do, how they arrive at their conclusion. Research is a process in need of reconstruction, and in writing its history it becomes apparent that institutions and the way in which science is organized is every bit as crucial as individual cleverness. In the case of the formulation of paleontology as a program for research, there is good reason to turn to the home institution of men like Cuvier and Brongniart, and others beside them, who all contributed to the formulation paleontology; the Paris Museum for Natural History.

MUSEUM SCIENCE

The French revolution had suddenly cast science and the scientist in a new role as servants of the people through the state, rather than of the king. Napoleon mobilized French scientists by appealing to their patriotic sense of the "gloire" of the nation. In 1798 he brought dozens of them along with his army to complement the military conquest of Egypt with a scientific one. These "savants" produced the multi-volume *La Description de l'Égypte*, detailing the archaeology, geography, and natural history of the country, sparking a craze for all things Egyptian, and inspiring the decorative style of Napoleon's reign, the "Empire" style.

One of the most important decisions of the revolutionaries was to nationalize the formerly Royal Garden, the Jardin du Roi, and various other natural history collections to form one gigantic museum dedicated to the science of natural history and for the edification of the people. The Paris Musée nationale de l'histoire naturelle (MNHN), the National Museum for Natural History, opened in 1793. It concentrated in one place the largest collections of natural history in the world, and these would be augmented by not always peaceful means as Napoleon's armies marched across Europe in the decades to follow. Not only that, the staff must either be considered the unlikely chance–assembly of individual brilliance, or more likely the effect of the confluence of means and resources allowing a number of initially very young and inexperienced scientists to establish an entirely new science. It can be argued that this success was as much an effect of the museum as of the efforts of the scientists working there. The historian John Pickstone has argued that the museum was part of a general reorganization of science in the early nineteenth century, or as he puts it:

> If one wished to be dramatic, one could claim that 1793–95 in Paris was the museological moment. As part of the assertion of a new cultural identity, or driven by the exigencies of disorder, the political authorities of France took unto themselves the collections which had belonged to the monarchy (and to some extent the church); they laid them out as public or educational displays. (Pickstone, 1994, p. 118)

Pickstone identifies over a dozen fields of science, which he describes as "*analytical/comparative* or *museological/diagnostic*," including of course the comparative anatomy and paleontology of Cuvier (Pickstone, 1994, p. 111). This new group of professional, middle-class scientists wielded the means of the state in the pursuit of "pure" science, and their efforts were principally directed toward deconstruction of their objects of study in order to make classifications. In Cuvier's case the process could be reverse-engineered to produce reconstructions from incomplete materials. The important recognition is that these sciences were institutionally and organizationally tied to the museum form.

The effects of scale of the new kind of institutions were dramatic: "The operative question [. . .] was not 'What have I got?,' as it might be for private collectors and their staff, but rather 'What is there?'" (Pickstone, 1994, p. 119). The confluence of this universal and totalizing ambition and the ability (at least seemingly so) to make it a reality is important. It is part of the make up of modern science, but only the state was able to guarantee the kind of resources necessary to attempt it in the early nineteenth century, and revolutionary France led the way. Looking specifically at the Paris Museum for Natural History, it became the model to emulate. It was an effect of both the advantages of scale, which allowed people like Cuvier and Geoffroy Saint-Hilaire to make breakthrough observations within anatomy, and the prestige of the both theoretical and practical achievements of the French scientists. The guns of the Napoleonic wars had barely silenced before representatives of the British Museum made the pilgrimage to Paris to learn on-site how to do it. The principalities of politically fragmented Germany began eagerly competing with one another to try to bolster their museums and rival Paris. Lacking the resources of centralized France this largely eluded them until the German reunification in 1871. France held a comfortable lead in sciences like anatomy, physiology, and zoology in the first half of the nineteenth century as part of the success of these analytical and diagnostic sciences, to use Pickstone's vocabulary. Only in the latter half of the century was French biology overshadowed by British and German.

As already stated, the staff of the new museum turned out to be a pretty extraordinary collection. Left over from the former administration, from Buffon's garden, was among others the professor of zoology Jean-Baptiste Lamarck, forty-nine at the establishment of the museum. Joining him in 1793 were then twenty-three-year-old Brongniart and Étienne Geoffroy Saint-Hilaire, just twenty-one, who was appointed professor of zoology alongside Lamarck. The surprising youth of some of the professorial staff has been attributed to the breakdown of the old, aristocratic system of patronage. The revolution had reversed the roles of seniority and minority, and someone like Geoffroy, with no scientific publication to his name at all, received his position for busting his illustrious teachers the anatomist Daubenton and the crystallographer Haüy (also on the staff) out of revolutionary prison, probably saving their lives. Joining the staff of the museum in 1795, at the age of twenty-six, Georges Frédéric Cuvier went on to overshadow them all.

The intellectual vigor displayed by the staff of the museum was quite extraordinary. It was here that Lamarck late in life developed the first true theory of organic evolution in his *Philosophie zoologique*. It was vigorously opposed by for instance Cuvier, who based his science on the ability to discern how presumably fixed species replaced each other within the fossil record, with no development of one into another. This opposition has been interpreted with varying degrees of bad faith. There is even a "black myth" of Cuvier as a dark genius machinating to destroy those with which he had scientific disagreements, including Lamarck. The more modern and measured interpretations rather indicate that the museum was in fact an intellectually heterogeneous place, where Lamarck and Cuvier publicly debated the issue of evolution. Lamarck's books sold well and he had little trouble in securing audiences even with Napoleon himself.

Eventually Lamarck's theory of evolution, his "transformisme," was championed by another member of the staff, his fellow professor of zoology Étienne Geoffroy Saint-Hilaire. Despite the curious conditions of his appointment, Geoffroy's contribution to the very science of paleontology, which Cuvier was about to found, was first rate, and alone would secure his place in the history of biology. True to his early form of a man of action, Geoffroy was one of the "savants" who accompanied Napoleon to Egypt. Eventually he found himself in a position to tell Lord Nelson that either he was shipping his collection back to the Paris museum, or he would chuck the lot in the Nile rather than see the British take it. Nelson backed down.

His great anatomical discovery was however to correctly identify "homologies" in vertebrate anatomy. A homology is an anatomical structure that may have been considerably changed through adaptive evolution. The important realization was that all vertebrates were in fact put together according to the same "blueprint." Even when very divergent anatomical structures were analyzed, one could still point out, bone for bone, that the plan was the same, with all structures accounted for, only at times changed to a point of almost unrecognizability. This was a key discovery for the later interpretation of the fossil record. At the time of its discovery it could be interpreted in a static fashion, by Cuvier most important, as confirmation of the general structural blue-prints of all organisms. But Geoffroy himself already viewed it as a means of reconstructing a process of development.

To give an example, the fingers of a human hand and the elongated fingers with skin between them that forms a bat's wing is a homology. It indicates a common descent, unlike an analogy, which are anatomical structures looking and working the same, but the outcome of two independent processes of evolution. The tails of fish and whales are a biological analogy. To be entirely clear it must be said that Geoffroy talked about "analogies" where modern science uses "homologies." The change of nomenclature was made by the British idealist paleontologist, and adversary of Darwin, Richard Owen later in the century. Owen made the identification of analogies in the modern sense. In any case, the discovery went straight into the scientific endeavor under formulation by Geoffroy's colleague Cuvier.

As professor of comparative anatomy Cuvier had benefited from hitherto unseen effects of scale in the analysis of animal anatomy. The Paris museum was a combination of a number of collections and Cuvier could literally scan a larger material faster than any anatomist in prior history. On the basis of this Cuvier undertook a complete revision of the classification of the animal kingdom. Linnaeus, and Ray before him, had used binominal systems, identifying class, genera, and species. Cuvier now realized that there were just a few basic structures of all living organisms. Taking these as the basis of a system of classification he finally arrived at what was a natural system of classification. It still holds up rather well as a lot of his conclusions could eventually be reinterpreted in the light of Darwinian evolution.

Cuvier recognized four basic plans for all animals: the *vertebrata* (animals with an internal skeleton and a spine, everything from fish to man), the *mollusca* (mollusks not possessing an internal skeleton), the *articulata* (arthropods, insects, shellfish, etc.), and the *radiata* (radially shaped starfish, sea-urchins, etc.). He referred to these as *embranchéments*, the major branches of life. To this realization was added two very important principles for the reconstruction of extinct animals: "the correlation of parts" and "the subordination of characters." The first established that all parts of the anatomy of an animal must be adapted to the lifestyle of the animal. Comparisons made in the vast material at his disposal gave at hand that all hoofed animals also possess molars suitable for grinding vegetable matter, while animals of prey have legs adapted to both running and grasping, with a set of teeth suitable to cutting and tearing flesh. The second rule means that those anatomical structures most important for the classification of an animal also display the least modification from adaptation to different ways of life, i.e., any one part of an animal is a reflection of the whole. Cuvier was sure enough of these things to claim the ability to make a general classification of any animal based on a single bone. As a basic rule of thumb this makes perfect sense. Cuvier went on to establish the prehistoric existence, and inferred extinction, of a large number of prehistoric mammals and vertebrates of all kinds; 110 mammals and 47 reptiles and batrachians.

It requires however that the anatomist is dealing with animals which roughly conform to what he has experience of. Cuvier knew the large vertebrate mammals, aside from the lizards and such, of the modern fauna, and spent most of his time working on likewise large prehistoric mammals of the tertiary strata. Paleontologists soon found fossils of animals that would leave them stumped when simply relying of Cuvier's methods. It was the English geologist William Buckland who in 1824 described and named the first known "dinosaur," or "terrible lizard," and paleontologist now started to assemble all these bizarre and unexpected lizard-like creatures found in the secondary formations. The problems of working out how dinosaurs were put together required finding well-preserved specimens. The first attempts identified them as lizards, which is at least half right, but actually confused the matter as the structural differences in the anatomies of lizards and dinosaurs are quite distinct. The first dinosaur discovered, the *iguanodon*, is a case in point. The first bones were discovered in 1822 by Mrs. Gideon Mantell, formerly Miss Mary Ann Woodhouse, of Tillgate, England. Her husband Gideon Mantell became fascinated with the bones and spent several years collecting more fragments. In 1825 he published a treatise naming the new beast *iguanodon*. The structure of the teeth made him draw the conclusion, in the fashion of Cuvier, that it was an enormous prehistoric iguana, a lizard. For decades reconstructions of the dinosaurs found in increasing number remained very lizard-like. It was not until Cuvier's indirect disciple Richard Owen realized in the 1840s that the dinosaurs anatomically more resemble mammals than true lizards. As for *iguanodon* it was first realized that this was an animal standing erect on its hind legs, with two spiky thumbs, hitherto interpreted as nasal horns, when entire well-preserved skeletons were discovered in 1878 in the Belgian coal mine of Benissart.

To be entirely fair, after pointing out the key role of the Paris Museum of Natural History as the venue where the science of paleontology was formulated, and from where theories and methods were disseminated, as one can see British scientists were not far behind their French counterparts. Most of the spectacular finds of secondary strata land-living reptiles were made by British men of science. Cuvier was first to describe and name the huge aquatic dinosaur the *mososaur* (though he thought it was land-living, and it is in fact a lizard), but it was British scientists who identified the *ichtyosaurs* (fish-lizards) and swan-necked *plesiosaurs*, of the long-gone shallow warm seas which covered most of Europe at their time, with strange flying lizards, *pterodactyls* (so named by Cuvier), wheeling in the air above it.

MUSEUM SCIENCE VERSUS FIELDWORK SCIENCE

As paleontology in a sense grew out of the comparative anatomy of Cuvier it acquired this aspect of a museum science. But as we initially saw, Cuvier and Brongniart also found it necessary to leave the museum to scour the fossil-carrying rocks and strata. The aspect of museum science was always offset by the necessity of fieldwork. Here the insistence on the necessity of the museum for the research process is partly due to the tendency of modern science to underestimate and misunderstand its key role. Modern scientists have little use for museum displays for the research process, but they were central for their nineteenth-century precursors. At the same time fieldwork is as indispensable as ever. This at times leads to the postulation of an incomplete dichotomy between the production of scientific facts as one between "the field" (the scientists on top of the rocks) and "the theory" (published articles and books). In the nineteenth century the necessary switching yard between these levels of scientific inquiry was the museum. Correct documentation of all circumstances surrounding the find and exhumation of a fossil was in fact necessary knowledge to guarantee its authenticity. Fakes were a known and common problem already in the eighteenth century, and talented taxidermists assembled everything from seven-headed hydras to furry fish in attempts to make a dishonest buck from wealthy collectors to the ire of the naturalists. However, in the museum a strange alchemy took place whereby all specimens were displayed shorn of the particular circumstances of their original burial and recent exhumation. The museum allowed the singular specimen to stand in for the species it represented as a collective. As Lee Rust Brown has put it, dealing with Ralph Waldo Emerson's reaction to the Paris museum:

> On the smallest scale—[. . .] of the single living plant or animal, the mineral sample, the fossil, even the heart or skull shown in the Comparative Anatomy cabinet—the representative aspect of the thing superseded the thing itself. So instead of particular creatures the visitor beheld "specimens," the representatives of species. (Brown, 1992, p. 62)

All the while fieldwork was of course important. Fieldwork was easily the best way for young men of science to make a contribution and get the attention of the movers and shakers of science and a leg up on the first step of the career ladder all through the nineteenth century. The fossils exhumed and collected through fieldwork were also for obvious reasons indispensable. The real crux might in fact have been who was going to do the digging and what scientific status their work was supposed to be accorded. It is a fact that early paleontology benefited immensely from amateur participation (see for instance a title like *Terrible Lizard*, 2001). While the museum-based scientists in principle recommended that fieldwork should at least be directed by members of their cadre of professionals, this was for most of the nineteenth century a perfectly unworkable proposition. What was at stake was the question which was the privileged place of observation; the field-naturalist's firsthand experience of things in some far-off and exotic but peripheral locality, or the measured analysis of the man working with the increasingly vast collections assembled in one of the great capital cities of Europe, simultaneously at the political center of nineteenth-century imperialism and of science. Dorinda Outram has pointed out how this opposition was summed up in a review by Cuvier himself of the work of one of his greatest rivals for the scientific limelight at the time, the German scientific South America traveler Alexander von Humboldt, in his day a true superstar of science:

> The field naturalist passes through, at greater or lesser speed, a great number of different areas, and is struck, one after the other, by a great number of interesting objects and living things. [. . .] But he can only give a few instants of time to each of them, time which he often cannot prolong as long as he would like. He is thus deprived of the possibility of comparing each being with those like it, of rigorously describing its characteristics, and is often deprived even of books which would tell him who had seen

> the same thing before him. [. . .] The sedentary naturalist, it is true, only knows living beings from distant countries through reported information subject to greater or lesser degrees of error, and through samples which have suffered greater or lesser degrees of damage. [. . .] Yet the drawbacks have also their corresponding compensations. If the sedentary naturalist does not see nature in action, he can yet survey all its products spread before him. He can compare them with each other as often as is necessary to reach reliable conclusions. He chooses and defines his own problems; he can examine them at his leisure. He can bring together the relevant facts from anywhere he needs to. The traveler can only travel one road; it is only really in one's study (*cabinet*) that one can roam freely through the universe. (Outram, 1996, pp. 259–261)

The opposition between fieldwork and museum, or collection in general, was—not unexpectedly—never resolved. Sciences like paleontology rested on both of these legs. But serious attempts were made to make a hierarchical division of roles whereby the traveling naturalists were expected to defer to the centrally placed scientists. Any number of "Instructions for travelers" were published in most major European languages to simultaneously try to ensure that a body of reasonably homogenous observations could be made by a mixed bag of missionaries, traders, big-game hunters, adventurers, and the occasional professional scientist. One of the first ones was in fact penned by Cuvier himself as early as in 1799, for a French expedition designed to circumnavigate the world; three compact pages dedicated to anthropological measurements of the natives encountered.

These attempts to subordinate and remote control the activities of scientific travelers occasionally drew the ire of those among them most conscious of their own self-worth. Thus the British Africa explorer and co-discoverer of the sources of the Nile, Richard Burton, protested: "We are told somewhat peremptorily that it is our duty to gather actualities not inferences—to see and not to think, in fact, to transmit the rough material collected by us, that it may be worked into shape by professional learned at home" (Driver, 2001, p. 67).

Of course in a science like paleontology fieldwork would always be of primary importance, but the general structuring of the science as a social hierarchy was implemented. It was easier to send young men eager to make their mark off into the remote, often desert areas that were prime venues for fossil hunting, than established senior researchers who played the role of recipients of the fossils. While paleontology started in the European heartland in the early nineteenth century, a lot of the early fossils came out of the Low Countries. As it progressed it became increasingly clear that the best material was coming out of first North America, with huge desert tracts becoming opened to settlement and exploitation, but also for scientific work. After the turn of the twentieth century a similar wave of spectacular finds started coming out of eastern China and Central Asia as expeditions were dispatched there.

The general structure of how science was organized is in evidence in one of the great scientific rivalrics of the young North American Republic, between the university-trained Othniel C. Marsh and the self-taught Edward Drinker Cope. Between them these two men augmented the knowledge of the fossil record by leaps and bounds with spectacular finds, the most famous one being that of the *tyrannosaurus rex*, the "lizard tyrant king." Their bitter personal and professional rivalry played out as part of the nineteenth-century story of how North America became The Place for fossil hunting. Not least the huge railway-building projects advanced the cause of paleontology in general, and paleontology as a particular American science. Since both men directed their expeditions out into the badlands of the United States, the means they employed were heavy-handed in a matching fashion, occasionally even resorting to violence to keep the competitor out. They battled both in the field and through their scientific publication, won by the autodidact Cope on points; 1400 publications to Marsh's 270, but Marsh arguably still came across as the greater scientist. Their rivalry unearthed a slew of localities filled with spectacular fossils. The first were located on the East Coast, in New York state, but attention shifted westward as fossils were located first in the Badlands of Nebraska already prior to Marsh's and Cope's rivalry, then Colorado (both); Como Bluff, Wyoming (Marsh); Judith River, Montana (Cope); and eventually in the 1890s the Carnegie Quarry, Utah, was found, later to become the National Dinosaur Monument.

THE PREHISTORY OF FOSSILS IN SCIENCE

We know that fossils have been found by humans for as long as there is recorded history. The most spectacular kind, the huge bones of various gigantic extinct vertebrates, has always drawn the most attention. The ancient Greeks dug them up and interpreted them as either the remains of those giants, Titans or Cyclops, their mythology was full of, or as the bones of mythical heroes from the past. This notion that great men of yore were in fact exceedingly strapping fellows carried over into medieval times. There are a considerable number of relics in the form of huge bones attributed to various saints in churches all over Europe, which on closer inspection in modern times have been reclassified as pertaining to one of the larger species of prehistoric mammals.

As far as fossil bones were attributed to giants, heroes, or holy men, it was at least clear that they were the remains of once-living beings. Things stood a little less clearly with plant fossils, or the many and curiously shaped invertebrate fossils discovered, especially the *petrificata*, the "petrified" varieties. In the sixteenth century the German botanist polymath Conrad Gessner went as far as to describe these kinds of fossil simply as the "sport" of nature, *ludi naturae*, perhaps God's little joke if one would like. There was also the alternative interpretation that these seemingly organic but petrified structures were in fact of demonic origin; the outcome of the Devil's own attempts at rivaling the All-highest by doing a bit of creation on his own. His efforts were of course immediately aborted, producing these curious fossils.

At the time the most widely spread fossil, of unclear origin, was the *petroglossa*, the "stone teeth." When found embedded in stone, these were found in great numbers. Their origin was obscure beyond the fact that they were obviously very sharp teeth, and as they caught people's imagination they were also imbued with magic powers. The *petroglossa* were thought to have special powers of negating poison, and became something of a fashion item in the courts and noble houses of the Renaissance. It was common to find *petroglossa* mounted in precious metals and stones as centerpieces of banquet tables, as part of an insurance against poisoning for the guests. These circles were hotbeds for fears, rumors, and accusations about poisonings. On balance food poisonings and infections due to deficient hygiene were greater culprits by far, but the suddenness of such illness left people eyeing their political rivals. This situation ensured that these petrified teeth were the most high-profiled fossils of their day. They are in fact fossilized shark teeth, and these can be found in abundance, indicating an abundance of sharks, rapidly shedding teeth in the prehistoric oceans.

This origin was finally elucidated in the seventeenth century by the Danish naturalist Nicolaus Steno, or Niels Stensen in his native Danish. Working in Italy, Steno in 1669 had the opportunity of comparing some *petroglossa* to the teeth of a head of a newly caught shark and made a positive identification of them as such. That they were sharks' teeth had been suggested earlier, but Steno's claim to fame rests not only on that, but on the fact that he went one step further and took an interest in the circumstances under which these fossils were found. He identified the fossil-carrying rocks as sedimentary, originally soft but hardening over time. More importantly he grasped the sequence of deposition, which means that the lower sedimentary strata represent an older age than upper sedimentary strata. Steno also realized that the fossils found in these strata represented long-dead animals, and that the rocks provided a relative chronology of the order in which they had been deposited. The biblical deluge was a handy mechanism for explaining these findings, but they stimulated discussion and further observation about the processes of sedimentation and deposition.

As was discussed already in the earlier chapter "Evolution as a Paradigm," geology is of key importance for evolutionary thinking in general and paleontology in particular. Steno provided an important first step toward the historical geology more fully developed in the eighteenth century, which also formed a prerequisite for paleontology. When Alexander Brongniart in 1812 identified "index fossils" as a means of establishing a relative chronology of prehistory, it was in a sense a logical extension of conclusions already drawn by Steno. At this point, instead of partly reiterating the history of geological thinking that paleontology extended into the organic world as discussed in the previous chapter, I would like to take the opportunity of briefly discussing fossils, geological periods, and the uses these can been put to by paleontologists.

THE MEANING OF "FOSSIL?"

To start from the very beginning, *fossil* itself is a very old Latin word. Based on the verb *fodere*, "to dig," fossil was originally the adjective applied to anything dug up. Only gradually was the adjective transformed into a noun. It continued to be applied to anything curious dug out of the earth, and that was originally what the remnants of prehistoric organic life were considered as, something curious, belonging in the cabinets of curiosity alongside everything else that was strange and wonderful. From the point of view of professional history there is nothing illegitimate about these older conceptions of fossils. Every period needs to be understood in its terms, as far as this can be attempted. At the same time, it is obvious that the history of the rise paleontology is to a great extent the history of the establishment of our present frame of reference of fossils. (But far from the only one, as there is a social history of professionalization, which is just as important.) They are the impressions left by the remains of once living organism, plants, animals, or simple cellular life, living in their environment. The definition of what may constitute a fossil these days becomes slightly tortuous because the range of variety of the kinds of fossils scientists have been unearthing, especially in the twentieth century, is staggering.

Fossils vary incredibly in age. The oldest fossils are the impressions left in rocks by unicellular life, *stromatolites* not dissimilar from modern blue-green algae, some 3.5 billion years ago during the period known as the Precambrian. Some of the youngest fossils are the prehistoric mammoths frozen in ice some 10,000 years ago, unearthed in Siberia, with their soft tissue preserved more or less intact. As for different kinds of fossils the already discussed *index fossils* continue to be a useful tool. As they have an already-known position in the stratigraphy of the local rocks, they are helpful in placing new and hitherto unknown organisms in their right position in the relative chronology of prehistory. Another peculiar kind of fossil are the *trace fossils*, the study of which is known as *ichnology* ("the science of footprints" in Greek). Not just footprints of large dinosaurs, though these rock faces that were once a sandy beach a dinosaur walked across capture the imagination, a trace fossil can be a wormhole through a piece of dirt later petrified. It might even be the pockmarks in the earth of a prehistoric heavy rain. There is also a further specialization within the trace fossils by those scientists who specifically study the prehistoric pollen, *palynology* ("the science of 'sprinkles'" in Greek). What they do have in common, and is part of their particular fascination, is that they are the traces left of animals, or some part of their environment, going about their daily business. The objective after all is to try to establish what the prehistoric world was like, to try to infer animal behavior from the evidence of the fossils record in comparison with the comportment of modern animals. Specializations of paleontological inquiry here include "paleoecology," how extinct species interacted with their environment, and "paleobotany."

This is unlike the vast bulk of fossils, which pretty much represent the refuse heap and charnel house of prehistoric life. Within modern paleontology there is even a special appellation for the scientific specialization of working out how and under what circumstances something was buried so that it could eventually reappear again as a fossil; the principles of *taphonomy*, derived from the Greek *taphos*, "burial" (as in epitaph), and *nomos*, "law," i.e., *the laws of burial*. Typically taphonomy concerns itself with attempts to reconstruct entire chains of events from why an animal found itself where it did, how it came to a sticky end, the sequence of its burial and the forces working on the body, its transition from dead animal to imprint, and finally the circumstances of its exhumation. In the modern sense "fossilization" refers to this process of turning dead organisms into hard structures, or imprints, in rocks.

A typical process of fossilization reads something like this: The medium, the stuff something becomes buried in, is referred to as the "embedding matrix." The properties of this medium will have consequences for the kind of fossils one can expect. The hard parts of the organism can simply lie buried within the matrix. For older fossils it can be assumed that the matrix might fill any internal cavities. Unless the conditions of burial dissolve the original hard parts as well, we are left with a cavity which however can carry a perfect imprint of the original organism. Then there is the possibility of some replacing, "precipitated" matter, seeping through the matrix, eventually replacing all or part of even the hard parts of the original organism. It is often some kind of silica, resulting in "petrification." It might fill internal cavities, it might entirely fill a cavity after the original organism disintegrates, and it might simply replace the original bones or shells within the surrounding matrix.

It is rare but possible to find fossils showing the preserved outline of soft tissue. It can happen though, for instance when a desiccated natural mummy forms, which then goes on to be embedded in some material which gradually replaces the non-decomposing tissue. In the case of plants, with sturdier cell membranes, it is possible to find fossils where the cellular structure can be analyzed microscopically.

Downright weird are two kinds of rare fossils; the first are of insects trapped in globules of resin, which when buried can solidify into amber; the second are a kind of fossil found outside Australia of Cretaceous aquatic lizards. The cavities left by their bodies have been entirely filled with opal, and there are virtually none to be seen in a museum. The cash value of these is simply too great. For a long time those examples seemed like the extreme of fossil weirdness. And then again, the fossil record continues to turn up surprises, and far from all seems to be known about it. Renowned *Science Magazine* in 2005 reported a find of 68-million-year-old still-pliable soft tissue being discovered inside a fossilized T-rex bone, with the possibility of more to find provided paleontologists start looking for it. The paleontologists did not make the claim that the tissue was original, but that some form of hitherto unknown process of fossilization on a molecular level is possible. Original tissue may have been replaced with long molecule chains—a (thermostatic) polymer, a kind of natural "plastic"—though recent developments do indicate that some genuine organic T-rex is sticking in there as well. After all, the oldest organic material whatsoever are some amino acids preserved in fossil shells formed 360 million years ago.

GEOLOGY AND THE FOSSIL RECORD

As has already been discussed in a previous chapter, the ability to make the study of geological formations and the study of prehistoric life proceed in lockstep is crucial. It is the geological classification of rocks and strata into "Eras," "Periods," "Epochs," and "Ages" which form the limitations in time and space of the organisms studied by paleontology. It makes them "four-dimensional" sciences; beyond the normal three dimensions of space they deal in time as well. A quick guide to the present ordering of geological periods, of geochronology, gives the sequence:

Era	Period	(Epoch)	Begins	Event
	Quaternary	Present/Recent,	0-10,000	Large mammals going extinct
		Pleistocene	10,000-2.5 mil.	Ice Age(s)
		Pliocene	6 mil.	First humans
Cenozoic	Tertiary	Miocene	26 mil.	

		Oligocene	38 mil.	Mammals specialize and modernize
		Eocene	55 mil.	
		Paleocence	65 mil.	Archaic mammals take over
	Cretaceous		135 mil.	Dinosaurs extinct, flowering plants
	Jurassic		190 mil.	First birds
Mesozoic	Triassic		225 mil.	First mammals
	Permian		280 mil.	Extinction event for many evertebrates
	Carboniferous		345 mil.	First reptiles
	Devonian		395 mil.	First amphibians, first forests, fish proliferate
Paleozoic	Silurian		430 mil.	Air-breathing invented, first land plants
	Ordovician		500 mil.	First vertebrates
	Cambrian		570 mil.	Spread of marine invertebrates
			700 mil.	First animals
Precambrian			3.4 bill.	Possible first bacteria, blue green-algae, and first multicellular organisms
			4.6 bill.	Origin of Earth

You can expect some shifting around of the dates of the geological periods. While geology and paleontology started out only able to provide a relative chronology of what beings appeared and what the sequence of events was roughly. In recent decades physics has lent a hand through the discovery of radiation. To be very brief it is possible to measure the half-life of almost any radioactive substance with a fixed rate of decomposition, of which there are a number naturally appearing in both rocks and fossils. Measuring the amount of radioactivity left allows you to calculate for how long the emission has been going on, which is since a strata was formed. If it contains fossils, these are of the same age. The first and probably still best known is the C[arbon]-14 method, which is not very accurate (or perhaps rather the application was not always rigorous enough), and has the great drawback of requiring organic matter to work. It doesn't bring you back much further than a few thousand years as well. For geological purposes there are a lot of commonly occurring substances with longer half-lives. Uranium-235 (^{235}U) decays into lead-207 (^{207}Pb) in 713 million years, and it takes U^{238} all of 4.510 billion years to turn into Pb^{206}. Other modern techniques include such things as thermo-luminescence.

The history of the establishment of the geochronological sequence, not least the naming practices, is a large subject in the history of geology. The recognized periods were originally much fewer, and of the present names only "tertiary" (Giovanni Arduino, 1760) and "quartenary" (Charles Desnoyers, 1829) are in use from the first geochronological schemes proposed in the eighteenth century. Most of the period names derive from specific parts of geography. For instance, the names "Cambrian" (Sedgewik, 1835, and by logical extension also "Precambrian"), "Ordovician" (Lapworth, 1879) and "Silurian" (Murchison, 1835) are all British in origin, Welsh to be exact. Behind them stick the province of Cumbria and the ancient Brython tribes of Silures and Ordovices inhabiting Wales in Roman times. Devonian (Sedgewick and Murchison, 1839) is of course named after the county of Devon, indicating to what a great extent this aspect of geology was a peculiar British concern. They did not have it all their way though, and the Permian is named after the Russian province of Perm (Murchison, 1841), while the Jurassic derived its name from the French Jura mountains, given this name for a section of the stratigraphic record by Alexandre Brongniart in 1829. Of course these are very general descriptions of the geochronology of the earth. All kinds of local varieties of geological sequencing are of paramount importance for paleontologists engaged in the study of the fossil record. It might still be useful for North American paleontologists to keep track of such things as "Mississippian" and "Pennsylvanian." The situation is similar elsewhere.

This activity of identification and naming was after all very much a collaborative effort, which also highlights the character of science as consensus-building through argumentative conflicts. A study which is a modern classic of the history of science is Martin Rudwick's *The Great Devonian Controversy*. It is an analysis of the integrated social and scientific process of the definition of the Devonian strata as representative of a separate geologic period in the history of the earth, possessing a specific fossil fauna of its own. The fact that geology and paleontology are intertwined to the extent they are gives the potential field of operations for the paleontologists a daunting scale. Your ordinary zoologist or botanist can keep himself perfectly occupied with the organisms currently living on earth, and which look and behave in a similar manner as they have been for the last couple of million years. The paleontologists have a beat which encompasses not just all of the world, but during the entire history of life on it. The simple fact is that all species sooner or later go extinct, and the extinct organisms in the history of life vastly outnumber the living. What makes the task manageable, in a sense, is the fact that we are dealing with a limited material of study. The fossils that have passed down to us are very much the accidental and fortunate remains of past worlds. Still, the material assembled is immense. Already in the 1820s the French paleontologist in the generation after Cuvier, Alcide d'Orbigny, in the Paris Museum for Natural History had assembled 100,000 fossils. The fossil record is now infinitely vaster and more varied. The nineteenth century dealt mostly with vertebrates, plants, hard-shelled aquatic invertebrates, and so forth, while the twentieth century has yielded the traces of soft-bodied organisms and not least the fossils of microorganisms, extending the range of paleontological reconstruction much further. Still, it is a song with modern paleontologists that "We need more fossils." More fossils are also forthcoming in large quantities, but our knowledge of the history of life on earth from them is nevertheless very incomplete.

Already from the outset the objective of paleontology has been to reconstruct past worlds and a reasonable shortcut is to use inferences from modern zoology, biogeography, and ecology. This means that the ambition is to try to work out as near as possible what life was like, how organisms acted and interacted with each other, and the environment at any given time and place. For practical reasons scientists are left with the ability to re-create local situations where happy circumstance has left them with something to work with. But apart from the time factor developments within, geology itself has somewhat changed the ground rules of paleontology through the introduction of plate tectonics—the realization that the ground beneath our feet is not fixed, but a set of plates "floating" and grating on each other at the edges. The classic paleontology of the nineteenth century assumed that continents were fixed in place. Species of land animals in order to travel from one continent or climate to another had to walk (or hop, or fly). As the fossil record indicated that certain animals had spread between continents with no present link it became a favorite game to postulate now sunk *ad hoc* land bridges or even entire continents to explain these movements. As they were assumed to be stuck in place, it was a bit of a problem how come certain continents with a land connection had divergent prehistoric faunas. Plate tectonics, formulated by Wegener in the 1920 and hooted down then only to be vindicated in the 1950s, allows for the fauna to stay in place while the continents move, breaking up and crashing into each other at geological speed. It adds yet another factor to be considered: movement in time not just by the flora and fauna but of the geography itself.

This brings up the work in recent years on great and cataclysmic events in prehistory. Far from a throwback to the nineteenth century discussions on principle between "cataclysmists" and "steady-state" theorists, there have been a number of realizations about specific large-scale dramatic events in the prehistory of the world. One such thing is the somewhat disturbing but fairly regular mass-extinction of large numbers of species. There have been several, but the most recent and most recently published is of course the event that wiped out the dinosaurs (non-avian), often referred to as the K/T-event, which is by some now directly linked with a huge meteor impact in the Gulf of Mexico. Earlier events might have had similar causes, but remain even more obscure for the time being.

A BETTER WORKING THEORY OF EVOLUTION?

Some aspects of the debates over the theory of evolution in recent years have brought out the difference between micro- and macro-evolution. Science has gained increasing insights into the inner workings of genes and cells with regards to heredity. Techniques for genetic analysis have allowed scientists to work out a genetic history of life at least as far back as we have access to organic remains with some viable genetic material. This also severely limits us. While *Jurassic Park* is inspired by science in a sense, it is still a far cry from the research done by Svante Pääbo at the Max Planck Institute on the genetics of Neanderthal Man. Rejuvenating fossilized dinosaur DNA is still as much science fiction as simply analyzing it. The possible "soft tissue" fossils, which have recently turned up, may give access to molecular structures of tissue, but the polymer-like structures involved are a lot sturdier than the fleeting DNA necessary. We are still obliged to work very much within the true and tested format of paleontology to reconstruct macro-evolutionary prehistoric life on earth.

It is hardly a coincidence that it is paleontology which historically has consistently been the branch of science displaying the most skepticism toward fixed theories of evolution. Paleontology is only interested in something like natural selection as a mechanism of evolution to the extent that it helps explaining the finds and observations. This is in fact still the case. The modern synthesis is widely accepted, but specifically within paleontology and macro-evolutionary reconstructions there have been recent irreverent attempts at improving it, to make the theory of evolution work better in face of the observations. If these attempts are simply attempts at a reform of Darwinism or a fully fledged alternative able to replace it would seem to depend on what part of the debate one latches on to.

One kind of challenge to evolution through natural selection came in the 1950s as the German entomologist Willi Henning latched on to the Darwinian concept of branching evolution. He formulated a research project to ensure the correct systematic reconstruction of evolution, calling it "cladistics" (after the Greek *klados*, branch). It is in fact a very consistent methodology for reconstructing and describing branching evolution. The challenge to the "modern synthesis" of evolution stems from the fact that no species can really be identified as a halfway point in an evolutionary transition from one to another, as predicted by Darwinian evolution. All species are equal within cladistics, to be described and inserted at their proper place in the branching structure. It's a very complex attempt at consistent systematization where all individuals in all groups in the system preferably should have a common descent. The problem with this attempt is that a lot of seemingly consistent groupings in nature have little to do with common descent. In a sense the old nineteenth-century division between analogies and homologies (see the section on *Museum Science* above for the distinction) keep cropping up, and working out which is which is far from simple. To give some examples mammals, so named after their mammary glands, share a common ancestry and by virtue of this physical trait can be described as *synapmorphic* (Greek *syn*-common, *apos*-apex, *morphe*-form) as it sets them apart from other species lacking these. The opposite are *symplesiomorphic* traits (Greek *syn*-common, *plesio*-close, *morphe*-form), which unites all species within a grouping. But many seemingly logical synapmorphic groups are problematic. Warm-blooded animals, *hemotherma*, for instance are beyond all doubt the effect of different evolutionary processes, with the mammals and the dinosaurs and birds having gone through that development independently, making such a group *polyphyletic* (*poly*-many, *phyla*-branches). The huge order of fish is for instance a very mixed bag. At the opposite end there is the recent decision to group dinosaurs and birds together as a synapmorphic taxonomic unit. This was a novel and controversial idea still in the 1980s. The warm-blooded birds were felt to differ very significantly from the dinosaurs, which according to the common consensus of the day were still regarded as cold-blooded reptiles. As the interpretation of the evidence swung toward warm-blooded dinosaurs, the interpretation of these became increasingly bird-like until today it is possible for paleontologists to speak of "non-avian dinosaurs," dinosaurs that are not birds. By rights dinosaurs and birds should hence be treated as a unit. In practice however the present 9,000 or so species of birds are such a well-defined taxonomic unit it is treated as a *parafyletic* unit, i.e., a subdivision not covering all relevant species, with the *dinosauria* making up the remainder.

To make matters still more confusing, cladistics based on morphological comparisons of structure and those based on genetic inquiries tend to differ in certain areas. The rise of molecular biology in the decades following World War II was of great importance also for paleontology and the research on fossils. The limitations imposed by the need to preferably have access to the DNA structure of the cells are apparent.

One might think it should be impossible to overestimate the role of heredity in evolution. After all, working out the structure of the DNA molecule, "the double helix," in Cambridge in the 1950s has rendered Francis Crick and James Watson the status of scientific heroes of the modern age. There is however, an aspect of evolution and heredity that has caused controversy within the sciences, not least paleontology in recent decades; the extent of adaptation through natural selection.

The 1970s saw a series of brilliantly argued works interpreting human behavior in terms of genetics and adaptation. The seminal one was Richard Dawkin's 1976 *The Selfish Gene*, standing our normal conception of ourselves on its head by arguing that humans, like all organisms, are basically the vehicles of replication for our genes. Others followed in a similar vein, like John Maynard Smith and Robert Trivers, who in 1985 published the very influential *Social Evolution*. It is clear that these scientists all advanced the case for the relevance of natural selection and evolution into the spheres that have always been highly charged, those areas that directly involve the personality, free will, and complex social behaviors and institutions of humans themselves. It was a very successful argumentation in favor of adaptation through natural selection as formulated within the modern synthesis of the theory of evolution.

But adaptation through natural selection had never been the panacea for everything, historically in particular within paleontology. By the late 1970s the above mentioned group, and other adherents, were polemically labeled "adaptationists" by the population geneticist Richard Lewontin and the young paleontologist Stephen Jay Gould. What they pointed out was that the assumption of the "adaptationists" that all observed phenomenon could and should be interpreted as the outcome of a slow process of adaptation through natural selection was in fact unproven. Just because an organ or a behavior today has a certain function, it does not by necessity follow that it was originally developed for that specific purpose. As an alternative to "adaptation" Gould, with his colleague Elisabeth Vrba, even proposed an alternative but theoretically just as possible process of "exaptation," specifically to describe how something developed for one purpose at a certain point might acquire a new, completely unforeseen, but highly beneficial function.

The "ex-" indicates that while the development is "apt," it is coming from out of left field, while "ad-" indicates that the development is a purposeful response to a situation. Specifically with reference to man several of the more complex features of human consciousness would seem to be of this order. It is fairly clear that what we are engaged in here, me writing, you reading, are not things that are the direct outcome of a slow process of adaptation, but rather some form of exaptation.

Paleontologists like Gould are interested in making the theory of evolution better explain the fossil record. This means they tend to look at the development of entire *taxa* or species over long periods of time. Where "adaptationists" are mostly concerned with micro-evolution, how DNA is transferred and advantages exploited on the level of individual organisms, paleontologists tend to be aware of inter-species competition. From a paleontologist's historical point of view it is of the utmost importance that the dinosaurs did go extinct, leaving the field relatively free for mammals to develop. There seems to have been nothing regular or law-regulated in that cataclysmic event, just a one-off disaster. And that is perhaps the major point of paleontological reconstructions, like all history-writing when dealing with the sketchy record of past events the story offers little in the way of inevitability. The dinosaurs might never have gone extinct, and the same follows for the other episodes of mass extinctions in prehistory. Working on this Gould, again, this time alongside Niles Eldredge, in 1972 launched the concept of "punctuated equilibrium" as a model for better describing some aspects of prehistory as evidenced by the fossil record. They put forth their model in opposition to a more common adaptationist view, which they labeled "phyletic gradualism," according to which new species arise from slow, even, gradual adaptation through natural selection, and this occurs to the majority of a population over a vast area. "Punctuated equilibrium" instead holds that changes can be very uneven, very rapid, and involve only a small population in a limited geographical frame, which then goes on to rapidly assert itself due to the new competitive abilities gained. This description of the workings of evolution does in fact tally with many observations of the fossil record. There are long stretches in the fossil record which gives the impression of a "lull" in activity, and natural selection would seem to work in such a fashion as to keep species relatively constant as the environment remains constant. There are on the other hand periods of intense activity, like the so called "Cambrian explosion" some 500 million years ago, which seemingly quickly threw up a great number of the basic forms of organisms still in existence today.

Whether the dogged insistence of paleontologists like Gould, Vrba, and Eldredge on pushing the theoretical interpretations of the fossil record beyond natural selection spells the dethronement of Darwinian evolution, or at least the "modern synthesis" of it, or not, is a matter of opinion. The insistence of someone like Gould is that the objective is to improve the theory of evolution as to form a better tool for understanding the history of life on earth. Paleontology has spent the last two centuries on the task, and did so already prior to the writings of Charles Darwin. No matter what direction the theory of evolution takes, the fossil record remains as the research material for this science.

FURTHER READING

Bowler, Peter J. *Fossils and Progress: Paleontology and the Idea of Progressive Evolution in the 19th Century* (New York: SHP, 1976).

Cadbury, Deborah. *Terrible Lizard: The First Dinosaur Hunters and the Birth of a New Science* (New York: Holt, 2001).

Cohen, Claudine. "Stratégies et rhétoriques de la preuve dans les Recherches sur les ossements fossiles de quadrupèdes" [Strategies and Rhetorics of Proof in the Recherches], in Blanckaert, Cohen, Corsi, and Fischer (eds.), *Le Muséum au premier siècle de son histoire* (Paris: Editions du Muséum d'histoire naturelle, 1997).

Copans, Jean, Jamin, Jean, *Aux origines de l'anthropologie française* (Paris: J. M. Place, 1994).

Jardine, N., Secord, J.A., Spary, E.C. (eds.), *Cultures of Natural History* (Cambridge, MA: Cambridge University Press, 1996).

Pratt, Mary Louise. *Imperial Eyes: Travel Writing and Transculturation* (London: Routledge, 1992).

Rudwick, Martin J.S. *The Great Devonian Controversy: The Shaping of Scientific Knowledge Among Gentlemanly Specialists* (Chicago: University of Chicago Press, 1985).

Simpson, George Gaylord. *Fossils and the History of Life* (New York: Scientific American Library, 1983).

Uddenberg, Nils. *Idéer om livet: En biologihistoria* [Ideas About Life: A History of Biology], vol. 1–2 (Stockholm: Natur och Kultur, 2003).

Geologic Time Scale and Radiometric Dating

Bowdoin Van Riper

We are accustomed, as a species, to reckoning the passage of time in years, decades, and centuries. The passage of 4.6 *billion* years—the time from the formation of the earth to the present day—is literally beyond our comprehension. The geologic time scale is a tool for making that vast span of time comprehensible by dividing and subdividing it into named segments. The names of those segments serve as a convenient shorthand for specifying particular segments of earth history. "The Devonian Period" (or, in casual usage, just "the Devonian") substitutes for the more cumbersome and less memorable expression "416 to 359 million years ago," and "the Holocene Epoch" for "the last 10,000 years." Some names also evoke images of how the earth—or, to be accurate, parts of the earth—looked during the corresponding slice of geologic time. "Carboniferous" brings to mind lush swamps teeming with giant insects, "Jurassic" a world ruled by dinosaurs, and "Pleistocene" a glacial wasteland populated by cave bears, woolly mammoths, and fur-clad human hunters wielding primitive stone-tipped spears.

The intellectual foundations of the geologic time scale were laid in the decades around 1700. The time scale itself emerged between 1800 and 1850, when geologists established and popularized the hierarchical system of named eras, periods, and epochs still in use today. Geologists could, at the time, determine the relative ages of rocks and fossils (A is older than B, but younger than C) but not their absolute ages (A is 10 million years old). The geologic time scale was designed around these realities. Its system of divisions and subdivisions, with agreed-upon names and agreed-upon boundaries marked by geological and biological events, was designed to give geologists from different regions a common language in which to discuss issues of relative age. Absolute ages were added to the geologic time scale beginning in the 1950s, as geologists discovered how to use the decay rates of radioactive elements as a clock, and so invented "radiometric dating."

The idea that earth and its living inhabitants have changed over time has been central to the science of geology since at least the 1680s. Earth's rocks and fossils record a long series of such changes, stretching from the most distant past to the most immediate present, and estimates of the length of that series have grown steadily since the early 1700s. Geological and biological change over time do not *demand* evolution as an explanation (Georges Cuvier, Richard Owen, Louis Agassiz, and other eminent scientists of the early nineteenth century explained them by other means), but the abundant rock and fossil evidence of such changes is powerful evidence for the reality of evolution. Charles Darwin understood that power and used it to his advantage in *On the Origin of Species*, the first (1859) edition of which included two complete chapters totaling sixty-five pages on the subject. In our day as it was in Darwin's, the geologic time scale is a powerful tool for organizing evidence of geological and biological change over time present in the earth's rocks

CENOZOIC ERA		
Period	***Epoch***	***Years before present***
Quaternary	Recent	11,000 – today
	Pleistocene	1.5 – 2 million
Tertiary	Pliocene	13 million
	Miocene	25 million
	Oligocene	36 million
	Eocene	58 million
	Paleocene	63 million
MESOZOIC ERA		
Cretaceous		135 million
Jurassic		181 million
Triassic		230 million
PALEOZOIC ERA		
Permian		280 million
Carboniferous		345 million
Devonian		405 million
Silurian		425 million
Ordovician		500 million
Cambrian		600 million
PRE-CAMBRIAN ERA		
Back to the formation of the Earth as a celestial body, possibly 3 billion plus years ago.		

The Table of Geologic Time

and fossils. Augmented by radiometric dating, it is also essential for analyzing the tempo—and, by extension, the nature—of evolutionary change. Perhaps most important, the time scale provides the framework within which the geological and paleontological evidence for evolution is presented to the public.

The geologic time scale's utility to scientists is not, however, what makes it an icon. It is an icon because it is inextricably linked to the popular image of evolution as a "march through time" from the biologically primitive past to the biologically advanced present. The public perceives the geologic time scale not as hierarchical, but as linear. The names of its periods (from Cambrian to Cretaceous) and epochs (Pliocene to Holocene) serve the public as "captions" for a "slide show" of images familiar from hundreds of books, posters, museum exhibits, and films depicting the geologic past. The image labeled "Cambrian," for example, shows trilobites. The "Devonian" image shows primitive fishes, the "Carboniferous" image towering land plants, and so on through various dinosaurs, early birds, the ancestors of the horse, and finally—in the image labeled "Pleistocene"—cavemen hunting mammoths. Taken as a whole, and coupled with these iconic images, the iconic sequence of period and epoch names reinforces the central idea in the modern evolutionary worldview: Earth changes over time, and life changes with it.

THE EXPANDING PAST, 1650–1800

James Ussher, archbishop of Armagh, confidently proclaimed in 1654 that the world had been created in 4004 B.C.E. Ussher's date was regarded, at the time, as authoritative: the result of applying first-rate textual analysis to the complex genealogical passages in the book of Genesis. Ussher started from two assumptions widely held at the time: that the Bible was divinely inspired and thus free from error, and that it was best read in a straightforward, literal sense. Assured of the quality of his data, he applied logic and some basic mathematics to the lives of the patriarchs and calculated the time that had elapsed between the creation, the flood, and the present day. Ussher's date of 4004 B.C.E. for the creation was printed in the margins of English bibles for centuries to come, but by the mid-eighteenth century scientists' faith in it was fading rapidly.

Use of the Bible as an authoritative source of information about earth history was not limited to Ussher and other scholarly churchmen. Nicholas Steno, the most important geological theorist of the mid-seventeenth century, noted the correspondences between his ideas and biblical texts. Thomas Burnet, author of the *Sacred Theory of the Earth* (1684), was the first of many late-century writers who wove together speculative histories of the earth from a mixture of science and scripture. The writers in question—scientists, though the word did not yet exist and they would have called themselves "natural philosophers"—turned to Genesis not out of fear but out of the conviction that it was the best available source of information about the early history of the earth. How could it not be, given that it was an eyewitness report that, by definition, could not be wrong?

Taking Genesis seriously as a source of scientific information shaped scientists' understanding of earth history in important ways. First, it fixed the age of the human race at a modest six thousand years. Second, it clearly established that earth history was virtually concurrent with human history, having begun only a few days sooner. Finally, it provided descriptions of the earth shortly after its creation and the precise date of a geological event of worldwide scope and enormous significance: the flood. Genesis thus provided a solid chronological framework within which geologists could interpret the geological history of the earth. It defined the length of that history and provided a convenient chronological scale (recorded human history) for it.

Estimates of length of earth history steadily expanded over the course of the eighteenth century, straining and then breaking the once-close connection between geology and Genesis. France's preeminent naturalist—Georges Louis Leclerc, Count Buffon—proposed at mid-century that earth was not thousands but tens of thousands of years old. Benoit de Maillet, in a book published at mid-century but conceived decades earlier, suggested that millions or even billions of years was more likely. Scottish polymath James Hutton wrote, in the 1780s, that all traces of the earth's early history had long since been eroded away. There was, he argued in his *Theory of the Earth* (1797), no way to reliably estimate the earth's age—except to say that it must be unimaginably vast.

The rapid expansion of earth history in the eighteenth century presented geologists with two serious problems. One was how to make chronological sense of the time between the origin of the earth and the origin of the human race, to which neither sacred nor secular records referred. The other was how to determine how old the earth actually was. The earliest versions of the geologic time scale, which appeared in the 1750s, were attempts to solve the first problem. Solutions for the second would not emerge for another century.

PRIMARY, SECONDARY, AND TERTIARY (1750–1800)

Steno's Laws, the intellectual foundation on which much of modern geology is built, are four principles for interpreting the relationships between rock layers (or "strata"). Like many revolutionary ideas they seem glaringly obvious in retrospect:

1. Any stratum in a formation is younger than those below and older than those above it.
2. All strata, whatever the current orientation, were horizontal when first deposited.
3. Strata are horizontally continuous when deposited, unless interrupted by a solid object.
4. Features cutting across multiple strata must be younger than any of those strata.

"Steno" was Nicolaus Steno, the Latinized name of a naturalist and priest who had been born born Niels Stensen in Denmark. What made his laws revolutionary was the idea that underlay all four of them: that it was possible to translate geometry into history, and reconstruct the chains of geological events from the rock layers they left behind.

Steno's laws implicitly equated layers of rock with slices of time. Each layer represented, in physical form, the length of time it took for the sediment making up that layer—silt, mud, sand, gravel, or chemical precipitate—had been deposited. It also represented a particular *moment* in earth history. No matter how wide an area a given rock formation covered, it must have been deposited over that entire area during the same period of time. Johann Gottlob Lehmann and Giovanni Arduino applied Steno's ideas on a grand scale in the 1750s. Lehmann (working in Germany) and Arduino (working in Italy) simultaneously but independently investigated the geological structure of mountain ranges and discovered that it followed a clearly defined pattern. The core of any given mountain range was made up, they both concluded, of steep-sided masses of non-layered, fossil-poor rocks like granite. A deposit of layered rocks, flat or slightly tilted and rich in fossils, overlapped the base of this core and were, in turn, overlapped by one (according to Lehmann) or two (according to Arduino) other deposits. The uppermost layer in both systems consisted of patchy deposits of loose sand and gravel, called "alluvium." The sequence of formations was always the same—an observation later confirmed by Pierre Palassou in France, Peter Pallas in Russia, and Alexander von Humboldt in South America.

Lehmann and Arduino each used the Law of Superposition to conclude that the deposits making up mountain ranges always formed in the same order, beginning with the granite core and ending with the alluvium. Arduino's names for the four types of deposits he recognized—primary, secondary, tertiary, and quaternary—reflected their shared belief in a universal sequence. Both men, however, went even further. They concluded, again independently, that the primary rocks (Lehmann termed them "primitive rocks") in all the world's mountains had been formed at the same time. "Primary, secondary, tertiary, quaternary" thus came to refer not just to a universal sequence of rock formations but to the eras in earth history when they had been formed.

Abraham Gottlob Werner, writing in the late 1780s, broadened the idea from mountains to rocks in general and added a physical explanation. The earth, he theorized, had once been entirely covered by water. "Primitive" rocks like granite, gneiss, basalt, and slate had been laid down first, forming by chemical precipitation on the sea floor. "Transition" rocks were formed when life was new and small amounts of land were beginning to appear as the water receded. They contained limited numbers of fossils, therefore, and were formed primarily from chemical precipitate along with some material washed off the new land surface. The still younger *floetz* ("stratified") rocks, products of a time when the sea had receded still further, were rich in fossils and primarily formed by erosion debris, with some chemical deposits. The "alluvium" consisted of loose material swept off the land by running water or spewed onto it by volcanic eruptions. Werner's categories of rocks did not correspond precisely to either Lehmann's or Arduino's, but he shared their belief in "universal formations" that had been deposited worldwide roughly simultaneously and characterized these four types of deposits as "universal formations." Each one, he argued, had been laid down worldwide at a specific time in earth history.

The systems devised by Lehmann, Arduino, and Werner, which divided up earth history by using distinctive sets of rock formations as time markers, remained popular through the eighteenth century. Werner's continued to be taught well into the nineteenth century, and would-be medical student Charles Darwin encountered it at the University of Edinburgh in 1825–1827. By 1800, however, the use of distinctive fossils as time markers was beginning to come into its own.

The use of fossils as time markers rested on the realization that species have finite life spans and that countless species have appeared and disappeared from the face of the earth. The fossil species found in a particular rock formation are thus a sample of the living species that existed at the time and place it was deposited. The presence of the same set of fossils in two formations implies that the formations are of the same age, regardless of the differences in composition or the geographic distance that might separate them. French naturalists Alexandre Brogniart and Georges Cuvier applied these principles to the rocks around Paris, publishing the results in 1808. William Smith, a surveyor and civil engineer, did the same in southern England, producing a large scale geologic map of the area in 1815.

The use of fossils as time markers allowed geologists greater precision in dividing up the geologic past. Fossil assemblages—groups of contemporaneous species—could be used to define large segments of geologic time, just as Arduino's and Werner's "universal formations" could. Unlike rock-based systems of dividing time, however, fossil-based systems could be easily scaled down. A single species with a short lifespan could be used to identify a specific formation, and so a narrow slice of earth history. The geologic time scale that took shape in the 1820s and 1830s was a product of both approaches.

SYSTEMS AND PERIODS (1800–1840s)

European geologists had, by the first decade after 1800, established both the coarse and the fine structure of what would become the geologic time scale. The coarse structure was provided by the sweeping four- and five-fold divisions of earth history formulated by Lehmann, Arduino, and Werner. The fine structure was provided by a growing body of geologic maps, cross-sections, and papers that placed the rocks and fossils of particular localities in chronological order. What was missing was an intermediate view: a classification system whose basic units were not individual formations but groups of related formations. These intermediate units became known as "systems."

Alexander von Humboldt defined the first one in 1799, when he coined the name "Jurassic" to refer to an extensive set of limestone formations found in the Jura Mountains along the Swiss-French border. The Jurassic System was, for more than twenty years, the only grouping of its kind. Then, in 1822, Belgian geologist Omalius d'Halloy proposed that the distinctive set of chalky limestone formations that overlay the hard Jurassic limestones in western France be defined as the Cretaceous System. The same year, British geologists William Daniel Conybeare and William Phillips coined the terms Upper and Lower Carboniferous Systems for, respectively, the coal beds of north-central England and the group of limestone and sandstone formations immediately beneath them. They also explored deposits known, in England, as the Oolite and the Lias, which would in the 1830s be reclassified as part of the Jurassic.

The definition of the Cretaceous and Carboniferous in 1822 opened the floodgates. Over the next twenty years, European geologists grouped virtually all of Europe's fossil-bearing rocks into systems. Arduino's term "tertiary" had, for decades, been loosely applied to the interleaved formations of sand, clay, limestone, and gypsum that filled the Paris Basin. It became a system more by longevity and default than by formal definition. The establishment of the Cretaceous in 1822 defined its lower boundary, and in 1829 French geologist Jules Desnoyers defined its upper boundary by proposing the name Quaternary System (another borrowing from Arduino) for the loose alluvial deposits overlying it. Five years later, in 1834, Friedrich von Alberti coined the term "Triassic System" for a characteristic trio of clay, limestone, and reddish-brown sandstone formations from the Rhine River valley into the Triassic System.

Meanwhile, two of Britain's leading geologists were busy unraveling the complex series of formations that lay between the bottom of the Carboniferous and the top of the massive crystalline rocks that Arduino called "primary" and Werner called "primitive." Over the course of the 1830s, Adam Sedgwick and Roderick Murchison defined three distinct systems: the Cambrian and the Silurian in 1835, and (in company with many other British geologists) the Devonian in 1840. Murchison—whose energy was exceeded only by his ambition—capped this extraordinary decade of work by defining the Permian System (above the Carboniferous and below the Triassic) on a visit to Russia in 1841. The establishment of the Permian filled the last major gap in the new classification scheme. An unbroken chain of systems now encompassed everything from the oldest fossil-bearing rocks (the Cambrian) to the most recent sediments deposited by running water (the Quaternary).

Systems, like the formations that make them up, are units of rock that also stand for units of time. A formation represents, visually, the span of geologic time required for it to form. A system represents, in the same way, the longer span of geologic time required for all the formations within it to form. Nineteenth-century geologists, recognizing this, used the term "period" to refer to the time required to form the rocks of a given system. Used this way, however, "period" took on a second, crucially important meaning. It referred not just to a *length* of time but to a specific *segment* of the past: a specific episode in the history of the earth. "The Devonian Period" or simply "the Devonian" was thus shorthand for "the period during which the rocks of the Devonian System were formed." Some geologists—William Buckland and W.D. Conybeare in England, Louis Figuer in France, Franz Unger in Germany—went further, using the rocks and fossils associated with particular periods to reconstruct what a typical landscape of that period might have looked like. Idealized representations of "the Silurian" or "the Devonian" became standard illustrations in popular books on the history of the earth and the history of life by the middle of the nineteenth century.

The concept of a generic Silurian or Devonian landscape made sense because periods were, by definition, universal. The "Devonian Period" took place at exactly the same time in Europe as it did in Africa, North America, or Antarctica. The Devonian fauna of one locality might have differed somewhat from the Devonian fauna of another, but to say that a fossil was of "Devonian age" was to locate it, with considerable though not absolute precision, in earth history. The establishment of periods imposed a measure of order on the vast and (in the nineteenth century) still-growing stretches of time between the first fossils and the first humans. The modern geologic time scale was born during the system-defining flurry of the 1820s and 1830s, and periods have remained its heart and core ever since.

ERAS AND EPOCHS (1820s–1850s)

The use of index fossils became standard geological practice in Europe during the 1830s. The Silurian, Devonian, and Permian systems were all defined in terms of their characteristic fossils as well as their characteristic rock types. Older systems that had been defined by rock type alone—the Triassic and Cretaceous, for example—were gradually redefined, in the 1830s and 1840s, in terms of their own characteristic fossils. Fossils also proved useful, during this period, in adding additional levels of organization to the geologic time scale. On one hand, they were used to group periods into larger "eras." On the other hand, they were used to subdivide periods into smaller "epochs."

The grouping of systems into larger "supersystems" and (by extension) periods into "eras" was proposed by Adam Sedgwick in 1838 and by John Phillips, the nephew of William Smith and a respected geologist in his own right, in 1840. Phillips's system, adopted in 1841 and still in use today, consisted of three such groupings for which he coined the terms "Paleozoic," "Mesozoic," and "Cenozoic" (Ancient Life, Middle Life, and Recent Life). The Paleozoic covered the Cambrian, Silurian, Devonian, Carboniferous, and Permian. The Mesozoic included the Triassic, Jurassic, and Cretaceous. The Cenozoic encompassed the Tertiary and Quaternary. The boundaries of systems were, by 1840, being adjusted to correspond to major changes in the fossil record, such as the appearance or disappearance of particular groups of species. The boundaries of the new units were chosen according to this pattern. The Mesozoic began at the Permian-Triassic boundary and ended at the Cretaceous-Tertiary boundary, the sites of what were—even in 1840—recognized as massive extinctions of species.

A number of geologists used fossils to subdivide existing systems and, therefore, existing periods. Murchison, for example, distinguished between the Upper and Lower Silurian even as he defined the system. Leopold von Buch, after carefully studying its fossils, divided the Jurassic into Upper, Middle, and Lower divisions in 1839 and then further subdivided each of those. Alcide d'Orbigny broke the Cretaceous into a number of small "stages" in the 1840s, again using fossils to do so. Charles Lyell's subdivision of the Tertiary in the early 1830s was the most spectacular of all.

Having traveled through France and Italy with Murchison in 1827–1828 and on his own in 1828–1829, Lyell knew the Tertiary strata of southern Europe well. Recognizing that the Tertiary deposits were rich in the fossils of mollusks and other marine invertebrates, he decided to subdivide the system according to the proportion of extinct species to still-living species in each of its formations. Stopping in Paris on his way home to London in February 1829, he formed an alliance with Gerard Deshayes, an expert in fossil mollusks who had been considering a similar project. Over the next two years Deshayes analyzed 40,000 fossil mollusks representing 8,000 species and assigned each to one of three tentative subdivisions. Applying his own interpretation to Deshayes's work, Lyell turned the three subdivisions into four: the Newer Pliocene, in which 90 percent of the fossils represented living species; the Older Pliocene, in which a third to a half of the fossils had living representatives; the Miocene, in which less than 18 percent did; and the Eocene, in which roughly 3 percent did.

Assigning the two most recent periods variations of the name "Pliocene" had been a diplomatic nod to Deshayes, who had contributed much to the project and who believed that only three epochs could be distinguished in the Tertiary. By the end of the decade, Lyell had thought better of the idea. Both halves of the Pliocene were acquiring subdivisions of their own, and their names ("Lower Newer Pliocene") were becoming unwieldy. Beginning in 1839, in an appendix to the French translation of his *Elements of Geology*, he renamed them. "Lower Pliocene" became simply "Pliocene," and "Upper Pliocene" became "Pleistocene." Twenty years later, in the fifth edition of *Elements*, he expanded the Pleistocene to include deposits containing (only) still-living mollusks and the remains of extinct land animals—deposits usually included in the Quaternary rather than the Tertiary. The issue was further complicated by Edward Forbes's 1846 proposal that "Pleistocene" be associated with the Ice Age, the existence of which (in the Quaternary) had been established less than a decade before.

Other modifications of the Tertiary epochs were considerably less complicated. In 1854, for example, Heinrich Ernst von Beyrich took slices from the end of the Eocene and the beginning of the Miocene to form a new epoch called the Oligocene. It included formations that were poorly represented (and seemingly inconsequential) in the areas of France and Italy that Lyell had studied, but well-represented and significant in the parts of Belgium and northern Germany that von Beyrich studied. Twenty years later, W.P. Schimper split off the lower Eocene and renamed it the Paleocene, noting that it possessed a distinctive set of fossil flowers that the rest of the Eocene did not.

By the time Lyell died in 1875, the subdivisions of the Tertiary had stabilized into the five still in use today: from oldest to newest, the Paleocene, Eocene, Oligocene, Miocene, and Pliocene. The Pleistocene was still complicated, with some geologists preferring to define it by its climate and others by its fossils, but both groups agreed that it belonged in the Quaternary. The six Tertiary/Quaternary epochs are, along with Phillips's three eras, the only divisions of the geologic time scale *other* than periods that are familiar to most non-geologists.

GEOLOGIC TIME AND HUMAN ANTIQUITY (1850s–1885)

Geologists' estimates of the age of the earth expanded significantly between 1650 and 1850, but their estimates of the age of humankind did not. The majority of career geologists saw humankind as 6,000 to 10,000 years old, and assigned them to the rarely defined, seldom-discussed Recent Epoch at the very end of the Quaternary Period. Two things defined the Recent: the presence of humans, and the absence of any species that were now extinct. Its flora, fauna, and climate were indistinguishable from those of the present, which made it—from a geologist's standpoint—dull and uninteresting. Indeed, for many geologists, the Recent lay just beyond the proper boundaries of the field.

These chronological arrangements reflected, in part, a lack of clear and compelling evidence that humans *were* older than 6 to 10,000 years. The oldest written records of life in Western Europe reached back roughly 2,000 years and the oldest written records of life *anywhere* reached back no more than 5,000 years, leaving plenty of room to accommodate the few ancient artifacts that were discovered. Reports of human artifacts and remains found mixed with the bones of now-extinct animals appeared occasionally during the 1820s, 1830s, and 1840s, but most were ambiguous, poorly documented, or both. Geologists, to whom the investigation of such claims fell, dismissed them out of hand or declared them insufficient to force reconsideration of established ideas. Geologists' ingrained skepticism on the issue extended even to colleagues with established reputations, such as Jules Desnoyers in France and Robert Godwin-Austen in England.

The skepticism was reinforced by the fact that the established ideas were intellectually comfortable. A strong progressionist strain ran through nineteenth-century geology, and many geologists saw the history of the earth as a steady march toward its current state and the history of life as a steady march toward human beings. Having both processes "finish" at the same time was, in this context, satisfyingly tidy. Having humans appear after the earth had taken on its present form also resonated with Christian teachings that earth had been providentially designed with humans in mind. Why, if it was made for humans, would God allow humans to appear before it was complete?

A series of fresh discoveries made in 1858 and 1859 firmly established that early humans *had* coexisted with mammoths and other now-extinct animals. By 1863 humans had been firmly established in the Pleistocene, and the Pleistocene-Recent boundary had lost some of its stratigraphic and most of its symbolic significance. Over the next twenty years, books by Charles Lyell, James Geikie, William Boyd Dawkins, Edouard Lartet, and Gabriel de Mortillet seamlessly combined prehistoric archaeology and Pleistocene geology. Most geologists remained content to leave the study of early humans to archaeologists, but the wall separating them from the geological past had been decisively and permanently breached. The merger of earth history and human prehistory advanced further in 1867, when Paul Gervais proposed the name Holocene Epoch (defined by stratigraphy rather than the presence of humans) as a replacement for the older Recent. The Holocene was formally adopted by the International Geological Congress in 1885, and has remained part of the geologic time scale ever since.

THE PROBLEM OF ABSOLUTE AGE (1860–1900)

Nineteenth-century (and earlier) geologists had no way of measuring the ages of rocks and fossils directly and precisely. They could tell that one formation was older than another, but not by how much. Nor could they tell how many years before the present a given formation had been laid down. The only way to find numerical ages for geological events—to calibrate the geological time scale—was to estimate the rate at which geological processes operated. It was at that point that the problem of absolute age became, again, virtually insoluble.

Questions about the tempo of geologic change swirled through geology in the eighteenth and nineteenth centuries. Two distinct sets of answers emerged. One (sometimes called "uniformitarianism") assumed that all geological changes, no matter how large, could be explained by small forces acting over long periods of time. The other (sometimes called "catastrophism") assumed that, though small changes could be explained by small forces, large changes required proportionally large forces. A uniformitarian and a catastrophist could, therefore, look at the same set of formations and draw radically different conclusions about the time it had taken to form them.

Incomplete and uncertain data compounded the problem. When John Phillips tried, in 1860, to calculate the approximate age of the earth, he used the deposition of sediment on the sea floor as his "clock." Noting that the basin of the Ganges River, in India, was eroding at the rate of 1 foot every 1,332 years, he used the same figure for erosion worldwide. Calculating the total thickness of all the fossil-bearing rocks on earth (earliest Cambrian to latest Pleistocene) at 72,000 feet, he multiplied that figure by 1,332 and concluded that the history of life on Earth was about 96 million years long. Phillips acknowledged, however, that his calculations were loose at best. If the Ganges eroded more quickly than the rest of the earth's surface, if the sediment was not distributed uniformly, or if large thicknesses of fossil-bearing rocks had been eroded away without a trace, the calculation would be too low. Phillips (or any other geologist of the time) had no way to know.

Other geologists, using different assumptions or different methods of calculation, came to different conclusions. C. Lloyd Morgan adjusted Phillips's model to allow for varying rates of erosion and sedimentation, and, in 1882, concluded that it had taken 82.5 million years to form the world's fossil-bearing rocks. Alfred Russell Wallace, co-discoverer of evolution by natural selection, made his 1880 calculation on the assumption that all the eroded sediment from the continents is deposited within thirty miles of their coast. This increased the rate of deposition (at least near the coast) by a factor of 19, to 1 foot every 158 years. Wallace thus shrank the time since the early Cambrian (during which all the fossils then known had been formed) down to a mere 28 million years.

The 1893 calculations of William J. McGee, an American, moved sharply in the other direction. Citing the existence of *pre*-Cambrian stratified rocks around Lake Superior, he increased the thickness of sediment that had to be accounted for to a total of fifty miles. Using the rate at which the Mississippi River dumped sediment into the Gulf of Mexico, he allowed 30 million years for the accumulation of a single mile of sediment. To compensate for unreliable data, he introduced a varying "factor of safety" into his calculations—64 for the Cenozoic, 256 for the Mesozoic, 1024 for the Paleozoic—and provided maximum, minimum, and mean estimates for each. His mean estimate for the length of the Cenozoic, for example, was 90 million years, which made the minimum estimate 1.4 million (90,000,000 ÷ 64) and the maximum 5.7 billion (90,000,000 x 64).

Sedimentation rates were not the only tool that geologists used in their efforts to calculate absolute age. One alternative method revolved around the changing chemistry of the oceans, and particularly their salinity. The idea of using salinity to estimate the age of the earth goes back at least to 1715, when it was proposed by British astronomer (and comet namesake) Edmond Halley. Reasoning that dissolved salts were carried into the ocean by rivers and then had no way to escape, Halley concluded that the oceans must be getting more salty over time. The question was: How quickly? Halley, lacking the data he needed to answer that question, abandoned work on the problem. It was picked up again in the 1870s by geologist T. Mellard Reade, who approached it by trying to estimate the amounts of certain salts carried by the world's rivers annually. Assuming that those rates remained constant and the oceans started as fresh water, Reade concluded that it would take roughly 25 million years for an ocean to build up the concentrations of calcium sulfate and magnesium sulfate that it now has. Reade's estimate for the time necessary to reach modern concentrations of sodium chloride was an order of magnitude larger: 200 million years.

John Joly, an Irish-born geologist who taught the subject at Trinity College in Dublin, proposed an even more elegant solution in his 1899 paper "An Estimate of the Geological Age of the Earth." He proposed taking the weight of sodium now in the world's oceans and dividing it by the weight of sodium carried into them by the world's rivers in a year. The result would be the age of the earth. Joly's original (1899) answer was 99.4 million years, which he revised down to 89.3 million years to account for sodium already present in the ocean at the beginning of the process. He revised the estimate upward to 90.8 million years in 1900, and upward again to 150 million years in 1909. William J. Sollas, professor of geology at Oxford, drew similar conclusions—80 to 150 million years old—from the same method. George Becker, of the U.S. Geological Survey, argued for an age of 50 to (more likely) 70 million in 1910, stating that the rate of sodium accumulation was diminishing because there were fewer and fewer exposed crystalline rocks to supply the sodium in the first place.

A third method involved estimating the age of the earth based on the rate at which it had cooled from its initial, presumably molten, state. Originally attempted by Buffon in the mid-1700s, it was resurrected in the 1860s by William Thomson, a leading British physicist and a pioneer in the field of thermodynamics. Thomson—later given the title Lord Kelvin, by which he is better known—assumed that earth had originally been a ball of molten rock, and had been cooling steadily ever since. He calculated the time necessary for a cool, solid crust to form and the time that must have elapsed between then and the present. He concluded, in an 1862 paper, that the earth was no less than 20 million and no more than 400 million years old. Returning to the numbers in 1868, he narrowed the range considerably, dropping the upper limit of the range and settling on 100 million as a plausible figure for the earth's actual age. By 1897, citing the availability of new data that had been lacking thirty years before, he had narrowed the range even further, keeping the lower limit at 20 million but reducing the upper limit to 40 million. The real figure, Kelvin then argued, was likely closer to the lower number than to the higher one. Clarence King, a leading American geologist, had come to a similar conclusion in 1893. Working from assumptions about the initial temperature and interior structure of the earth and data on the melting point of diabase, a common igneous rock, King arrived at a figure of 24 million years for the age of the earth.

Kelvin's (and later King's) figures were widely and readily accepted by most scientists. Physics was regarded, in the late nineteenth century, as the "queen of the sciences." The mantle of certainty and precision that

Newton and his successors had conferred on the discipline in the seventeenth and eighteenth centuries (and that Bohr, Einstein, and Heisenberg would leave in tatters by the mid-twentieth) remained in place. Geological ideas, by contrast, were associated with neither certainty nor precision in the late nineteenth century. Physics was a quantitative science and, though the rise of geophysics and experimental geochemistry was beginning to shift the balance, geology had the reputation of being a qualitative (and thus, in some sense, inferior) science. The elaborate mathematical apparatus behind Kelvin's and King's ideas about the age of the earth reinforced the seeming invincibility of their conclusions. So, too, did Kelvin's reputation as one of the world's most eminent physicists. Most observers found it inconceivable that someone of his stature, speaking on a subject so familiar to him, could be wrong.

The weight of authority behind Kelvin and King's calculations ensured that the few who did challenge them tended to be eminent authorities in their own right. T.H. Huxley—whose expertise, eloquence, and boundless self-confidence made him Darwin's leading public advocate in the years after *On the Origin of Species* (1859)—confronted Kelvin head-on a decade later. The problem, he argued, was not Kelvin's mathematics but his geology. The assumptions underlying his calculations were simply too speculative, Huxley argued, for the whole exercise to be of any real value. "Mathematics," he wrote, "may be compared to a mill of exquisite workmanship which grinds you stuff of any degree of fineness; but, nevertheless, what you get out depends on what you put in; and as the grandest mill in the world will not extract wheat flour from peascods, so pages of formulae will not get a definite result out of loose data" (Huxley, 1869, p. xxxviii). John Perry, a former assistant of Kelvin's who became a leading physicist in his own right, took a similar position in 1895. He showed that calculations identical to Kelvin's and King's could, if based on different but equally plausible initial assumptions, show that earth was 29 billion years old—a figure more than 1,000 times greater than either Kelvin's or King's. Thomas C. Chamberlain, then professor of geology at the University of Chicago, went further. Kelvin had, he argued, based all his calculations on a particular model of the formation of the earth and failed even to consider any of the possible alternatives. Several of the alternatives, including the one that Chamberlain preferred, implied that the earth had cooled far more slowly, and thus was far older, than Kelvin assumed.

Huxley, Perry, and Chamberlain were all less concerned with defending particular estimates of the age of the earth than with criticizing the supposed certainty of Kelvin and King's estimates. The essence of their argument was that far too little was known about the earth's interior structure and early history to draw conclusions about its age with any certainty. The critics were, in this sense, arguing that Kelvin and King's estimates were just as dependent on imprecise data and unverifiable assumptions—and thus no better than—those built around sedimentation rates or ocean chemistry.

For most working geologists, all three methods of estimating absolute age had a further drawback: They answered the wrong question. Calculations based on the salinity of the oceans or the cooling of the earth could give a rough estimate of the overall age of the earth, but they were useless for calculating the age of specific geological events. Calculations based on sedimentation rates could provide rough estimates of the duration of particular periods and epochs, but even then they were imprecise. Calculating the age of a specific geological event based on the rate at which the overlying sediment had accumulated was theoretically possible, but not attractive (or widely used) as a practical tool. Geologists left the nineteenth century as they had entered: without a practical method of measuring absolute age. Only a few years into the twentieth century, however, the discovery of radioactive decay provided the basis for just such a method.

THE RADIOMETRIC REVOLUTION (1900–1940s)

The discovery of radioactivity, in the first years of the twentieth century, changed geologists' understanding of the geologic time scale in two critical ways. First, Ernest Rutherford and Frederick Soddy's 1903 discovery that radioactive decay generated heat demolished most of Kelvin and King's calculations of the age of the earth. If the interior of the earth contained radioactive minerals, then Kelvin's assumption that no new heat had been added to the earth's interior since its formation was wrong. The heat generated by radioactive decay would offset at least some of the heat loss from cooling, so calculations based on a steady loss of heat would by definition be too short. Second, the discovery that radioactive decay proceeded at a constant rate suggested that it might provide a precise, reliable mechanism for determining the absolute ages of rocks. Radioactive elements decayed by emitting alpha particles (helium nuclei, consisting of two protons and two neutrons) or beta particles (electrons), transforming as they did so into other, lighter elements. As the process continued, more of the radioactive "parent element" would be converted into the stable "daughter element." Carefully measured, the ratio between them should reveal the age of the rock.

British chemist John William Strutt (later Lord Rayleigh) made the first rough attempts at radiometric dating in 1905, by measuring the quantity of helium in samples of radium. Based on these measurements and the rate at which radium decayed, he calculated that the radium sample was roughly 2 billion years old. American chemist Bertram Boltwood measured the uranium-to-lead ratios of forty-three different mineral samples, theorizing that the radioactive decay of uranium transformed it into lead. The results ranged from 400 million to 2.2 billion years. The lowest of Boltwood's numbers, in other words, matched Kelvin's most liberal 1862 estimate of the age of the earth.

Geologists welcomed the numbers cautiously, aware that the Strutt and Boltwood's calculations—like Kelvin's and, for that matter, their own—contained a measure of uncertainty. They also wanted to withhold judgment until the absolute ages derived from radioactivity could be checked against the relative ages derived by traditional geological methods. In 1911, Strutt's assistant Arthur Holmes carried out just such a check. Dating specimens from a many different parts of the Paleozoic Era, he found that they fell neatly into the expected chronological order. The Carboniferous samples had the lowest ages (340 million years), followed in order of increasing age by the Devonian (370 million) and Silurian (430 million) samples. The Precambrian samples turned out to be significantly older than the rest: 1.02 to 1.64 billion years.

Holmes went on to make the development of radiometric dating his life's work, refining the methods used to do the dating and building up a body of rock samples whose radiometric dates and position in the geologic time scale were both known. Holmes used twenty-three such samples to draw up, in 1927, the first version of the geologic time scale to be calibrated using radiometric dating. Most of the samples came from the midpoints of systems and thus of periods. Holmes's dates for the boundaries between periods depended, in most cases, on estimations made using traditional methods for calculating absolute age. Revising his calibrated time scale as more and better dates became available, Holmes published new versions in 1937 and 1947.

One reason why better dates were becoming available in the 1930s and 1940s was a growing awareness of how radioactive decay worked. The rates at which parent elements decayed into daughter elements were measured with growing precision, and the intermediate steps they passed through were defined more carefully. Frederick Soddy showed, in 1912, that most elements existed in several forms ("isotopes") that were visually indistinguishable but behaved differently at the atomic level because they possessed different numbers of neutrons. The development of mass spectrometers in the 1920s allowed chemists to easily determine the proportions of different isotopes in a sample of material, and to add new dating methods to the traditional uranium-helium and uranium-lead ones. The array of dating methods available to modern geologists is extensive and varied: potassium-argon, argon-argon (two isotopes of the same element), rubidium-strontium, samarium-neodymium, lutetium-hafnium, rhenium-osmium, thorium-lead, and lead-lead.

Holmes was, fittingly, a member of a multidisciplinary committee formed in 1931 by the National Academy of Sciences to investigate the age of the earth. His contribution to the committee's report was a detailed summary of what radiometric dating had thus far contributed to the question. Based on the best available data, he argued, the earth was probably at least 1.6 billion but less than 3 billion years old. Another section of the report repeated, with fresh numbers, a traditional calculation of the time necessary to lay down all the formations deposited on earth since the beginning of the Cambrian. The results were consistent—to the author's considerable surprise—with the 500-million-year figure arrived at through radiometric dating. Yet another section of the report re-evaluated Joly's (and others') methods for using the salinity of the ocean to calculate the age of the earth, and concluded that it was too imprecise to be of value. This, too, was a victory for radiometric dating, since calculations based on salinity had consistently yielded much lower numbers than those based on radioactive decay.

The precision and reliability of radiometric dating was firmly established by the end of World War II. The increasingly numerous, increasingly precise dates published in the late 1940s were received, by most geologists, the way descriptions of new fossils had been received by their nineteenth-century counterparts: with occasional quibbles over details, but acceptance of the underlying scientific principles. The idea of assigning absolute ages—that is, numbers—to the age of the earth, to the endpoints of particular eras or periods, and to specific geological events had become an accepted part of geology.

Beginning in the early 1950s, radiometric dating also became an integral part of archaeology, and of those branches of geology that dealt with the most recent portion of earth history. Working at the University of Chicago in 1949, Willard Libby and his colleagues developed carbon-14 (or "radiocarbon") dating. As its names suggest, the new method was built around carbon-14, a rare (one part in a trillion) but widely distributed radioactive isotope of carbon. Carbon-14 is continually produced by the action of cosmic rays on atoms of nitrogen-14 in the middle layers (30–50,000 feet) of the atmosphere. Once formed, it combines with oxygen to form carbon dioxide, which is absorbed by the oceans and by plants during photosynthesis. Living plants and animals continually exchange carbon-14 with their environment. The ratio of radioactive to non-radioactive carbon in their tissues remains the same, therefore, as the ratio in the environment as a whole (one part in a trillion). When the organism dies, the exchange stops, and the carbon-14 in its tissues begins to spontaneously decay into nitrogen-14. The organism's death thus starts the radiometric "clock" in the same way that the formation of a rock does in other methods. The more time has elapsed on the clock, the less radioactive element there will be in the rock, or (for carbon-14 dating) a sample of the organism's tissue. Techniques for measuring the amount of carbon-14 in a sample initially involved measuring how often an atom of it decayed, and extrapolating the number of atoms from the average time between events. Modern techniques, which rely on mass spectrometers that enable researchers to count individual atoms of carbon-14, give more precise results with smaller samples.

Even more than other radiometric dating methods, carbon-14 dating requires careful calibration. Though nominally constant, the levels of cosmic radiation striking earth's atmosphere actually fluctuate slightly over time and so cause the levels of carbon-14 to vary as well. Aquatic plants and the animals that eat them absorb some of their carbon from ancient, slowly dissolving carbonate rocks that contain lower-than-usual levels of carbon-14. Cold War–era testing of nuclear bombs, carried out frequently in the northern hemisphere but rarely in the southern hemisphere, made carbon-14 levels higher above the equator than below it. Developing appropriate corrections for these variables—called "calibration curves"—was a critical element in the refinement of carbon-14 dating.

Radiocarbon dating can be used on virtually any organic material—wood, bone, plant material—but, because of the rate at which carbon-14 decays, is effective only on samples less than 50,000 years old. Its impact on geology has, therefore, been profound but limited. It works only on un-fossilized organic matter, and only on organic matter from the (geologically speaking) very recent past. Despite these limitations, however, it has done for the study of the geologically recent past what other forms of radiometric dating have done for study of the distant past. It provides a reliable way of working out sequences of events that—if geologists had to rely solely on the spatial relations of organic remains and strata—might otherwise remain ambiguous. The impact of radiocarbon dating on prehistoric archaeology has been even more spectacular: equivalent both in form and in depth to the initial impact of radiometric dating on geology in the 1920s and 1930s.

THE VERY YOUNG AND VERY OLD (1950s–PRESENT)

Radiometric dating gave geologists, for the first time, the ability to define segments of the geologic time scale in purely numerical terms. Radiometric dating made it possible to associate the boundary between two periods (for example, the Cretaceous-Tertiary boundary) not with just with a geological change (the shift from chalk to sands and clays in southern Europe) or a biological one (the disappearance of dinosaurs), but with a round number (65 million years ago). The effects of this change are most visible at the extreme ends of the geologic time scale.

The Holocene is the last of the Cenozoic epochs, the shortest (by a factor of 100), and the only one with a boundary defined by a radiometric date. The upper boundary of the Holocene is the present; the lower boundary is a point 10,000 years before the present. Setting the boundary at 10,000 years makes it memorable and allows it to mesh smoothly with the archaeological record. The entire Paleolithic (Old Stone Age) Era falls on the Pleistocene side of the boundary, and the more technologically sophisticated Mesolithic and Neolithic (Middle and New Stone Age) Eras on the Holocene side. Despite these advantages, the radiocarbon-based definition of the Holocene breaks a long-established pattern of using biological events to define epoch boundaries. It also means that the Pleistocene-Holocene boundary falls at an essentially random point near the end of the most recent glacial period, rather than *at* the end of the glacial period or at some other climatic turning point.

The location of the Pleistocene-Holocene boundary is also a minor element in a larger, ongoing debate about the boundaries of the Neogene Period. Coined by Moritz Hörnes in 1856, "Neogene" originally referred to what was then known as the Upper Tertiary: the Miocene, Pliocene, and Pleistocene epoch. Geologists working on Quaternary terrestrial deposits have supported the continued use of this definition, arguing that the boundary between the Pliocene and Pleistocene (that is, the Tertiary and Quaternary periods) is marked by significant biological and geological changes. Specialists in marine geology, who see few differences between Pliocene and Pleistocene sea-floor deposits, argue that the Neogene should be extended to the present. The Cenozoic would then be divided into the Paleogene Period (Paleocene, Eocene, and Oligocene epochs) and the Neogene Period (Miocene, Pliocene, Pleistocene, and Holocene epochs), with the old Tertiary and Quaternary redefined as "sub-eras" of the Cenozoic or abolished entirely. Specialists in the geology of the glacial period, while supporting a separate Quaternary Period, argue that its lower boundary should be redefined. Setting the boundary at 2.58 million years would make it correspond to the onset of the glacial period, and put the entire glacial period (which now crosses the Tertiary-Quaternary boundary) in the Quaternary period.

At the opposite end of the geologic time scale, controversy also surrounds the division of the 88 percent of earth history that falls before the Cambrian Period. The "Precambrian," with few hard-shelled (and thus readily identifiable) fossils, remained essentially undivided until the development of radiometric dating. When it *was* divided into eons (groups of eras), eras, and periods, the boundaries between them were set at multiples of hundreds of millions of years. The Proterozoic Eon, for example, begins 2,500 million years ago. Its three eras—the Paleo-, Meso-, and Neoproterozoic—begin 2,500, 1,600, and 1,000 million years ago. The four periods that make up the Mesoproterozoic begin at evenly spaced intervals of 200 million years: 1,600, 1,400, 1,200, and 1,000 million years ago. The Archaean Eon, which immediately precedes the Proterozoic, is subdivided into eras but not periods, and its lower boundary remains imprecisely defined at "circa 3,850 million years." The use of the term "Hadean Eon" to designate the time between the formation of the earth and the beginning of the Archaean, though first proposed (by geologist Preston Cloud) as early as 1972, remains unofficial.

Critics of the current Precambrian dating system—most of whom are specialists on the rocks and sparse fossils of the period—contend that the system is incomplete and that its boundaries are arbitrary. The dates of those boundaries, they argue, bear no relation to the geological and biological changes that actually took place during the relevant periods of earth history. Finally, they point, the use of perfectly round figures (even hundreds of millions) creates a spurious image of absolute certainty and precision. Actual radiometric dates from the Precambrian invariably involve a margin of error, ranging from +/– 6.5 million years at 2,500 million years ago to +/– 10 million years at 4,000 million years ago. The false precision of the current date-based boundaries should, the critics conclude, give way to boundaries defined by actual geological events—defined, that is, the way geologists have defined chronological boundaries since the 1750s.

CONCLUSION: IMAGINING THE PAST

The proper way to subdivide Precambrian time and the proper time to begin the Pleistocene Epoch are, of course, far from the everyday thoughts of anyone but specialists in those areas. The geologic time scale remains, for most scientists and all lay people, a stable and uncontroversial subject. Like the periodic table of elements, it simply *is*. The geologic time scale is different from the periodic table, however, in that it is more than simply a scientific tool. Images of it shape the way we think about geologic time, and about the processes for which it serves as a backdrop.

The standard image of the geologic time scale, especially in publications aimed at popular audiences, is a single vertical column divided into segments that represent the various periods. Variations—a horizontally oriented bar instead of a column, or a double column that provides room to indicate the names of eras—exist, but are comparatively rare. The Precambrian is almost invariably at the bottom of the column (or the left end of the bar) and the Quaternary almost invariably at the top (or on the right). The boxes are almost never drawn to a consistent scale, with their relative heights (or widths, in a horizontal layout) representing their relative durations. Embellishments of individual segments with drawings representing the dominant (or most interesting) species of the period are common but by no means universal. The ultimate expressions of this last convention are timeline-murals like Rudolph Zallinger's "Age of Reptiles" and "Age of Mammals" at Yale University's Peabody Museum of Natural History and John Gurche's "Tower of Time" at the Smithsonian.

There are excellent reasons for these conventions. Having the illustration "read" chronologically from left to right mirrors the way we read text; having it read chronologically from bottom to top mirrors the arrangement of strata and fossils in the real world. Constructing diagrams of the time scale *to* scale would make them visually awkward and pedagogically compromised. The 3,225 million years of the Proterozoic and Archaean Eons (about which we know comparatively little) would dwarf the 625 million years of the Phanerozoic (about which we know a great deal) if the whole thing were drawn to scale. Individual eras and periods within the Phanerozoic—what non-scientists think of as the geologic time scale—would be smaller still. Illustrations of animals add visual interest, and help to reinforce the difference between (say) the Permian and the Pleistocene. Useful and sensible as they are, however, these visual conventions have side effects.

The most visible of these side effects is the subtle but constant reinforcement of an outmoded image of evolution. The idea that evolution, if it occurs, is linear and progressive is deeply rooted in Western culture. It was a central element in the theories of Jean-Baptiste Lamarck and most other pre-Darwinian evolutionary theorists, and remained popular among post-Darwinian theorists (sometimes described as "Neo-Lamarckian") who accepted evolution but not natural selection. The essence of the Lamarckian and Neo-Lamarckian views was that evolution followed a straight path that lead, by clearly defined stages, from simple organisms to complex ones. Evolution involves simple species literally *becoming* more complex over the course of thousands of generations and millions of years, with fish (for example) being succeeded by amphibians, reptiles, and mammals as Monday is succeeded by Tuesday, Wednesday, and Thursday. Departures from the path—where they exist at all—are "dead ends" or "blind alleys." The stages along the main line are the one *true* path to evolutionary success, and humans—the last stage—are, by definition, evolutionary success incarnate.

Depicting geologic time as a narrow column with trilobites (or some other simple life form) at the bottom of the Cambrian and humans at the top of the Holocene reinforces precisely this view. It implies that evolution is like a ladder in which each new rung represents a step toward greater complexity, rather than (as Darwinian models suggest) an endlessly branching bush in which each twig represents an adaptation to a particular set of environmental conditions. The use of one or two familiar animal images to illustrate each period encourages the viewer to conclude that "to all things there is a season." The decline of the dinosaurs and the rise of mammals at the Cretaceous-Tertiary boundary becomes a matter of inevitability rather than accident. Above all, linear images of the geologic time scale cater (as non-Darwinian theories of evolution do) to viewers' taxonomic vanity. *Homo sapiens*, they hint, is not just one contemporary species among many. It is the climax of the evolutionary process: *the* goal that the long climb from trilobites to mollusks to fish, amphibians, reptiles, birds, and mammals was intended to reach.

Using images of particular plants or (more often) animals to decorate and designate particular subdivisions of the geologic time scale also has unintended consequences. First, because the chosen images are nearly always of species that *appeared* during the time interval in question, it emphasizes novelty over continuity. Describing, and visually depicting, the Devonian Period as the "Age of Fishes" is a convenient shorthand, but it also means that fishes are rarely acknowledged in later stages of the diagram. Flowering plants are, in the same way, rarely seen after the Cretaceous, nor small mammals after the Paleocene, where they provide a visual counterpoint to the inevitable *Tyrannosaurus rex* and *Triceratops* used to illustrate the Cretaceous. Second, the use of one or two iconic species to represent each period subtly downplays both the diversity of life and the accumulation of small but significant changes *within* any given period. The history of life, as sketched in depictions of the geologic time scale, implicitly involves long periods of total stasis separated by the appearance of radically new and different species—often without any apparent biological precedent.

The lack of scale in most depictions of the geologic time scale has, perhaps, the most pervasive effect of all. First, it is nearly universal, as common in technical depictions aimed at professional geologists as in stylized ones aimed at the general public. Second, it affects not just the way we think about the history of life, but the way we think about geologic time itself. Third, it reinforces our innate inability to grasp vast spans of time with any precision. Fourth—and perhaps most important—there is no practical way of getting around it. Drawing the geologic time scale to a constant scale would mean making intolerable choices: Make the Cenozoic (say) so small that its epochs would become invisible, or make the Proterozoic so large that it would take up most of a double-page spread all by itself? Reduce the size of the Holocene a modest quarter-inch, and the Pleistocene alone (if drawn to scale) balloons to a little over two feet. Depictions of the geologic time scale thus distort, out of necessity, one of the critical facts that they should, in an ideal world, convey: the extent to which the era of complex life on earth (much less the era of humanoids) is utterly dwarfed by everything that went before.

Metaphors for Deep Time

The geologic time scale is a useful tool for subdividing earth history but not for conveying its vastness. The latter task is better accomplished through metaphors, most of which involve either time or distance and all of which are designed to highlight the chronological insignificance of human history when compared to Earth history.

Charles Lyell, in *Principles of Geology*, compared the immense vistas of time revealed by geology to the immense vistas of space revealed by Galileo and Newton: "Worlds are seen beyond worlds immeasurably distant from each other, and beyond them all innumerable other systems are faintly traced on the confines of the visible universe" (Lyell, v. 1, p. 63). John McPhee, who coined the term "deep time" in his 1980 book *Basin and Range*, also coined the best-known linear metaphor for it. Imagine, he wrote, that the span of your outstretched arms represents the entire history of the earth. "The Cambrian begins in the wrist, and the Permian Extinction is at the outer end of the palm. All of the Cenozoic is in a fingerprint, and in a single stroke with a medium-grained nail file you could eradicate human history" (McPhee, p. 126).

The most famous metaphor of all, first presented by astronomer Carl Sagan in his 1977 book *The Dragons of Eden*, maps the history of the entire universe onto a single calendar year. The Big Bang (15 billion years ago) occurs just after midnight on New Year's Day. Our galaxy is formed on 1 May, our solar system on 9 September, and earth itself on 14 September (4.5 billion years ago). The oldest known rocks are formed around 1 October, and the oldest fossils shortly afterward. The Cambrian Period and the first flowering of complex life begin on 17 December, and the heyday of the dinosaurs covers the three days after Christmas. Human beings appear late on New Year's Eve, and invent agriculture roughly forty seconds before midnight. All of recorded human history—every book, every building, every idea, and every individual we can name—unfolds in the last ten seconds of the cosmic year.

All of this matters, ironically, *because* the geologic time scale is so vital to our understanding of earth history, and of the history of life for which it forms a backdrop. It allows us to deal with the vastness of earth history in manageable chunks, and gives us guideposts against which to measure what we see. If, as L.P. Hartley wrote, "the past is a foreign country," then the geologic time scale is our map and—with its period-defining rock formations and groups of fossils—our gazetteer as well.

FURTHER READING

Albritton, Claude C. *The Abyss of Time* (San Francisco: Freeman, 1980).

Berry, W. B. N. *The Growth of a Prehistoric Time Scale* (San Francisco: Freeman, 1968).

Burchfield, Joe D. *Lord Kelvin and the Age of the Earth* (Chicago: University of Chicago Press, 1990).

Dalrymple, G. Brent. *Ancient Earth, Ancient Skies* (Stanford: Stanford University Press, 2004).

Jackson, Patrick Wyse. *The Chronologer's Quest: The Search for the Age of the Earth* (Cambridge: Cambridge University Press, 2006).

Rudwick, Martin J. S. *Bursting the Limits of Time* (Chicago: University of Chicago Press, 2005).

Piltdown Man and the Almost Men

Brian Regal

The fossils of this chapter are evolutionary icons of a different sort. They come less from the earth than from the darker aspects of the human psyche in the form of duplicity, over-eagerness, and self-delusion. At the same time they represent the strengths of science with its continuous testing of facts and its pursuit of knowledge about the workings of the universe. The late nineteenth and early twentieth centuries were the first golden age in the discovery of fossil humans. A number of these finds, however, were not what they were originally thought to be: They were almost men, but not quite. They were misidentifications of things that at first sight seemed to be ancient, but turned out to be modern, or seemed to be human when in fact they were not. Piltdown Man stands out from this group because it was not an honest mistake of misidentification, but a carefully calculated hoax. It was a lie meant to fool and injure. Like the story of Jack the Ripper, the culprit remains unknown, though there are several popular suspects. The person or persons who perpetrated the hoax took advantage of a time and place where a select group of scientists—who are after all human like the rest of us—were ready to see what they wanted to see in the field of human origins research.

Piltdown Man was not alone. The fossil that came to be called Nebraska Man was another find that generated quite a bit of excitement. Some thought it proof of the presence of early hominids in the Americas. Unfortunately, they were wrong. After some fanfare, it was discovered that a while a genuine fossil, Nebraska Man turned out to

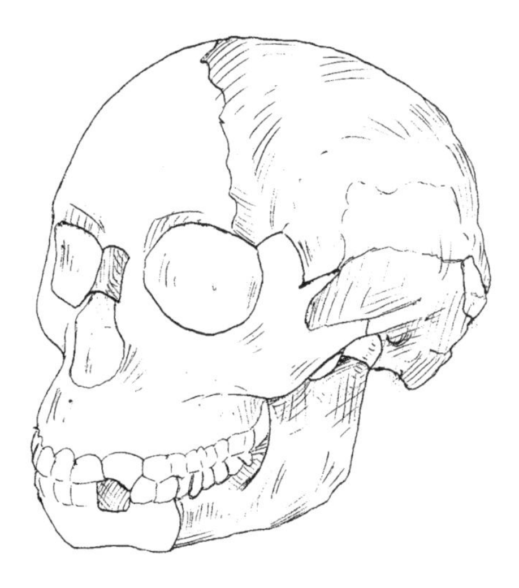

The Piltdown Skull as it was reconstructed. Only those parts that are shaded were the original material, all the rest is built up from plaster.

be a misidentified pig. The fossil, a single tooth, had been worn by use and erosion in such a way that it looked superficially like an anthropoid tooth. Researchers, one in particular, who were sure such a fossil would turn up, embraced it quickly. Only after careful and sober study was it found to be something other than a human ancestor.

These stories are held up by detractors of science and evolution as an example of the hubris, conceit, and arrogance of science and scientists and how duplicitous and gullible they can be. It is in reality a story of the corrective nature of science and how scientists themselves are the first ones to question new discoveries and claims and that this system is what makes scientific data reliable and the best way of discovering facts about the universe.

Sasquatch

Like Piltdown Man, Bigfoot, or Sasquatch, has been a popular topic in the realm of hoaxes and possible hoaxes. Generally dismissed by mainstream scientists as at best a product of folklore and at worst a hoax, Bigfoot has yet to be definitively proven or disproven. It can be argued that Piltdown was relatively easy to unmask as it constituted a small collection of testable artifacts. The tracks of Bigfoot are plaster casts of remains no longer available which cannot be tested directly. Believers have thousands of eyewitness reports—as well as all those plaster footprints—to base their contentions on, but those too are difficult to verify. The famous Patterson film of 1967 is a unique and fascinating artifact, but the blurry and jumpy nature of the footage renders it resistant to analysis, despite the many attempts to do so. An interesting aspect of the Bigfoot evidence is that unlike Piltdown, even hardcore believers acknowledge that some of the evidence has been hoaxed: though not all of it. One lesson that might be learned here is that the more concrete the evidence put forward for a hoax the easier it is to unmask, while vague data resists scrutiny better.

From the beginning, the search for evidence of human antiquity was as much political and cultural as it was scientific. The fossils and artifacts being found—mostly in Germany and France—were used to advance political agendas along with working out the riddles of human origins. Like many human endeavors, the search for our origins involved those who were looking to claim the first, or the oldest, or the most complete remains, or to say that their fossil showed them superior to other people. As this was all new, naturalists had not yet learned quite how to tell the difference between human fossils and other things. Early examples of misidentifying fossils included some which looked disturbingly human. Christened *Homo diluvinii testis* by the Swiss naturalist Johann Jacob Scheuchzer (1672–1733), these fossils had a human-like head and eye sockets, backbone, and shoulder blades. It seemed proof of the antiquity of humans until the influential Frenchman Georges Cuvier (1769–1832) pointed out that it was not a human at all, but a fossil salamander. Cuvier—who pioneered the field of comparative anatomy—did not believe in transmutation so was happy to debunk these relics. In 1805 another human "fossil" turned up. Clearly a human skeleton, not a salamander, the Guadeloupe skeleton was locked in a block of stone like any other fossil. Eventually, Cuvier took a look at it and realized it was a human skeleton, but not a fossil. It had been encased in a type of sand that could harden quickly giving the appearance of great age.

While most cases like these were honest mistakes there were the occasional outright hoaxes. One of the earliest was an American hoax known as the Cardiff Giant. In November 1869 a farmer named Newell said he dug up a large, bizarre fossil on his land outside Cardiff, New York. The story quickly spread through the media to the outside world. It was a human figure almost nine feet tall which had turned to stone through some unknown but extraordinary process. People came from far and wide to gawk at the forlorn giant and speculate on what it was. To take advantage of the crowd of rubes who showed up the landowner put up a tent and charged visitors a fee. Protestant ministers seized on the Cardiff Giant as scientific support for the accuracy of scripture. A New York geologist, James Hall, also came to look. His statements about the thing being an intriguing find were taken out of context by newspapers as scientific support for its genuineness. Eventually, a young Yale University paleontologist early in a long, distinguished career, Othniel Charles Marsh (1831–1899), inspected the anomaly and quickly saw that it was a block of gypsum carved to look like a human body. Marsh's assessment that the Cardiff Giant was a hoax did not deter showman P.T. Barnum who, unable to buy the humbug outright, simply made one himself and put that on tour. A few years later a similar giant showed up in County Antrim, Ireland. Found by a Mr. Dye, the Irish Giant looked like the Cardiff Giant and was exhibited in Dublin and other Irish cities. It too, eventually drifted into obscurity.

By the late nineteenth century a number of genuinely ancient humans like the Neandertals and Cro-Magnons had been found and were forcing naturalists to reassess how humans fit into the wider scheme of life on earth. Attempts were made to organize these ancestral humans into an evolutionary framework. The straight-line approach simply put the different hominids into an order based solely on age: oldest at the bottom and youngest at the top. This way *Pithecanthropus* from Asia led to the Neandertals which led to the Cro-Magnons of Europe which led to modern humans. It was a nice, simple arrangement which seemed to make sense. The influential French anthropologist Marcellin Boule (1861–1942) devised a more bush-like branching tree configuration modeled on that of the German Ernst Haeckel. Part of Boule's reasoning for using this approach was that his study of the Neandertals led him to believe they were not human, despite superficial appearances. He thought them dark, bestial monsters, and he did not want them to be connected in any way to modern Europeans. Indeed, the common, though erroneous, image of the Neandertals as "cavemen" comes from Boule. He allowed his biases to affect his conclusions. By adhering to a branching tree scheme, Boule and others could push the Neandertals—as well as the primates and any other hominid that annoyed them—off on a dead-end side branch and argue that while superficially similar to modern humans, the Neandertals had no direct connection to them as ancestors. Other researchers adhered to the pre-sapiens theory. They argued that modern humans appeared much earlier than even *Pithecanthropus* and so modern people were not related to the other hominids or primates, but predated them all. The Pre-sapiens theory became popular among those who accepted human evolution but were uncomfortable with admitting they were related to any fossil hominids.

Politics as an integral part of the search for the history of humans was not limited to France. English anthropologists had politics—both personal and national—on their minds as well. Sir Arthur Keith (1866–1955) took over at the Royal College of Surgeons in 1908. Born in Aberdeenshire, England, Keith studied medicine and zoology. He went to Thailand in the early 1890s to work as a doctor at a mining company and became fascinated by primate anatomy. He did studies on the relationship of humans and apes before starting work at the London Hospital Medical School in 1895. At the Royal College of Surgeons he influenced many students with his views on human evolution. His approach to natural history, the acquisition of fossils, and his relationship to scientists of other countries suggest that he saw it all as a competitive game he wanted to win. He felt British science was being left behind by the discoveries being made in France and Germany. He wanted a fossil that would launch British paleoanthropology—and his own career—ahead of his continental rivals. Until 1911 Keith accepted the straight-line theory of human descent, but after reading the work done on the Neandertals by Marcellin Boule he pushed the Neandertals aside, and accepted the Pre-sapiens theory. In working out how a human ancestor should look Keith formed in his mind's eye a picture and focused his search on that image.

While many Europeans wanted to push the hominids as far away from them as possible, others liked the idea of being able to claim the oldest human fossils or cultures as their own. In the late nineteenth and early twentieth centuries the nations of Europe were locked in a growing battle for national supremacy and had engaged in an arms race: a situation which would lead to World War I. Patriotism ran high as each state looked for ways to outdo the other in any field possible. The sciences were not immune from this frenzy. Citizens and scientists alike in France, Germany, Belgium, and Britain often considered themselves superior to the others. These political and cultural rivalries spilled into the study of human antiquity. In 1863 the French amateur fossil hunter Jacques Boucher de Perthes (1788–1868) found a human jaw in a quarry near Abbeville at a place called Moulin Quignon. De Perthes was an early pioneer in the attempt to show the great antiquity of the human race. He discovered a wide range of objects he believed to be human artifacts in the Abbeville region. He was eager to find actual human remains and so put out a bounty to be awarded to any quarry worker who brought him some. The bounty quickly brought results when a jaw was produced. The discovery of the Moulin Quignon jaw caused a good deal of excitement as it was thought to date from the early Quaternary age. If true this would radically change ideas about how old the human race was. It was a source of pride for many French scientists. It even made an appearance in Jules Verne's novel *A Journey to the Center of the Earth* (1864). After finding a number of still living prehistoric monsters one of the characters in the story asks if there "were any of these men [represented by the Moulin Quignon find] of the abyss wandering around the deserted shores of this wondrous sea of the center of the earth?" The hero, Professor Lidenbrock, whose ideas about the age of the earth lean to the catastrophist, makes references to the earliest humans being Caucasians. Verne threw into his story virtually every topic of geologic and paleoanthropologic discussion of his day. Many French scientists accepted the Moulin Quignon find as genuine. The well-known anthropologist Jean-Louis Armand de Quatrefages (who is mentioned by name in *Journey to the Center of the Earth*) championed it. British anthropologists, however, were not so enamored of the jaw and voiced their concerns. They thought it was much younger than the French did. Political animosities between the two cultures came out in how people responded to the jaw. The British antiquarian John Evans thought the jaw and the flint tools found with it fakes. Offended by the British reaction Jacques Boucher de Perthes declared that his prehistoric Man was being assaulted not only by the political and scientific, but religious establishment of England. A few years later the roles reversed when the British found a human fossil on their home turf.

In 1888 along the banks of the Thames River near Kent, chalk workers uncovered a broken but fairly complete human skull. The skull was brought to the attention of amateur fossil hunter Robert Elliot who had previously asked the chalk workers to keep an eye out for such things. The skull was quickly dug out under the protests of local schoolmaster Matthew Keys, who was also interested in fossils. The late nineteenth century was a period where fossil-hunting techniques were only just starting to get sophisticated, but old ways die hard. There was still much haphazard collecting by amateurs done in ways that would horrify modern paleoanthropologists. Keys, however, was a bit more forward-looking and wanted to photograph the skull in the ground as a way of properly recording the find before digging. His protestations fell on deaf ears and it was wrenched from its burial spot. This act fed suspicions that Galley Hill Man, as it came to be known, might not be a genuinely ancient find.

The Galley Hill skull was found in ancient strata but looked very modern. This pleased Arthur Keith because this was just what the Pre-sapiens theory suggested. Keith accepted Galley Hill as genuine, but others (Marcellin Boule included) did not. In reality Galley Hill Man was not as ancient as Keith hoped. It was found to be a Bronze Age intrusion dug down into the strata of an earlier age, making the newer skeleton appear to be from the earlier time. Ancient people, much like modern ones, dug graves for their dead. These graves would sometimes intrude into layers that were much older than the body itself. Years later when the skeleton was rediscovered it could easily appear to researchers to be a modern skeleton in an ancient layer of earth. This is part of the reason modern archaeologists and paleoanthropologists are so careful when they excavate a site and why they go through great pains to document every step of the process so that the material recovered can be recorded accurately. Misinterpretation was what Matthew Keys hoped to avoid by photographing the fossil. Arthur Keith had allowed himself to be swayed by his desire to find an ancient British human fossil, something that would bridge the gap between the past and present: What Keith really wanted was a British missing link.

The concept of the "missing link" has been a powerful one in Western biological science. It was thought logical that if living things evolved from one form to another there must be an intermediate stage that was a combination of the old and new types. These intermediate forms would put an end to any haggling over whether evolution actually worked. Anthropologists like Keith argued that a form of human ancestor must have existed that had both ape-like and human-like characters. The human characteristic would surly have been a big brain, as that was seen as the most obvious marker of humanness. There were arguments about which came first. Did our ancestors walk upright first? Did they acquire big brains first? Did they evolve modern grasping hands first? Did they come down from the trees, or did they start off on the ground? A missing link might answer these questions. The missing link hypothesis fascinated scientists, theologians, and the public alike. It would also set the stage for one of the great hoaxes in science history.

PILTDOWN MAN

Arthur Keith was, like many paleoanthropologists of the day, looking for something that fit his expectation of what a human ancestor should be. Australian anatomist Grafton Elliot Smith (1871–1937), then working in England, had his own view. As a brain specialist, Smith believed that for early ancestors to progress to modern humans, their brains had to enlarge first. As a curator at the British Museum of Natural History working primarily on fish, Arthur Smith Woodward (1864–1944) was looking for any early ancestor that would boost his career. In 1912 all three men got what they were looking for in the same package.

Piltdown Man had everything an early-twentieth-century anthropologist and human evolutionist could want: It had a modern-design skull, an ape-like jaw, and it was found in an ancient layer of strata. It fit the missing-link hypothesis nicely. At the center of the story was Charles Dawson (1864–1916), a lawyer by training, a naturalist by desire. Dawson, likable, middle-aged, with a big mustache and big dreams, was an amateur who wanted to enter the rarefied world of British science as a fossil and artifact finder. To that end he roamed the English countryside searching for them and enlisting the services of workmen at quarries and construction sites. In 1908, as the story goes, workers digging gravel on the land of Barkham Manor near Piltdown, Sussex, where Dawson had a business connection to the local landowners, had used some strange bits of rock while repairing the road. Dawson picked them up asking the workers if they saw any more they should hold onto them, as he wanted them. He thought these rocks were eoliths (ancient man-made tools). Returning to the site at a later time, Dawson was presented with bits of rock that looked to him like fossil bone and which he determined were pieces of a human skull. By 1911 additional fragments were brought out of the dig site. At this point Dawson contacted an acquaintance of his at the British Museum.

Arthur Smith Woodward was a trained geologist and curator of fish at the museum who had struggled to lift himself up from his working-class roots. He joined the museum as a teenager. He put together an extensive catalog of fossil fishes and while not an anthropologist he had studied first hand the Java Man fossils as well as a number of Neandertals including Heidelberg Man. Where Dawson was big and blustery, Woodward took himself seriously and was somewhat reserved. The two men had known each other since 1891 when they described some fossil mammals Dawson had discovered. Woodward named them *Plagiaulax dawsoni* in honor of his friend. Dawson had a long history of fossil hunting so while an amateur he had considerable knowledge of fossils and artifacts and experience at fieldwork. Woodward saw that Dawson's energy could be made to work to his advantage, so he gave the lawyer the honorary title of "collector" for the museum. Woodward described Dawson as,

> one of those restless people, of inquiring mind, who take a curious interest in everything round them–from an unusual form of ancient rushlight-holder to the latest device in electric torches; from an old parchment deed to a horn-like growth on a horse's head; from the proverbial live toad in stone to an escape of inflammable gas from the ground; from fossils and minerals in the Wealden rocks to the tools and other leavings of prehistoric man. (Woodward, 1948)

Eager for the recognition, Dawson threw himself into excavating the Piltdown site in 1912 and sent everything he found to Woodward. Dawson employed a local man, Venus Hargreaves, to do the digging, but the kept the site secret. They eventually uncovered more pieces of a skull, a good portion of a jaw, and several teeth.

After seeing the fossils Woodward and Dawson along with paleontologist Pierre Teilhard de Chardin (1881–1955) visited the Piltdown site together. De Chardin was a French Catholic priest and scientist who had studied with Marcellin Boule. He was gaining a reputation as a paleontologist of scholarly ability and by 1905 had already taught physics and chemistry and had studied geological strata and fossils in the field. He came to the Jesuit College at Hastings, where Dawson lived, in 1908. The next year he and another priest began inspecting the quarry there for fossils. The local workmen alerted Dawson who had been collecting there since he was young and he rushed over. The two men met and became friends. Pierre Teilhard de Chardin would go on to a long, distinguished career writing empirical books on human evolution as well as deeply mystical works on the relationship of evolution, religion, and the meaning of life. His metaphysical writings on the existence of a "noosphere," or moral super-organism, brought him occasionally into conflict with the more traditional and conservative elements of the Catholic Church.

In June 1912 the eager, jovial lawyer, the stiff scientist, and the lanky priest headed out to Barkham Manor. They found a few promising mammal fossils and a few eoliths. The following November a newspaper story came out in the *Manchester Guardian* that a human ancestor, probably the oldest ever found, had been dug up at sleepy Piltdown, Sussex. Charles Dawson and Arthur Smith Woodward presented Piltdown

Man to a meeting of the Geological Society of London on 18 December 1912. Although the actual fossil comprised relatively few pieces, a reconstruction was put together that filled in all the missing parts with plaster. The fully formed skull and jaw were a striking sight. The upper skull and face were human and the jaw apelike. The brain, going by the size of the skull, was on the large size. In their presentation, Dawson put forward the geological and archaeological material while Woodward did the presentation on the anatomical aspects. Displaying a racist edge unfortunately all too common in that day, Woodward said the teeth while primate, were of a kind seen in modern humans of "low types."

Woodward christened the creature *Eoanthropus dawsoni*. Piltdown Man, as it was popularly called, had all the characteristics believers felt a missing link should have. At the meeting Woodward was persuasive and most of the crowd accepted his assessment. Arthur Keith—a bit miffed that he as an anthropologist should have been kept out of the discovery loop in favor of Woodward who was a fish expert—had some reservations, but leaned toward acceptance. One or two in the audience were not so sure. Professor David Waterston, anatomist from Kings College, did not think the skull and jaw went together. The chin and important parts of the jaw where it articulates to the skull were missing so it was difficult to be sure just how the two worked together. There was some haggling over the correct size of the skull, but in general comment was positive. With his greater knowledge of primate anatomy, Keith argued that the skull should be bigger than Woodward made it but accepted Piltdown as genuine because it fit his preconceived image of pre-sapiens. Piltdown was not the ugly beetle-browed Neandertal of dim wit and growling countenance, but had a graceful, even beautiful face. Though they disagreed on details, the three British anthropologists agreed that Piltdown Man was a human ancestor.

One thing all in the audience agreed upon was that if there had only been a canine tooth found that would help clear up a lot. A canine had been fashioned for the reconstruction to fit what was thought to be the right morphology for such a creature. During later excavations at the site de Chardin found a canine tooth. Like the skull and jaw it was exactly what was predicted. At the original meeting the scientists also commented that a second collection of bones would help confirm the first set and might help fill in some of the missing parts on the first one. As if by magic, in 1915 at a different, but nearby site Dawson found the remains of a second individual. The discovery of Piltdown II seemed to confirm that the first was neither a fluke nor an accidental mix of different parts. Shortly after that, however, Dawson fell ill and passed away in 1916. While Arthur Smith Woodward kept digging, no more material was ever found. When he retired, Woodward went to live near Piltdown in order to keep searching.

Aleš Hrdlička (1869–1943), a curator at the Smithsonian Institution in Washington, DC, was at first noncommittal about Piltdown. Shortly after the fossil was announced he wrote about it for the *Annual Report of the Smithsonian Institution* (1913). As "the specimen was not yet available for examination by outsiders" (p. 509), he said, he only had the published accounts to go by. As a result he simply relayed to his American audience what the British were saying. Gathering more information, however, he grew uncomfortable with the awkward fit of the skull and jaw. It made no sense from an evolutionary point of view. Originally from Bohemia, Hrdlička was a meticulous anatomist with an eye for finicky details. He examined every new fossil that was claimed to be a human ancestor and found most of them wanting. He was puzzled as to why Piltdown had a modern-size brain, but not a modern jaw that would allow it to employ that brain to talk? Primates like apes, chimps, and monkeys can make sounds, but they cannot engage in human speech. This is because of the shape of their jaws, the number of facial muscles they have, and the movement of their tongues. The mechanics of their jaws do not allow for the subtlety and nuance of movement that gives humans speech. Piltdown had a brain like a human but no ability to articulate complex language. It was like a creature that had flippers for swimming but no way to hold its breath underwater. Hrdlička said moderns appeared as a result of a slow, steady evolution from an as-yet-unknown pre-anthropoid ancestor, not a pre-sapiens. He held to a more straight-line hypothesis with the Neandertal as an ancestor.

A wide range of naturalists accepted Piltdown because it seemed to have all the elements that they thought it should have: part human, part ape, big brain, ancient strata, and worn molar teeth. If you looked hard enough you could see whatever you wanted. British scientists were particularly proud that this important creature was one of their own. They now had a specimen that ranked with those being discovered in Europe. Even Marcellin Boule liked Piltdown because it was the kind of evidence that would help him keep the Neandertals out of the human line.

Several Americans were at first skeptical of Piltdown. In particular, William K. Gregory (1876–1970) and Henry Fairfield Osborn (1857–1935) of the American Museum of Natural History in New York hesitated to accept it. In 1917, however, pieces of the second skull were discovered. With this new development Osborn came around to a more sympathetic view of Piltdown. He was already in the pre-sapiens camp, and though Piltdown baffled him as it did Hrdlička, it was easier for Osborn to overlook the discrepancies because like his British colleagues, he was looking for something just like this: something that would prove human ancestry went back much further and did not include the brutish Neandertals.

There were other elements that helped Osborn come around. In 1855 what was thought to be a fossil human jaw was found at Foxhall, England, not far from Piltdown. While the jaw was a little dubious it seemed proof of a Stone Age culture in Britain. It was made public in an 1867 publication by an American doctor living in the area named Robert Collyer. He believed the jaw a genuine relic and the oldest such human fossil ever found. After making the rounds of many of the leading British anthropologists it fell to T.H. Huxley who pronounced it a modern skull from the Roman period of British history. Still believing he had a genuine fossil Collyer took it to France with similar results. Collyer may have been a student of the famous American anthropologist Samuel George Morton (1799–1851), the father of American physical anthropology and ethnology studies who wrote on human racial variation. Morton based his work on a monumental collection of human skulls from peoples all over the world. After his rejection by the British and French scientific communities, Collyer and his jaw dropped into obscurity until it was resurrected by British naturalist J. Reid Moir. In 1909 Moir tracked down the site and found human-worked flints in the same area. When Piltdown was made public in 1912 Moir argued that the flints found at Foxhall might have been made by Piltdown Man. Osborn agreed and felt these earlier discoveries placed Southeast Britain at the fore of human antiquity studies: something Woodward and Keith wanted all along. Osborn said, "We have here at last found in the Foxhall flints proofs of the existence of real *Tertiary man* [italics in original], of geologic age exceeding a million years" (Osborn, p. 35). He saw the Foxhall and Piltdown material as complimenting each other and was especially pleased with this because the British material supported the notion of a large-brained ancestor going well back in time.

Though of dubious antiquity, the Foxhall jaw helped pave the way for Piltdown's acceptance. It spurred J. Reid Moir to find the flints that were held up in support of Piltdown when it was found. In addition, the flints were from a geologic layer similar to the one in which flints were found in France by Jacques Boucher de Perthes. All these things helped convince Osborn and others that Piltdown was legitimate. As a result he became a major promoter of Piltdown in America. He wrote in his usual romantic style of Piltdown, "If there is Providence hanging over the affairs of prehistoric men, it certainly manifested itself in this case." (Osborn, p. 51). He used Piltdown in many of his publications including his bestselling *Men of the Old Stone Age* (1916) as well as *Man Rises to Parnassus* (1927). In 1925 Osborn opened his groundbreaking exhibit on human evolution, The Hall of the Age of Man. Piltdown figured prominently but oddly in the exhibit. Osborn said that the issue of the Piltdown skull and jaw going together were "settled" by the discovery of Piltdown II. In the exhibit Piltdown Man is shown in an elaborate human family tree as being on the line to modern humans but branching off in a way not easy to understand. The image seems to show either that Piltdown is a direct ancestor to moderns or a dead-end offshoot. The guidebook for the exhibit also figured Piltdown in a similar way. This small detail suggests that despite his enthusiasm for the fossil, Osborn was still not really sure what to do with it. The 1932 edition of the guide had much of the enthusiasm over Piltdown replaced by a straightforward description and a quick mention it might be from the Upper Pliocene Age. As an indication of the increasingly awkward position of Piltdown, by the time the 1938 edition of the Hall of the Age of Man exhibit handbook was released Piltdown had an even less enthusiastic description. Osborn had passed away in 1935 so his longtime colleague W.K. Gregory wrote the text for the new edition. Gregory had never liked Piltdown and so in the 1938 edition of the museum guide he described Piltdown as an anomaly not accepted as a human ancestor by most authorities. He says that if the jaw had been found separately from the skull, it would never have been thought of as associated with human evolution. The overall description of Piltdown in this edition of the guidebook leaves the reader with the feeling that the museum is giving a kind of shoulder shrugging over Piltdown: not discounting it, yet not really supporting it either. A nagging, vague disquiet led many anthropologists to drop Piltdown from their researches.

Piltdown Man suffered from the problem of fit. It was still unclear whether the jaw and skull went together. There was a problem of philosophical fit as well. Throughout the early part of the twenty-first century, other fossil humans were turning up. In addition to those found in Java a new batch had been discovered in China at a site outside Peking. Both Java Man and Peking Man—who were later recognized as both being *Homo erectus*—gave a different story to the course of human evolution than Piltdown did. Piltdown had been in line with earlier thinking about big brains developing first. Java and Peking suggested the opposite. Their development suggested that a large brain may have come relatively late in the process. In addition, the 1925 Taung find, while not convincing to many at the time began the move to view Africa as the cradle of the human race instead of Asia. As the twentieth century progressed Piltdown seemed more and more an anachronism. No more material was found and scientists were left with this small collection of intriguing bits that did not fit the growing view of how humans had evolved.

Over the years an enormous amount of literature was produced about Piltdown Man as scientists and theologians alike tried to work it into their theories with varying levels of success. By 1953, however, little in the discussion of human origins included Piltdown Man. Such was the case that when in that year a major human origins symposium was held in England Piltdown was not even on the official list of topics to be discussed. It was discussed informally though by a number of the attendees including Kenneth Oakley, Wilfrid Le Gros Clark, and Joseph Weiner. They discussed Piltdown's problem of fit and had many questions about how the fossils were found and why the material looked the way it did (these scientists had not been part of the generation which had been intimately involved with Piltdown and thus were not so wedded to it). A creature with the characteristics of Piltdown did not go with the current thinking on human evolution. These scientists were bothered by the fact that the skull was so human that it had no primate morphology. The jaw was almost totally simian, only the wear on the teeth was in the human realm. The Java and Peking fossils showed a blending of primate-like and early human morphology throughout their entire structure. Their traits were not isolated and segregated the way Piltdown's was. The question about Piltdown's big brain was where did it come from? Why when all other hominids being discovered had small brains why did this creature have a large one?

The scientists at the meeting tried to come up with explanations for how a human skull could come to be associated with such a primate jaw. It might have been a case of the two halves getting mixed together somehow. This sort of thing did happen, but when a second Piltdown was found that was ruled out as the chances of such a mix occurring in the same way at two different locations tested credulity. Oakley, Le Gros Clark, and Weiner decided to try to find out once and for all if Piltdown Man was what Arthur Smith Woodward always claimed it was. They went to the British Museum (Natural History) where the original fossils were kept with the plan of examining them with the latest investigatory techniques. Permission was granted and a series of chemical tests were performed. The battery of tests and experiments—including the new x-ray spectrograph—showed that the teeth and jaw had been altered by cutting and filing to show wear and to obscure anatomical details which would have given the fraud away. They had been stained with chromium as well. The jaw was that of an orangutan and the skull was an early modern human: Piltdown was a fake, built from the ground up from different parts like Frankenstein's monster.

In 1914, Charles Dawson found an artifact under a hedge at the Piltdown site that looked like an ancient tool of some kind. Dubbed the cricket bat—from the British game—this object was extraordinary. It was unlike anything ever found in England. The French anthropologist Abbe Bruiel who had done extensive work on the physical culture of Neandertals and Cro-Magnons thought the cricket bat the result of animals gnawing on an elephant bone, not human activity. Oakley made a copy of the cricket bat by cutting a bone with a knife and staining it. His efforts produced a replica that was identical to the original. Nothing about the Piltdown discoveries were right: not the skull, the teeth, the associated artifacts, or even the animal bones found with it. Everything must have been planted. The entire assemblage was hoaxed. J.S. Weiner put forward a possible explanation for how so many were fooled. "The great success of the Piltdown hoax," he said, "came from the clear conception on the part of the perpetrator that a man-ape of the right age appearing in the hitherto unknown gravel had a good chance of deceiving the paleontological world" (Weiner, p. 75). Once the excitement had calmed down, a sober examination of the evidence by researchers with no axes to grind had discovered just what Piltdown was.

The next problems to solve were who did it? Why did they do it? What did they gain from doing it? No one knows for sure. It could have been Charles Dawson using the "discovery" as a way up the ladder of scientific success or as a way to embarrass those who had already made it there. It has been argued that Dawson did not have the skill or anatomical knowledge to produce such a convincing forgery. In recent years some have questioned whether it really is a convincing forgery. One school of thought says Dawson did have the requisite knowledge from his university training; another blames Arthur Smith Woodward. The latter's motivation was advancement to the position of director of the British Museum, a post he coveted and was actively campaigning for. Woodward also had access to all the raw materials needed to make the forgery while Dawson had access to the site. The thirty-year relationship between Dawson and Woodward makes this scenario plausible. Later investigations showed that an alarming number of Dawson's pre-Piltdown finds—including a Roman figurine made of iron, Stone Age hand axes, and even *Plagiaulax dawsoni*—were also fakes. Charles Dawson was not the only candidate. Many have pointed to Arthur Smith Woodward, Arthur Kieth, and de Chardin. Several unusual suspects have also been named as the culprit, including Arthur Conan Doyle (creator of Sherlock Holmes) who lived near the Piltdown site. For years researchers have puzzled over and examined every scrap of evidence in an effort to find the hoaxer's identity.

To complicate the story, a discovery was made public in 1996 about the contents of a storage box. Originally stumbled across in the attic of the British Museum in the late 1970s, the box contained skeletal materials stained in the same way as the Piltdown fossils. The box belonged to Martin Hinton who was curator of zoology at the time of the Piltdown discovery. Woodward had turned down Hinton in 1910 for a research grant. The theory is that Hinton concocted the Piltdown forgery as revenge. He knew of Dawson's incompetence and Woodward's gullible pomposity and used their weaknesses to make them look ridiculous. Woodward was enamored of the idea of a British missing link, so Hinton led him down that trail. He even created the bogus cricket bat for Piltdown Man to play with (something that would have been especially enticing to an Anglo-centrist).

Since the discovery of the box of fragments, many have come to believe the case is closed, with Hinton the final guilty party. That seemed to end the story, but it didn't. There are various camps that champion one suspect or another, or support their man as not being the culprit. For example, E.T. Hall argues that while Professor Brian Gardiner accused Hinton there was the problem of manganese. Gardiner said that a chemical analysis of Hinton's specimens showed that they had manganese in them. He also said that early tests on Piltdown showed they too had manganese in them: this probably from some chemical treatment the hoaxer used to make the specimens seem old. Hall argued that the originals did not have manganese in them. So, Hall argues that if the two groups of specimens—the original Piltdown ones and Hinton's—did not contain the same chemical they could not be the same. In other words, Hinton's specimens were not part of the original Piltdown group, leaving Hinton off the hook as the perpetrator.

In his 2004 biography of Charles Dawson, Miles Russell undertook a major reassessment of Dawson's private papers by cross-referencing them with the correspondence of Woodward and others. This allowed him to formulate a more nuanced picture of these men's relationship. Russell's investigations led him right back to the original suspect. He argued that Dawson's desire for fame as a naturalist was the consuming passion of his life. Dawson wanted scholarly recognition from the British scientific community. He wanted to be elected to the Royal Society and maybe even win himself a knighthood, both prestigious signs of accomplishment. The large number of fraudulent discoveries Dawson had made in the years leading up to 1912 were a pattern that Russell saw as leading in a specific direction as part of a grand plan. Dawson was building up a catalog of finds designed to make him seem the equal of the superstars of British science. In this light, Piltdown can be seen, not as a single incidence of fraud, but as part of a long line of fraud designed to lead to the acknowledgement that Dawson was a man of importance. If Russell is correct his explanation lets all the other suspects off the hook with the exception of Hinton who may have been an accomplice to the scheme. It does, however, leave Arthur Smith Woodward, Arthur Keith, Henry Fairfield Osborn, and the others who supported Piltdown as dupes taken in by the cunning of the country lawyer with big ambitions.

NEBRASKA MAN

Not every false ancestor is a hoax like Piltdown. Many are simple misidentifications. The nineteenth and early twentieth centuries were a heady time for discovering fossil humans. England and Europe were not the only places fossils were found. A virtual parade of "men" appeared. While many were historic finds of great importance which opened up new avenues of thought about human evolution others were of peculiar countenance and dubious integrity. The British, of course, had Galley Hill Man and Piltdown. The Americans had Lansing Man in California, Vero Man in Florida, and Trenton Man in New Jersey, among others. In Argentina, paleontologist Florentino Ameghino discovered what he thought were early humans he called *Tetraprothomo* and *Diprothomo*. They all turned out to be misidentifications of modern skeletons.

There was a brief flurry of interest in a find from Oklahoma in 1918. An Associated Press story appeared in the *Tulsa Tribune* about a fantastic collection of finds from the rock quarry of A.H. Holloman. Workers there had unearthed artifacts from strata half a million years old. These included pottery shards, spear points, and tablets etched with a crude form of writing. There was also something that looked like a human hand and foot. The odd collection was distributed between the Smithsonian Institution and the University of Oklahoma Museum in Norman. It seemed to confirm an early human presence in the Americas. While they were not named, the article said that many scientists had visited the site. Despite this interest, the Oklahoma finds quickly dropped into obscurity.

A close second to Piltdown Man's infamy was a discovery from the Badlands of Nebraska. In September 1906, Richard Gilder, a Nebraska journalist and amateur fossil hunter, was searching for human artifacts near Indian burial mounds outside Omaha. Gilder had been investigating the local burial mounds of which there were quite a few. Searching near the Platte River he found where some boys had been digging out a rabbit hole and found human bones in the debris. Digging further he found the remnants of a fire pit and then a piece of human skull. He took the specimens to University of Nebraska professors Edwin Barbour and Henry Ward. Barbour thought them unusual but more modern than a Neandertal and while he thought they represented a type that was "low and savage" they were advanced enough to have developed religious rituals like burying the dead. Ward thought the bones unusually robust, the teeth well worn, and the brain case unusually small. Back in 1894 a group of explorers looking for the burial site of the Native American Chief Black Hawk also found some unusual skeletal material. When word of Gilder's discovery went out this earlier material was sent to him. An article about the discoveries then appeared in the *Omaha World Herald*, a copy of which was sent to Henry Fairfield Osborn in New York.

In these pre-Piltdown days Osborn was so excited by what he read that he immediately dropped everything he was doing and went to Nebraska to view the material firsthand. Osborn had only recently turned his attention to the specific question of human evolution (he had been working on general mammal evolution prior to this). He had followed the work of Marcellin Boule, Arthur Keith, and other physical anthropologists, but he had little hands-on human fossil experience himself. He had, however, a hunch that humans had made it into the Americas far earlier than anyone suspected. He was intrigued by the fact that the Nebraska find came from a locality he was already familiar with through his extensive study of fossil horses. Osborn excitedly examined the material—a skullcap, several long bones, and other assorted pieces. Heavy brow ridges on the skullcap and a general robustness of the bones suggested their great antiquity. Osborn published an article about the find in *Century* magazine the following month (there was in fact a whole flurry of articles in journals and magazines about the finds published by all the parties involved). He said that anthropologists were of two minds about the appearance of humans in the Americas. They believed they had appeared either very early or very late. Osborn believed in a very early entry date accepting the concept of a land bridge connecting the Americas to Asia that allowed for the migration of animals during the Pleistocene epoch. He concluded that early hominids must have crossed over in the same way. Osborn and other scientists had done much work on the wide range of fossil horses that had thrived in North America in ancient times. Horse fossils were abundant in the American West. The horse fossil series was a significant factor that helped prove the idea of the slow gradual change many believed the hallmark of evolution. He also knew that there was often evidence of human activity found in the same localities as horse fossils in other parts of the world. Osborn surmised that if these early horses had migrated from Asia to the Americas, humans would have followed them. Some heralded Nebraska Man as the oldest hominid known in the Americas. Osborn, while excited about the find was still cautious, but eager to have an American hominid. He had heard of similar material being found in the Western states but did not mention it in his articles because the reports were sketchy at best and he did not want to refer to anything that he did not know for sure.

Not everyone was so enthusiastic over the discovery. Sent by his boss at the Smithsonian Institution to view the bones, anthropologist Aleš Hrdlička examined all the Nebraska Man material with an eye toward anatomical detail but had disappointing news. His thorough study of the Nebraska material showed that the bones were not fossilized. His conclusion was that there was not even a remote chance the bones were what Osborn had hoped. The bones were the disinterred modern burial of a Native American. Edwin Barbour prophetically stated, "Unconsciously or otherwise an investigator is often influenced to see that which seems confirmatory rather than that which is contradictory to his conceptions and beliefs" (Barbour, 1907). It took Hrdlička, who did not believe in an early human presence in the Americas to see the material for what it was. As a result of Hrdlička's work, the first Nebraska Man vanished. Hrdlička did a survey of all the various fossils and bones found in the Americas which were purported to be human ancestors. He came to the conclusion that there were no ancient humans in the Americas. He did, in fact, help establish the first widely held scientific explanation that humans had entered the Americas very late, only within the last two thousand years before Columbus. This notion held until the 1930s when it was overturned by the discovery of human artifacts at Clovis, New Mexico. The Clovis hypothesis was that humans had entered the Americas around twelve thousand years before, earlier than Hrdlička but still much later than Osborn thought.

With Hrdlička's deconstruction of Nebraska Man, Osborn went about his other work. Though disappointed that the Nebraska material was not an ancient hominid he continued to hope one would be found. He felt all the right signs were there and that sooner or later an American fossil human would turn up. Osborn was given a second chance at an American hominid in February 1922. A paleontologist and fossil hunter from Nebraska named Harold J. Cook sent Osborn an unusual tooth he had uncovered. Upon receiving it, Osborn determined that it belonged to some kind of anthropoid. Cook found the tooth while searching the fossils beds near his family ranch on the Snake River. Cook and his father had associations with many leaders of American paleontology going back to the 1870s including such luminaries as Othniel Charles Marsh and Edward Drinker Cope. The Cook ranch was situated in the middle of a vast fossil field that routinely gave up many valuable and important specimens. The tooth was found along the Snake Creek, which was a tributary of the Niobrara. He and Osborn had a professional relationship going back some years from Osborn's work on the fossil horses of Nebraska. At Osborn's prompting Cook left Nebraska to come to New York to study paleontology with him, W.K. Gregory, and W.D. Matthew at Columbia. While a student at the University of Nebraska Cook's mentor was Edwin Barbour, who was involved in the first Nebraska Man of 1906.

Cook's first impression of this strange heavily worn tooth he sent Osborn in 1922 was that it came from an anthropoid, though he had no idea exactly what type of primate it was. He checked it against other known fossil teeth from the area and confident it was a primate of some sort sent it to Osborn. In New York Osborn was ecstatic at seeing the tooth. He concurred with Cook that it did seem to be a primate of some type and congratulated Cook on being the man to find the first American anthropoid. Unlike the first Nebraska hominid material, this was clearly fossilized and there was no chance of having confused it with a modern burial. Preferring to err on the side of caution, Osborn too sent the tooth out for double-checking. William K. Gregory and William D. Matthew, besides teaching at Columbia University were also on the staff at the American Museum of Natural History so they examined the tooth for Osborn. Matthew had done research in the very area where Cook had found the tooth and Gregory was well acquainted with fossil and modern primates. As Cook had done out in the field Gregory and Matthew compared it to other known fossils including, ironically a peccary, and saw no similarities. While not a human tooth it was not a recognizable primate either, though it seemed to have affinities with both groups. After conferring with Matthew and Gregory Osborn decided it was the first known anthropoid of the Americas so he christened it *Hesperopithecus*: the ape of the Western world. He declared the fossil in a paper in *Nature* and at a scientific meeting of the National Academy of Science in Washington. Though he said he had anticipated just such a find and was waiting for it to come along, he also called it the "greatest surprise" in the history of American paleontology.

Henry Fairfield Osborn believed that mammals had first evolved in Central Asia. His study of the distribution patterns of certain organisms—particularly horses and their extinct relatives the Titanotheres—from around the world suggested to him that mammals had originated in Asia then radiated out to other parts of the globe. As they did they adapted to their local conditions in Africa, Europe, and the Americas evolving into more specialized forms. Geologic evidence suggested to him that a thick, forested belt had once spanned a region from Europe and North Africa through Central Asia into the Americas allowing for a mass migration of animals. He also believed that primates and humans, being mammals, had also originated in Central Asia spreading out into Europe and Africa. Osborn saw in the fossil record that horses spread from Asia to North America using the Bering land bridge as a gateway. Osborn then postulated that if the horses had spread to the Americas from Asia it would be reasonable to conclude that primates and possibly human ancestors had as well. Osborn's human evolution theory—like many from the late nineteenth and early twentieth centuries—also had a good deal of racial and religious baggage attached to it. He saw Caucasians as the epitome of human evolution and, as a devout Christian, thought the fossil evidence proved the ultimate reason for evolution was to raise humans up to a point of spiritual salvation ordained by God. Osborn had first articulated this idea in a number of articles beginning in 1901. This helps account for his embrace of the first Nebraska Man, Piltdown Man, and the second Nebraska Man. Osborn did not base his enthusiasm for *Hesperopithecus* on the single tooth alone. His belief in an American hominid was predicated on a solid theoretical base (at least for what was known at the time). He had been building up circumstantial evidence for North American hominids for years. The tooth was just icing on a cake he had already baked.

Osborn shared his thoughts on *Hesperopithecus* (he never called it Nebraska Man) with his British anthropologist, friend Grafton Elliot Smith telling him he thought it fell between the anthropoid apes and humans. He told Smith that he was so excited about the find that he was having trouble concentrating on a book on the evolutionary history of elephants he was preparing. Smith was even more enthusiastic about *Hesperopithecus* than Osborn was. He asked Osborn to send him all the published material on *Hesperopithecus* then available: something Osborn immediately did. Smith started making mention of it in the British press and trumpeting it as a major find in the field of human evolution studies. Osborn was nervous, however, because the European press began referring to *Hesperopithecus* as a human ancestor. Osborn emphatically did not believe it was. The problem of Nebraska Man started with the 24 June 1922 edition of the *Illustrated London News*. They ran an article on *Hesperopithecus* with an illustration showing the creature as an archaic but clearly human-like creature. Using Grafton Elliot Smith as an authority, the *Illustrated London News* called the find "an astounding discovery of human remains." Osborn knew it would cause trouble because the paper was making statements not supported by the fossil record. He wrote Smith saying he could not agree with him that the creature be called an ape-man because there was still so little evidence to confirm it. Osborn argued that *Hesperopithecus* was neither an ape ancestor nor a direct human ancestor. Grafton Elliot Smith had jumped the gun on Osborn's pronouncements. (The Nebraska tooth did not particularly impress Arthur Smith Woodward.) The damage had been done, however; now it appeared as if Osborn was touting *Hesperopithecus* as a human ancestor.

While all the public pronouncements were being made, staff at the American Museum of Natural History reexamined the tooth. By 1925 Osborn's assistant, William K. Gregory, was having second thoughts about *Hesperopithecus*. Osborn had sent an expedition to the original site on the Snake River in order to find more remains. They did find more, unfortunately those pieces showed Nebraska Man was not a man, or an ape, but an extinct peccary, *Prosthennops*. Unmasked as a pig Nebraska Man had to be explained. Osborn had been preparing to travel to Dayton, Tennessee, to appear for the defense in the Scopes Monkey Trial in July 1925. He was planning on using *Hesperopithecus* as evidence supporting the concept of human evolution. He was also going to use it to make his rival William Jennings Bryan—who was appearing as co-prosecutor in the case—look out of touch with the latest scientific discoveries. Bryan and Osborn had been publicly jousting over evolution (Bryan was an outspoken Creationist) since 1922. In what at the time seemed like a delicious coincidence just waiting to be exploited, Bryan was from Nebraska as well. Osborn's plan was to go to Dayton and wave *Hesperopithecus* in Bryan's face. Lucky for Osborn his scientists discovered the reality of what the creature was before Osborn left. As a result Osborn did not go to Dayton to testify because now that it was clear the tooth was not from an anthropoid he assumed—probably rightly—that anti-evolutionists would be able to throw Osborn's fossil back in his own face as a hoax or an excellent example of the perfidy of scientists. Word of what *Hesperopithecus* really was began to filter out into the scientific community almost immediately and William K. Gregory made an official statement at a meeting of the New York Academy of Science and in an article in the December 1927 issue of the journal *Science*. The scientific world immediately dropped *Hesperopithecus* from its consciousness and moved on.

At about the same time as the Nebraska Man debacle another strange American primate was discovered, or so some claimed. François de Loys was a Swiss naturalist and explorer on an expedition scouting for oil deposits up the Tarra River along the border of Venezuela and Colombia. At a campsite, a pair of strange primates allegedly attacked de Loys and his party. The attack was so ferocious that de Loys's men fired on the creatures, killing one and driving the other off, after of course, the appropriate feces throwing. The creature was so unusual that de Loys propped it up on a box and took a photograph of it. He then skinned it and prepared the skeleton to bring back with him. According to de Loys, the hardships of the rest of the expedition were such that all of his equipment and specimens, including the creature's remains, were lost. All that was saved was the photograph, which caused a sensation. De Loys said the creature was almost five feet tall, tail-less, and weighed more than one hundred pounds. De Loys's friend geologist George Montandon thought de Loys had found an unknown American species. Montandon was a polygenist who also subscribed to white-supremacist Aryan theory. He seized on the photo as evidence of the separate evolutionary line for the aboriginal Americans. The British just sniffed at this Swiss monstrosity.

Arthur Keith suspected de Loys had fabricated the whole thing. He said it was nothing more than a large spider monkey with its tail cut off and that there was nothing to show that it was anywhere near the five feet tall de Loys claimed. However, in the photo, de Loys had set the dead animal on a Standard Oil Company packing crate, which was known to be a uniform size. This allowed other scientists to extrapolate the creature's height in comparison to the box. It came out to be a little less than five feet tall. With nothing to go on but the photograph, de Loys's ape never made it into the discussion of human or primate evolution. In the 1960s pioneering cryptozoologist (one who studies unknown animals) Ivan Sanderson reexamined the de Loys photo and came to the same conclusion as Keith. Had the creature been what de Loys claimed it was, it would have been doubly intriguing, as five-foot-tall, tail-less, upright anthropoids were not supposed to have ever existed in the Americas. Supporting the hoax claim, no other examples of de Loys's ape have since appeared.

THE REACTION

The lessons learned from Piltdown Man and Nebraska Man by the scientific community involved learning to be more cautious about accepting discoveries when they seemed a little too good to be true, and to be careful not to let the desire to find a particular answer blind them to other possibilities. These incidents also help reinforce the corrective nature of the scientific method. In other words, science works by the constant accumulation of material. No answer is ever final, no analysis the complete answer. Scientific ideas, theories, and even facts are only held as long as evidence is there to support them. Once new evidence is found old ideas are modified, changed, and even swept away entirely. Practitioners accept the ever-changing nature of scientific inquiry. Because of this attitude Piltdown Man and Nebraska Man have both faded to mere footnotes in the scientific literature, if they appear at all.

For the most part, the only reason these incidents are remembered today is because anti-evolutionists continue to dredge them up as evidence of the perfidy of scientists, the hollowness of science itself as a way to acquire knowledge about the universe, and of course that these incidents prove that evolution is a sham. Creationists argue that as evolution has no evidence supporting it scientists must manufacture it (there is a long line of reasoning which accuses scientists of making up essentially all evidence for evolution). With Piltdown a proven hoax, anti-evolutionists argue, all the evidence must be false. Piltdown and Nebraska Man also show how insidious and demonically guided scientists work tirelessly to keep the "truth" from the world. This concept, also very popular within the anti-evolution and anti-science world, is humorous considering it was scientists themselves who unmasked both Piltdown Man as a hoax and Nebraska Man as a misidentification. Anti-evolutionists would never have known Piltdown was a hoax unless scientists told them it was: because it was scientists who were the first to question it and to try to figure out what it really was.

A case study in the reaction to Nebraska Man is the writing of Ian Taylor. A corporate scientist with a degree in metallurgy from London University, Taylor, like many of the most ardent and outspoken anti-evolutionists, converted to Christianity relatively late in life. Following his conversion he left industry and managed to get himself involved in Christian broadcasting as a documentary producer. He then went on to be a Canadian religious radio talk-show host. His most well known work is *In the Minds of Men: Darwin and the New World Order* (1984). Taylor believes that evolution is the root of all the evil in the world, and does not particularly care for the American public school system either as it spreads belief in evolution. Public schools, he believes, only regurgitate falsehoods and teach children to conform to doctrine. Taylor also discusses Nebraska Man in this book. Promotional blurbs for *Minds of Men* said it was "thoroughly researched from original sources" yet he makes wildly inaccurate statements. Most of these inaccuracies could have been easily avoided had he actually worked from original sources. For example, Taylor refers to Henry Fairfield Osborn as a Marxist and a leading proponent of the American Civil Liberties Union. It was Osborn's Marxist tendencies, Taylor argues, that led him to oppose Christian Fundamentalism and to accept Nebraska Man as genuine. Anyone with even a passing knowledge of the life of Henry Fairfield Osborn would know that far from being a Marxist, Osborn was a staunch conservative and devout Presbyterian whose theories of human evolution were as much metaphysically based as they were empirically, and Osborn was not particularly thrilled with the ACLU. Part of the reason he did not go to Dayton to testify at the Scopes trial was that he was uncomfortable with Clarence Darrow's socialist affiliations, which as a conservative did not mesh well with Osborn's. In June 1925 Scopes came to New York to meet with American Civil Liberties Union officials. While not completely thrilled with the ACLU, Osborn liked Scopes and gave him advice as well as money to help tide him over. Besides the scientists, Scopes's defense team drew to it a circle of political liberals, socialists, and Progressives. It was a situation that annoyed the more conservative Osborn. A number of scientists Darrow had asked to testify bowed out because they did not want to be publicly associated with a radical, socialist, anti-eugenics agnostic like Darrow. Osborn confided to Scopes's advisor, G.W. Rappleyea, that he thought Darrow the wrong man to lead the defense.

Ronald Rainger, the author of the first major modern work on Osborn's career, rebutted Taylor through an Internet site that was carrying some of Taylor's remarks and corrected them publicly. As the author of the only other modern biography of Osborn I also contacted several Web sites that carried Taylor's words to correct them as well. Neither Professor Rainger's advice nor mine was heeded, because by the fifth edition of Taylor's book (2003) the same mistakes were still in the text. As late as 2006 the Creationist group Answers In Genesis had a Web page up on Nebraska Man that regurgitated Taylor's inaccuracies about the case as a reprint of Taylor's 1991 article for *Creation* magazine.

As with a number of anti-evolutionists who use Nebraska Man as a straw man, Taylor had a habit of misrepresenting original sources. For example, in chapter 8 of *In the Minds of Men* he takes the work of Osborn out of context with an almost giddy disregard for what the author is saying. In *The Origins and Evolution of Life* (1917) Osborn makes a statement that the ancient Greek philosophers were looking for more naturalistic explanations of the origins of life and were moving away from supernatural explanations. Osborn is clearly making a statement about what he thought the Greeks were doing not about his personal feelings. While Taylor could take issue with Osborn's analysis, he sidestepped this and instead interpreted Osborn's statement as being a personal philosophy. Taylor saw this as an example of how Osborn was drifting toward ungodliness, unable to differentiate between a statement made by someone about someone else and a statement of someone about their own feelings and beliefs. Taylor used this to show his audience of Christian Fundamentalists that anyone believing in evolution is suspect and an agent of darkness.

In chapter 15 Taylor suggests that the rise of eugenics, Hitler, and the Nazis were a direct result of the belief in and teaching of evolution (this is a popular vein constantly mined by anti-evolutionists both religious and political). Curiously Taylor says that Osborn was one of the people who influenced the Nazis. Osborn was the first American scientist awarded an honorary degree by a German university during the Nazi era, and privately thought the Nazis were doing a good job at rebuilding Germany after World War I. However, Osborn had been fooled by the Nazis' early embrace of science, and never really saw the true face of the Nazi regime before he passed away in 1935. The problem here is that Taylor had already accused Osborn of being a Marxist, and that his Marxist leanings contributed to Osborn's belief in evolution (despite the fact that Osborn was drawn to evolution because of his Christian beliefs). Taylor did not seem to know that the Nazis were major anti-communists. They considered Marxism a threat to the world and felt it was their duty to snuff it out. Taylor never explains why the Nazis would have embraced the work of a "Marxist" like Osborn. Taylor—and other anti-evolutionists—want it both ways; he wanted to equate evolution with Nazism and communism, but got his facts so tangled he only came off looking ill-informed. In reality it was Osborn's friend, the infamous eugenicist Madison Grant that was influential to the Nazis' extermination program.

One explanation for why Taylor insists on calling Osborn a Marxist when he was not has something to do with the American anti-evolution movement and its kinship to political conservatism. In the late twentieth and early twenty-first centuries political and religious conservatism converged more strongly than ever before. Kindred souls within the anti-evolution/conservative political nexus put forward the myth that all creationists and anti-evolutionists were political conservatives as well and vice versa. In opposition to this they manufactured an easily identifiable image which suggested anyone who supported evolution must be a communist, liberal, or atheist: probably all three. Promoting this idea helped solidify wider support for the anti-evolutionist cause which in turn became a plank in the conservative political agenda. This set up a simple "us versus them" dichotomy tailor-made for mass consumption. A line could be drawn where good people are on this side, bad people are on that side. Henry Fairfield Osborn—and others like him then and now—represent an anomaly for the anti-evolutionists. He was a man who was staunchly pro-evolution, yet at the same time was an arch political conservative and a devout Christian. He and many others like him throw into contention the idea that only liberals or atheists are for evolution and that all religious people oppose it. Osborn's career shows that the discussion of who does or does not support evolution is far more nuanced and complex than any simplistic black-and-white portrayal of it could ever be.

CONCLUSION

There are big differences between hoaxes and misidentifications. Scientists are not the only ones who fall prey to this. Anti-evolutionists have had their share of misidentifications and hoaxes as well. There have been modern hammers found locked in sedimentary rock, coins embedded in stone, human skulls under ancient strata all held up as evidence that evolution is a sham. They have all been found out to be faked or misidentified. Two well-known examples of this are the "Meister Print" and the Paluxy River "man tracks."

In 1968 fossil hunter William Meister found a strange artifact while prospecting for trilobites, in 500 million-year-old strata known as the Wheeler Formation. When he cracked open a slab he discovered what looked like a human shoe print. As if that weren't amazing enough the shoe had crushed a live trilobite under its heel. Creationists seized upon the Meister Print as evidence that humans had trod the earth during a time scientists said they did not exist. They held the Meister Print as proof that the fossil record, the geologic time scale, and the very notion of evolution were false. Studies showed the print was in reality an example of a common geologic occurrence known as spalling, in which slabs of rock break away from each other in flat plates. This particular case of spalling had created a simulacrum suggestive of a shoe print. After the determination of prosaic natural causes, some creationists stopped acknowledging the print in the 1980s, while others continue to do so.

An enduring belief within creationist circles has been that humans coexisted with the dinosaurs. To support the contention, many creationists pointed to the "man tracks" of Texas. They argued that what paleontologists called dinosaur tracks were in reality human footprints frozen forever in mud turned to stone. The tracks were first discovered in 1917 along the banks of the Paluxy River outside the town of Glen Rose, Texas. This region of Texas had been a floodplain 70 million years ago. Many of the tracks were clearly those of dinosaurs, but a few took on odd shapes because of the distortion caused by the animals walking across the wet mud. It was these tracks that were claimed to be human. Roland T. Bird (1899–1978), the American Museum of Natural History excavator who did the first extensive work at Glen Rose in the 1930s, tried to convince creationists that the prints were just deformed dinosaur tracks. Enthusiasts ignored him and happily measured, photographed, and discussed the tracks anyway insisting they were human.

When searching for the secrets of the universe and the life contained in it, searchers—whether scientists, historians, or theologians—must always be careful to see what is really there instead of what they want to be there. Like a character from a classical tragedy Piltdown Man stands behind us and with his toothy grin whispers, "be careful!"

FURTHER READING

Osborn, Henry Fairfield. *Man Rises to Parnassus* (Princeton: Princeton University Press, 1927).

Osborn, Henry Fairfield, W.K. Gregory, and George Pinkley. *The Hall of the Age of Man, Guide Leaflet Series* (New York: American Museum of Natural History, 1925–1938).

Regal, Brian. *Henry Fairfield Osborn: Race and the Search for the Origins of Man* (Aldershot, UK: Ashgate, 2002).

Regal, Brian. *Human Evolution: A Guide to the Debates* (Oxford: ABC-CLIO, 2004).

Russell, Miles. *Piltdown Man: The Secret Life of Charles Dawson* (Stroud, UK: Tempus Publishing, 2004).

Spencer, Frank. *Piltdown: A Scientific Forgery and The Piltdown Papers: 1908–1955*. (Oxford: Oxford University Press/Natural History Museum Publications, 1990).

Taylor, Ian. *In the Minds of Men: Darwin in the New World Order* (Zimmerman, MN: TFE Publishers, 1984).

Weiner, Joseph, Oakley, Kenneth, and Clark, Wilfred Le Gros. "The Solution of the Piltdown Problem," *Bulletin of the British Museum (Natural History) Geology* (November 1953): 139–146.

Fruit Flies

Jeffrey H. Schwartz

INTRODUCTION

"Fruit fly." What images that moniker conjures up. Pesky little insects that even to the naked eye appear an oxymoron of morphology. Chunky body. Small wings. How they become airborne and can avoid our rapid clapping of hands to demolish them is a mystery that plagues us from childhood on. They also seem to be capable of spontaneous generation, appearing as if out of nowhere the moment a bunch of bananas becomes ever so slightly overripe.

For many of us, "fruit fly" also brings back memories of experiments in high school and college biology classes in which we set out to learn the fundamental principles of heredity. One of the reasons fruit flies were chosen as the hapless players in these experiments was because hundreds of them are born each generation, you could breed hundreds of generations of them during a semester—which meant that you could get lots of data in a short period of time.

The principles of heredity that breeding experiments with fruit files were supposed to teach us were discovered in the mid-nineteenth century by the Czech monk, Gregor Mendel. Many of us also know that Mendel's organism of experimentation was not the fruit fly, but the common, edible garden pea—which raises the question of how a study of inheritance that was based on observing plants came to be applied to animals. What many of us probably do not know is that the underlying motivation for Mendel's experiments was not purely for the sake of science. To the contrary, it was primarily economic. Thus before we can get to the nitty-gritty of fruit flies themselves, we must first understand the history, personalities, and goals, scientific or otherwise, that preceded experimentation with these insects.

GREGOR MENDEL

Mendel was born in 1822 into a peasant family living in former Silesia, which lies in the northern part of what is now the Czech Republic. His given name was Johann. Rather than insist that he devote his life to the family farm, first Johann's father and then his sister supported his education. Upon completing his studies at the age of twenty-one in physics and mathematics at the Olmütz Philsophical Institute, his physics professor, Friedrich Franz, recommended that he continue his studies at the Augustinian monastery in Brünn (later called Brno). He changed his given name to Gregor when he entered the Altbrünn Monastery as a novice. At the age of twenty-six, Mendel was ordained into the priesthood, which both freed him of concerns about his daily needs and provided him with a means of entry into cultural and scientific circles. At the monastery, Mendel's primary task was teaching mathematics. Eventually, he also took over tending to the monastery's gardens, which afforded him the opportunity to experiment with plant hybridization.

According to the historian of science Robert Olby, during his education, Mendel probably learned about cells, fertilization, organismal (especially plant) growth, and early ideas on evolution. Together with his desire to standardize approaches to plant breeding, it was probably the latter intellectual endeavor that initially provoked Mendel to experiment with plant hybridization because, as is evident in his writings, he struggled with the question of whether species were stable in nature (as was advocated by the Church and part of the ideology of the Great Chain of Being), or whether variation was a natural phenomenon, with differences between individuals of a species being of a different kind than those that distinguished species from one another. Although Mendel knew from the efforts of plant breeders that his project would be daunting and would require thousands of breeding, crossbreeding, backbreeding experiments, he persevered. For it would only be through such painstaking work that he might be able to formulate "a general law governing the formation and development of hybrids" that would then allow him "to determine the number of different forms under which the offspring of hybrids appear, or to arrange these forms with certainty according to their separate generations, or definitely to ascertain their statistical relations" (Mendel, 1866, translation, p. 1965).

Because of his background in science, and physics in particular, Mendel brought to plant breeding an element that until then had been lacking: a prediction of the outcome of certain parental crosses. This was crucial, because while plant breeders knew that features of parental plants were passed on intact to their offspring in the succeeding hybrid generation, they also knew from observation that the expression of these traits was variable. But they didn't know if the expression of traits in the hybrid generation was random or governed by a general law of inheritance. Clearly, being able to predict the composition of future crops was critical in order for plant breeders to minimize waste and economically useless hybrids. His scientific training aside, it was fortunate that Mendel chose the common garden pea with which to experiment. For, unlike most plants, the features he studied—such as seed size, form, and color, position of flowers on the stem, and stem length—were not linked. Consequently, because the expression of one character in garden peas is independent of others, Mendel could keep track of how each character was inherited.

Before Mendel could perform his experiments, he first had to ensure that his parental stocks bred true for their features—and from 1856 to 1858, he did just that by selectively breeding and then cross-checking by crossbreeding offspring until every generation of each line was the same. Having satisfied himself that he did indeed have pure parental stocks, he embarked on his hybridization experiments. Although other plant breeders used the terms "active" and "latent" to represent the differential inheritance in hybrids of parental features, Mendel chose instead to refer to these characteristics of heritable material as, respectively, "dominant" and "recessive." As he described them, dominant features are those "which are transmitted entire, or almost unchanged in the hybridization, and therefore in themselves constitute the characters of the hybrid," while recessive features "become latent in the process . . . [and] withdraw or entirely disappear in the hybrids, but nevertheless reappear unchanged in their progeny" (Mendel, 1866, translation, p. 8). Mendel represented a dominant character state by a capital letter (e.g., A or B) and a recessive character state by a lowercase letter (e.g., a or b).

Although plant breeders had known for some time that features that were absent in one generation might appear in another, it was Mendel who demonstrated that when individuals of a hybrid generation (the result of breeding a parent purebred for the dominant character state with a parent purebred for the recessive character state) were crossbred, the dominant character state was expressed in their offspring essentially three times more frequently than the corresponding recessive character state. Furthermore, in this particular generation, the number of hybrids (individuals bearing one dominant state and a corresponding recessive state) was double the number of individuals that were pure for the dominant character state or pure for the recessive character state. That is, for every one purebred individual, whether purely dominant or purely recessive, there were two hybrid individuals.

Mendel presented the results of his experiments in 1865 at two different meetings of the Natural History Society of Brno (formerly Brünn) and published his findings in 1866 in the Society's *Proceedings*. Mendel sent one of the forty reprints he received upon his article's publication to

Karl Wilhelm von Nägeli, who was generally regarded as *the* botanist of the time. No doubt von Nägeli took issue with Mendel's conclusion that

> The offspring of the hybrids in which several essentially different characters are combined exhibit the terms of a series of combinations, in which the developmental series for each pair of differentiating characters are united . . . [and] that the relation of each pair of different characters in hybrid union is independent of the other differences in the two original parental stocks. (Mendel, 1866, translation, p. 19)

Rather than entertain Mendel's suggestion of stable forms von Nägeli clung to the Great Chain of Being notion of the intermediacy of forms as well as to his own theory of the "idioplasm" to explain their existence. Without von Nägeli's support, Mendel's experiments and insights were not accepted by the scientific establishment. It was as if he had never done this work. Only in 1900 were his "laws" supposedly independently rediscovered, or at least resurrected, by three different plant breeders.

DARWIN'S THEORY OF INHERITANCE: PANGENESIS

Early in 1860, not long after the publication of *Origin*, Charles Darwin began writing the draft of a two-volume opus that would be published eight years later under the title, *The Variation of Animals and Plants Under Domestication* (Darwin, 1868). In this work, Darwin formally articulated a theory of inheritance, which he called "pangenesis." Recalling what he had written years earlier in Notebook D (during the year 1838)—the "whole parent [must be] imbued with the change"—Darwin's theory of pangenesis was predicated on the notion that inheritance in offspring emerged through the blending of contributions from every part of each parent's body. As such, even though an individual was the sum total of all its parts, Darwin conceived of each part as independent. He, of course, had to assume this because, in order for his ideas of selection to work, selection had to be able to act on individual features.

Basic to the theory of pangenesis is the notion that through all phases of an individual's life, each part or organ throws off minuscule particles, which Darwin called gemmules. Gemmules dispersed via bodily fluids and were capable of reproducing themselves. Because gemmules embodied every stage of development of the organ or feature from which they derived, and because they could multiply in number, wounds could heal and some animals could regenerate a lost limb or tail. Most gemmules were active, but those that were dormant could also be inherited across generations. Darwin suggested that a later re-expression of dormant gemmules explained the emergence of reversions or atavistic features (features that were present in a distant ancestor or ancestors). In order to account for the lack of features in offspring of features possessed by one or the other parent, Darwin proposed that dominant gemmules could be suppressed.

In cases where animals could produce viable offspring via unfertilized eggs (such as ants), Darwin had to speculate that their sex cells would carry a full complement of gemmules. However, the sex cells of organisms that reproduce sexually (which would include a host of plants) would not carry the full complement of gemmules, which would allow gemmules from both parents to blend in order to produce the traits of their offspring. Nevertheless, in order for embryos to conform to the theory of pangenesis, they could not, strictly speaking, eventually develop into an adult. Rather, an individual would develop as a result of a succession of gemmules that represent different stages and ages on the way to becoming an adult. In addition, because different gemmules represent different phases of a parent's life—from embryo to adult—features that were acquired through use or disuse (because, according to the "Lamarckian" tenets that Darwin wholeheartedly embraced, an organism's desires can cause it to change) were also recorded in one's gemmules and could therefore be passed on to one's offspring. Darwin's theory of age-related gemmules also allowed him to account for features that emerge later in an individual's life, such as sexually dimorphic or secondary sexual features.

Among the scientists with whom Darwin shared his thoughts on pangenesis prior to publication of *Variation of Animals and Plants Under Domestication* was his cousin, the polymath, first biometrician, and father of eugenics Francis Galton. Galton took Darwin's notion of gemmules dispersing via bodily fluids to mean that they were carried via the blood or circulatory system. As such, in 1871 Galton conceived of and conducted a number of experiments to test the theory of pangenesis in which he injected or transfused blood from one variety or breed of rabbit into another to see if the features of one would appear in later generations of the other. Needless to say, the results were a disaster.

To be sure, Galton would have accepted Darwin's theory had the transfusion experiments had worked. But, as with those experiments in which only a few ounces of blood were injected into host animals, total transfusion from donor to host failed to produce the results pangenesis predicted. Consequently Galton was forced to conclude in a paper he read on 30 March 1871 at a meeting of the Royal Society of London that

> If the reproductive elements do not depend on the body and blood together, they must reside either in the solid structure of the gland, when they are set free by an ordinary process of growth, the blood merely affording nutriment to that growth, or else they [are] . . . temporary inhabitants of [the blood], given off by existing cells, either in a fully developed state or else in one so rudimentary that we could only ascertain their existence by inference.

His comments were published a few weeks later.

Although Galton had shown Darwin a draft of this manuscript prior to making it public, Darwin attacked his cousin in a letter to the journal *Nature* in which (contrary to what he actually wrote) he denied even implying that gemmules might be dispersed via the circulatory system. Although Galton believed that he had been betrayed by Darwin, he kept his response more restrained and less hostile. But it is probably because of his attempts to demonstrate pangenesis that Galton went on to propose his own theory of inheritance.

FRANCIS GALTON'S THEORY OF INHERITANCE: STIRPS

In 1875 Francis Galton lectured at the Anthropological Institute of London on his new theory of inheritance. The text of this lecture was published the same year. A radical difference between this theory and Darwin's theory of pangenesis was that Galton situated the "stuff" of heredity in the reproductive organs themselves; it did not result from the accumulation in reproductive organs of products from elsewhere in the body. Galton called his units of heredity "stirp," which is a term he derived from the Latin word for root, "stirpes." And because stirps were products of the reproductive organs they, unlike Darwin's gemmules, were exempt from the influence of use and disuse. Although Galton admitted that an insult from an external force might impact the sex organs, and provide the basis for inheriting an acquired characteristic, he considered the likelihood of this happening very slim.

The germ or basis for each body part was contained in stirps. Continuity from parent to offspring was maintained by contributions from parental stirps to the offspring's stirps. Stirps that were not involved with the development and growth of an individual became that individual's stirps, which were then passed on to the next generation. But here Galton broke with received wisdom of the time, in which it was believed that inheritance provided an identity or sameness from one generation to the next. Instead, Galton proposed that "the personal structure of the child is no more than an imperfect representation of his own stirp, and the personal structure of each of the parents is no more than an imperfect representation of each of their own stirps" (Galton, 1876). Because of this, stirps could be passed from parent to child while an individual could maintain a unique identity.

In his stirp theory Galton somewhat paralleled Darwin by going against received wisdom by suggesting that any notion of heredity had to go beyond consideration of parent and offspring alone. For Galton, the popular opinion was too constrained to account for reversals or the re-emergence of atavistic structures. Instead, he argued, heredity should be thought of as the inheritance of stirps and of the germ cells that derived from past ancestors. Since each parent can contribute to offspring only half of its stirps, there must be a competition that makes it possible for some but not all of the germs to be passed on to the next generation. Perhaps more important, however, rather than embracing the idea that germs could blend in offspring, Galton insisted that the only way in which a child could have features recognizable as belonging to one or the other parent would be if germs behaved as "particles"—discrete units that acted independently of each other—and in this latter way, Galton's stirp theory unwittingly mirrored Mendel's. By combining the notion of particular inheritance with other ideas of his, Galton was able to explain by the same naturally occurring mechanism how children and parents could be similar in various ways and yet also uniquely different.

The implications of Galton's stirp theory and its being grounded in the notion that units of heredity are discrete entities are two-fold. First, as mentioned earlier, it leads to the rejection of blending inheritance. Second, if the existence of discrete units of inheritance also implies that differences between individuals are discrete, then so too are differences between races or varieties of a species as well as differences between species. Clearly, this picture is diametrically opposed to Darwin's vision of insensible gradations between individuals and between species both in time and space, and of evolution being a continual, gradual process. And it is this intellectual dichotomy that will rear its head in 1900 with the rediscovery of Mendel's laws. In addition to the theoretical aspects of Galton's theory are the practical. Because by being mathematically inclined, Galton developed various formulae by which to calculate the percent of heritable material contributed to offspring not only by their parents, but also from preceding generations. And in doing so, Galton set the stage for the biometricians and population geneticists of the twentieth century.

HUGO DE VRIES: INTRACELLULAR PANGENESIS, MUTATION, AND THE REDISCOVERY OF MENDEL

In 1900, three plant physiologists and hybrid experimentalists claimed to have independently discovered Mendel's principles of heredity: the independence of units of inheritance and the three-to-one ratio of dominant versus recessive state expression in the hybrid offspring of pure line parents. All three botanists—Erich Tschermak von Seysenegg, Karl Erich Correns, and Hugo de Vries—had the opportunity to read Mendel's article of 1866, but when each did so is unknown. Of these three, only Correns, whom de Vries had upstaged by getting his article published first, gave credit in his publication to Mendel for the original discoveries. Regardless of the actual events surrounding these individuals, it is the case that of them, the Dutchman Hugo de Vries is the most relevant to the history of theories of inheritance.

In 1889 de Vries published a theory of inheritance that he called "intracellular pangenesis" in honor of Charles Darwin (de Vries, 1910a). But this honorific did not also signify that de Vries was convinced by everything Darwin proposed in his theory of pangenesis. Specifically, de Vries rejected the notion of use-disuse because by then, the German biologist August Weismann had demonstrated conclusively that the "stuff" of heredity lay in the nuclei of sex or germ cells—which proved the death knell to all theories of the inheritance of acquired characteristics, including Darwin's. De Vries was, however, impressed with the idea that an organism's characters are inherited via real entities (gemmules).

In his theory, de Vries referred to units of inheritance as "pangens" or "pangenes." Because he envisioned pangens as particles, he could explain the results of hybrid and crossbreeding experiments in terms of their expression after being passed on from each parent. Although like Mendel, de Vries associated a unit of heritance with a feature, he was not Mendelian in the sense that he did not envision that each parent contributed to a pair of such units. Nevertheless, because he envisioned pangens as discrete entities, de Vries, like Galton, rejected Darwinism and its emphasis on continuous variation between individuals and between species throughout time and space. Rather, just as pangens underlay the development of discrete morphologies, so too were differences between species discontinuous. With regard to the question of how variability arises, de Vries proposed two mechanisms: the emergence of a new pangen or a change in the number of pangens.

When his theory of intracellular pangenesis met with little enthusiasm from the scientific community, de Vries pursued experiments that might demonstrate its tenets. And for these he chose to focus on the so-called sports of nature or monstrosities that as early as 1886 he noticed would crop up without warning in his gardens of evening primrose. De Vries saw these novel plants, which appeared in the course of a single generation bearing novel features, as providing potential insight into how different features may come about in nature. Importantly, he demonstrated that although random and sudden in appearance, these novel features—such as a twisted stem or overly broad leaf—were not only heritable from one generation to the next, but that with selective breeding he could increase their numbers in successive generations. He referred to these suddenly appearing features as "mutations" and, in 1900, the same year in which his article on the discovery of Mendel's principles appeared in print, he also published the first volume of his "mutation theory" (de Vries, 1910b).

Just as the English saltationists, such as the comparative anatomists Thomas Henry Huxley and St. George Mivart, rejected Darwin's claim that organismal change could and does occur by the gradual accumulation of small variations, so too did de Vries reject this assumption. But while the saltationist model of change envisioned novel features as emerging rapidly as the result of changes that occurred during development, de Vries's theory proposed that a new species arises because of changes in pangens or mutations that occurred while pangens were in the process of replicating themselves; it was then upon these new pangens or mutations that natural selection acted. Since de Vries had seen in his gardens of evening primrose and other plants (he also experimented with snapdragons) that more than one individual with the same novelty would emerge in a single generation, he believed that this phenomenon—multiple instances of the same change—was common in nature. Since, as de Vries saw it, mutations in pangens occurred randomly and by chance, they had no adaptive significance.

Again in stark contrast to Darwin, de Vries distinguished between the kinds of changes that produce novel features, and thus new species, and those that underlie individual variation. For de Vries, change in the *number* of already present pangens produces what we see as individual variation. But while de Vries's theory may sound somewhat Darwinian in that changes in numbers of pangens are brought about by changes in environmental conditions, for the Dutchman environmental shifts and thus shifts in expression of pangens were transitory phenomena. Even if a shift in the expression of a particular number of features in a subgroup of a species was sufficient to produce what a taxonomist might identify as a "race," "variety," or "subspecies," its existence was not stable. It was transient.

De Vries's mutation theory served as a powerful antidote to Darwin's selectionist theory. Rather than emphasizing a continuum of gradually accumulating change as the basis for the origin of new species in which virtually every feature of an organism has adaptive significance, de Vries promoted rapid and random change via mutations that produced "sports of nature," whose discretely defined features made them immediately different from one another. For example, physicists, of whom Lord Kelvin was the most prominent, could not keep pace with Darwin's plea in each edition of *Origin* for an increasingly older earth that would be needed to accommodate his model of gradually accumulating change. De Vries's theory was much more compatible with their estimates for the age of the earth.

But de Vries's theory also attracted scientists who had difficulty with Darwin's emphasis on natural selection as an agent that chose from among variably advantageous features, because this notion demanded that a struggle for existence was an inherent property of life. And it was an emphasis on a struggle for existence, with selection choosing increasingly advantageous features as it directed the course of evolution, that served as the basis for what Herbert Spencer would call "Social Darwinism," which promoted racial hierarchy, colonization, and the control of so-called savage races by supposedly civilized ones. For de Vries, species alone arise from mutations and thus from evolutionary relevant processes; as for natural selection, if it acts at all, it does so on variants of features that are already present. Perhaps most important and in contradistinction to Darwin's idea of selection, de Vries's notion of selection typically envisioned selection against disadvantageous features, not the selection of the most advantageous.

WILLIAM BATESON: A THEORY OF REPEATED PARTS AND THE BROADER APPLICATION OF MENDELISM

For support of his theory of pangens or separate units of heredity and therefore of the reality of discontinuous variation in nature (rather than Darwin's notion of insensibly graded, continuous variation), de Vries turned to the English zoologist William Bateson. For although he may be best remembered among historians of science as the first to publish a translation of Mendel's article from German into English and for expanding Mendel's principles from plants to include animals, Bateson had earlier made a name for himself as a radical evolutionary thinker.

While at St. John's College, Cambridge University, in the early 1880s Bateson distinguished himself by studying the embryology of a worm-like creature of the genus *Balanoglossus*, whose odd anatomy had stymied taxonomists' attempts to pigeonhole it into any classification (Bateson, 1928). Bateson demonstrated that *Balanoglossus* was a chordate (an animal with a supporting rod or vertebrae along its back), although, as an unsegmented animal, it was a very primitive chordate indeed. But instead of pursuing this avenue of research further, Bateson turned to the question of individual variation and species difference: What is it? How did it arise? And how does variation really fit with a tenable theory of evolutionary change?

Bateson recognized that although Darwin had based his entire theory of evolution by means of natural selection on the *availability* of variation between individuals, he had not tackled the thorny matter of the *origin* of variation and its supposed adaptive correlation with changes in an organism's environmental surroundings. Consequently, Bateson took it upon himself to find support for Darwin's claim of a connection between environmental and organismal change—that, indeed, organismal change does "track" environmental change—and chose the constantly fluctuating lakes and basins of dead and dying lakes of Western Central Asia in which to situate his research. From the spring of 1886 through the fall of 1887 he collected data on local environmental conditions, including lake depth and water salinity and density. He followed his stint in Western Central Asia with a shorter one in northern Egypt, where he conducted another study of brackish water-dwelling creatures.

In the end, however, Bateson could not demonstrate any correlation between change in organisms and the environmental conditions in which they lived. Sometimes it appeared as if a pattern were emerging but, more often than not, all Bateson really observed was a shift in the representation of variations of features that were already present in species. He was forced to conclude that there was no support for Darwin's belief that small environmental change can "nudge" new variations into existence that then become fodder for natural selection. To the contrary, if environment-based selection acted at all, it did so within the constraints of the range of expressed variation of features that organisms already possessed as it pushed the bell curve of individual variation first one way and then another. This was not, Bateson recognized, a process that would lead to evolutionary change and the origin of new species.

But while some scientists may have thrown up their hands in dismay and returned to other, more fruitful, intellectual pursuits, Bateson became obsessed with studying variation, including "abnormal" variation, in all life forms. He became so devoured by this pursuit that he wrote to his sister Anna late in 1888: "My brain boils with Evolution. . . . It is becoming a perfect nightmare to me" (Bateson, 1928). Some six years later, Bateson published the results of his studies in *Materials for the Study of Evolution*, in which he documented no less than 886 examples of discontinuous, not continuous, variation. He drew upon his studies of organisms and lake salinity to clarify the conflation that Darwin had mistakenly made: that while "most of the elements of the physical environment are continuous in their gradations . . . as a rule, the forms of life are discontinuous" (Bateson, 1894). Like Galton before and de Vries soon thereafter, Bateson questioned seeking relevance for understanding evolutionary change in individual variation rather than through an appreciation of the discontinuity between characters. And since recognition of the discontinuity between features led naturally to a theory of rapid evolutionary change, it is not surprising that Bateson's ideas were embraced not only by Galton, but also by the saltationists, including Thomas Huxley.

One consequence of Bateson's study of variation was recognizing that differences not only between individuals of the same species, but also between individuals of different species, were often expressed in differences in number of repeated or meristic parts. If you think about it, this makes sense. Various animals, such as starfish and octopuses, have their repeated parts arranged around a central axis, like petals of a flower. Bateson found that when starfish vary, they do so in terms of more or fewer "arms." The same applies to variation in flower petal number. Many animals are segmented (*Balanoglossus* being an exception), for example, in having distinct muscle bundles or rings (such as worms), vertebrae, teeth, bones of limbs, joints of antennae or appendages, or distinct body segments. Even the fronds of ferns are "segmented" in that their "ribs" form a series of repeated parts. Variation between individuals or differences between species is often in the form of more or fewer segmented or repeated parts.

Unlike Darwin, who rejected "sports of nature" or "monstrosities" as providing a window on nature and on evolution, Bateson mirrored the saltationists by seeing individuals that differed from the common or expected "body plan" as being of enormous significance for understanding biological, and thus evolutionary, processes. He saw in the repeated parts of animals and plants analogy with a rhythmic patterning similar to the motion of tidal waves: Like the ripple effect of waves on sand, variations between individuals or differences between species were produced by developmental waves of differing intensities.

The year of publication of *Materials for the Study of Variation*, 1894, was also the year that the Royal Society of London followed Galton's suggestion to establish a committee charged with analyzing statistically measurable features of plants and animals. In 1897 Bateson joined this committee, which shortly thereafter was named the Evolution Committee, and with modest financial backing turned to breeding plants and animals, in particular poultry. The grounds surrounding his house were soon occupied to the limit with students performing all sorts of hybridization and crossbreeding experiments with plants and animals. In 1899, he presented to the Royal Society the outline of a plan for determining through experimentation and statistical methods the laws of heredity in both plants and animals. Even if Bateson's wife, Beatrice, exaggerated in her biography of him that he was then on his way to discovering Mendel's principles, he was certainly ready to embrace them.

Also according to Beatrice, Bateson first learned of Mendel's article by way of de Vries's, Correns's, and Tschermak's publications. Bateson then acquired a reprint of it, which he supposedly read on 8 May 1900 while traveling by train to lecture before the British Horticultural Society. Mendel's ideas made sense to Bateson in terms of the breeding experiments that he and his students were conducting. But perhaps more important, Mendel's ideas of particulate inheritance via discrete units of heredity resonated with Bateson's studies on discontinuous variation. Before the train trip ended, Bateson had added to his lecture a summary of Mendel's experiments and results.

Bateson began his lecture thus:

> An exact determination of the laws of heredity will probably work more change in man's outlook on the world, and in his power over nature, than any other advance in natural knowledge that can be foreseen. . . . There is

> no doubt whatever that these laws can be determined . . . No one has better opportunities of pursuing such work than horticulturalists.

Before proceeding to his review of Mendel's work, Bateson made the point first articulated by Galton that one cannot study the process of heredity merely by analyzing the single generation from parent to offspring. Rather, it was only through application of Galton's "law of ancestral heredity" that one could calculate how much heritable material was passed on from parent to offspring. After summarizing Mendel's experiments and detailing the "three-to-one ratio" Bateson commented: "[Mendel's] account of [his experiments] will certainly play a conspicuous part in all future discussion of evolutionary problems. . . . It is not a little remarkable that Mendel's work should have escaped notice, and been so long forgotten" (Bateson, 1928, p. 177).

Although it would be some years until Mendel's work was broadly accepted, in 1902 Bateson had the opportunity to expand the monk's principles from plants to animals. In collaboration with Rebecca Saunders, who had been conducting plant-breeding experiments on his property, Bateson published *Reports to the Evolution Committee of the Royal Society*. There, for the first time, not only the inheritance of "normal" features (such as number of toes or bones in them), but also the inheritance of "abnormal" features (such as too many or too few toes or bones in them) were addressed using the same language with which Mendel had discussed the inheritance of wrinkled versus round seed coats in common garden peas. It was a lesson in the elegance of simplicity. In one fell swoop, Bateson and Saunders presented what would become the foundation of the nascent field of genetics: character states for a particular feature (such as seed-coat configuration or toe-bone number) come in ("antagonistic") pairs of alternative (dominant versus recessive) character states. From that seemingly innocuous realization one could figure out most of the important aspects of heredity.

Bateson and Saunders also brought to the lexicon of genetics (a term that Bateson coined in a letter to the Cambridge University Professor of Zoology, Adam Sedgwick, dated 18 April 1905 [Bateson, 1928]) nouns that until then were used only by embryologists and cytologists. Instead of, for example, "germ cell" and "fertilized egg," these scientists used, respectively, the terms "gamete" and "zygote," and Bateson and Saunders adopted them. Bateson and Saunders themselves suggested that "pairs of antagonistic characters" should be referred to as "allelomorphs" (a term that was shortened in the twentieth century to "alleles"). When the pair of allelomorphs represented each state—one dominant and the other recessive—the zygote was said to be "heterozygote." When both allelomorphs of a pair were in the same state—either dominant or recessive—the zygote was defined as a "homozygote."

But Bateson and Saunders did not stop with these terms. They devised a standardized system by which one could refer to successive generations of related individuals that, they believed, would allow academics and professional breeders alike to communicate among themselves as well as across disciplines. They suggested that the letter "P" should denote the parental generation (the generation with which the experiment began). The letter "F" indicated subsequent or filial generations, with each successive generation represented by the subscripted number of its generation (for example, F_1, F_2, F_3). Generations preceding the parental generation were designated, for example, as P_2, P_3, P_4, etc.; the original parent generation as P_1.

Although Bateson and Saunders could have ended their report to the Royal Society here, they chose not to do so. Rather, following Bateson's own internal wrestling with the question, they raised the matter of just how one defines a species. As Bateson wrote in 1894 (p. 2) in *Materials*:

> No definition of a Specific [= species-specific] Difference has been found. . . . But the forms of living things, taken at a given moment, do nevertheless most certainly form a discontinuous series not a continuous one. This is true of the world as we see it now, and there is no good reason for thinking that it has ever been otherwise. So much is being said of the mutability of species that this, which is the central fact of Natural History, is almost lost sight of, but if ever the problem is to be solved this fact must be boldly faced. [comment added]

Bateson also understood that the differences taxonomists used to distinguish species were those without apparent adaptive significance. Often these features seemed trivial. Bateson also apparently suggested that new species can arise by the splitting of existing species because his "undulatory theory" precluded an evolution in which one species becomes gradually transformed into another. From his studies on variation and environmental change, it seems that the "force" behind the origin of new species is neither the environment nor natural selection. But regardless of its source, discontinuous variation held for Bateson the key to the origin of new species: "On this hypothesis, therefore, [discontinuous] Variation, whatever its cause, and however it may be limited, is the essential phenomenon of Evolution. [Discontinuous] Variation, in fact, is Evolution" (Bateson, 1894, p. 7).

With regard to trying to achieve a practical identification of species, Bateson and Saunders ventured, "some degree of sterility on crossing . . . is one of the divers properties which may be associated with Specific [= species] difference." But while this might sound "modern"—individuals of different species, if mated, would not be able to produce fertile offspring—Bateson and Saunders also conceived of the possibility that individuals that presented themselves (in whatever way, morphological or behavioral) as belonging to different species might not be so different genetically that they could not mate and produce reproductively viable offspring. Unfortunately, this insight, as with others Bateson, de Vries, and the saltationists proposed, was banished from the version of neo-Darwinism that the architects of the evolutionary synthesis of the 1940s adopted.

In 1909 Bateson published *Mendel's Principles of Heredity*. From his breeding experiments and those of colleagues he went beyond the simple dichotomy of "dominant versus recessive" traits to more complicated cases, such as incomplete dominance (common in various strains of chicken), in which the heterozygote for a dominant feature cannot be distinguished from the homozygote for it. Bateson also identified features whose inheritance appeared to be linked to one sex or the other, as in male baldness or color blindness, in which the allele is actually inherited by the son from his mother. He referred to these sex-linked features as "sex-limited" features.

In *Mendel's Principles* Bateson brought to genetics the familiar "Punnett square." (Bateson had asked Reginald Crundall Punnett, who did much work on breeding poultry on Bateson's property, to produce a visual representation of an offspring's inheritance of alleles from each parent.) Punnett had the idea of listing in a box containing four squares the four possible alleles (in the base of hybrid parents, two dominant and two recessive, say T, T, t, and t) that each parent could (theoretically) pass on to offspring. When the two boxes were placed one atop the other, the result was a box with four squares, each of which bore the result of the mating (in the case of hybrid parents, the resultant offspring would represent Mendel's three-to-one ratio: TT, Tt, Tt, tt).

Toward the end of *Mendel's Principles* Bateson expanded his ideas on repeated parts. Being quite taken by the process of cell division, in the production both of body or somatic cells and of sex cells or gametes, Bateson suggested that morphological novelty—which, he documented, often affects the number and spacing of meristic structures—should be understood in terms of "divisions by which similar parts are divided from each other, and differentiating division s by which parts with distinct characters and properties are separated" (Bateson, 1909).

Bateson's most salient error, however, was not entertaining the possibility that the chromosomes in the nuclei of cells might be the bearers of the "stuff" of inheritance. Instead, Bateson rejected the "chromosome theory" in favor of the cell's cytoplasm being the substance in which the basis of heredity lay. Bateson also accepted the mistaken identification of different numbers of chromosomes in males and females. Because the small male Y chromosome had not yet been detected microscopically, females were believed to have $2n$ chromosomes and males $2n-1$ chromosomes. And since females always had $2n$ chromosomes (supposedly the result of two female-producing gametes coming together), while males always had *2n-1* chromosomes (presumably the result of one female-producing gamete and one male-producing gamete coming together), Bateson incorrectly believed that it was the male-producing gamete that was tied to sex determination. Nevertheless, even though Bateson's ingenious conclusion was based on commonly held misinformation, it is to his credit that he recognized an underlying genetic basis for femaleness versus maleness.

THOMAS HUNT MORGAN: A SCIENTIST OF CONTRADICTIONS

While Edmund Beecher Wilson, the Da Costa Professor of Zoology at Columbia University in New York City, and his colleague, Dr. Nettie Maria Stevens, were occupied with identifying sex-related chromosomes in various insect (primarily beetle) species and trying to figure out their relationship to what were then being called "factors" rather than "units" of inheritance, Thomas Hunt Morgan, a professor of embryology at Bryn Mawr College for Women in Pennsylvania was extolling the virtues of Mendelism. In 1903, in *Evolution and Adaptation*, he proclaimed that "[t]he theoretical interpretation that Mendel has put upon his results is so extremely simple that there can be little doubt that he has hit on the real explanation" and that "there can remain little doubt that Mendel has discovered on of the fundamental laws of heredity" (Morgan, 1903).

Morgan's early acceptance of Mendelism is interesting in light of his turn of heart a mere six years later in an article he published in the *Proceedings of the American Breeder's Association*, titled "What are 'factors' in Mendelian explanations?" There, Morgan made clear his belief that Mendelian inheritance was of no significance especially for understanding evolution. As he wrote:

> In the modern interpretation of Mendelism, facts are being transformed into factors at a rapid rate. If one factor will not explain the facts then two are invoked; if two prove insufficient, three will sometimes work out. The superior jugglery sometimes necessary to account for the results may blind us, if taken too naively, to the common-place that the results are often so excellently "explained" because the explanation was invented to explain them. We work backwards from the fact to the factors, and then, presto! Explain the facts by the very factors that we invented to account for them. . . . I cannot but fear that we are rapidly developing a sort of Mendelian ritual by which to explain the extraordinary facts of alternative inheritance. (Morgan, 1909)

What is also interesting about Morgan—the individual who would eventually meld Mendelism and Darwinism into the model that would inform the evolutionary synthesis—is that, in *Evolution and Adaptation*, he did not embrace Darwinism. Like Huxley, Mivart, Bateson, and de Vries, Morgan could not accept Darwin's vision of evolution by means of natural selection. He rejected the utilitarian notion of adaptation that was so crucial to Darwin's view that natural selection will choose those individuals whose characteristics make them better adapted, or more fit, than other individuals. Neither could Morgan accept Darwin's analogy between artificial and natural selection because, he believed, "new species comparable in all respects to wild ones have not been formed [by breeders], even in those cases in which the variation [of the new breed] has been carried farthest" [comments added]. Whereas Darwin formulated his theory of sexual selection to account for features that were not adaptive for the organism, Morgan asked the question: How can we decide what is and what is not adaptive? On the other hand, he suggested, "[i]f . . . we assume that the *origin* of the responses has nothing to do with their value to the organism, we meet with no difficulty in those cases in which the response is of little or no use to the organism" (Morgan, 1903).

Because he largely embraced Hugo de Vries's mutation theory, Morgan rejected Darwin's belief that evolution occurs gradually, through a process in which natural selection picks and chooses from among myriad small or "fluctuating" variations. Instead, Morgan argued, if a new feature can arise via a single mutation, so, too, can a new species. In support of this view, Morgan cited various examples of the sudden origin of individuals that served as the basis for new breeds of sheep (ancon, for example) and cattle (niata)—the very examples that Darwin had dismissed as providing evolutionarily relevant evidence. With regard to niata cattle:

> In Paraguay, during the last century (1770), a bull was born without horns, although his ancestry was well provided with these appendages, and his progeny was also hornless, although at first he was mated with horned cows. If the horned and the hornless were met in a fossil state, we would certainly wonder at not finding specimens provided with semi-degenerate horns, and representing the link between both, and if we were told that the hornless variety may have arisen suddenly, we should not believe it and we should be wrong. (Morgan, 1903, p. 315)

Further like de Vries (as well as the saltationists and Bateson), Morgan decoupled the origin of novel features, and therefore the origin of the individuals of new species with these features, from the survival of these individuals and the persistence of their novel features. Again echoing de Vries and Bateson, Morgan argued that if natural selection was in any way relevant in nature, it was only after evolutionary novelties (and new species) had emerged through a process of mutation.

Morgan also rejected the Darwinian practice of lining up known forms—whether extant, fossil, or fossil and extant—from simple to complex as a demonstration of a continuum of gradual transformation. "[I]t does not follow," he wrote, "because we can arrange such series without any large gaps in its continuity, that the more complex conditions have been gradually formed in exactly this way from the simplest conditions" (Morgan, 1903).

Perhaps the biggest hurdle to Morgan's acceptance of Darwin's ideas was the latter scholar's increasing emphasis through the six editions of *Origin* on a struggle for existence. As Darwin had written in 1859:

> All we can do, is to keep steadily in mind that each organic being is striving to increase at a geometrical ratio; that each at some period of its life, during some season of the year, during each generation or at intervals, has to struggle for life, and to suffer great destruction. When we reflect on this struggle, we may console ourselves with the full belief, that the war of nature is not incessant, that no fear is felt, that death is generally prompt, and that the vigorous, the healthy, and the happy survive and multiply.

To which in *Evolution and Adaptation* Morgan (p. 116) responded:

> The kindliness of heart that prompted the concluding sentence may arouse our admiration for the humanity of the writer, but need not, therefore, dull our criticism of his theory. For whether no fear is felt, and whether death is prompt or slow, has no bearing on the question at issue—except as it prepares the gentle reader to accept the dreadful calamity of nature, pictured in this battle for existence, and make more contented with their lot "the vigorous, the healthy, and the happy."

We can summarize thusly Morgan's position in 1903. Animals and plants do not change in order to become better adapted to their surroundings and, in fact, some species are actually not well adapted at all. But if "a struggle for existence" does exist in nature, individuals would not be imperfectly adapted. On the other hand, some organisms appear to be overly adapted to their surroundings. Consequently, it is not appropriate to evaluate features and adaptation in utilitarian terms. Indeed, the existence of features that may seem overly or poorly perfected is best appreciated without the constraints of natural selection, selection pressures, and a struggle for existence (Schwartz, 1999). He concluded *Evolution and Adaptation* (p. 464):

> If we suppose that new mutations and "definitely" inherited variations suddenly appear, some of which will find an environment to which they are more or less well fitted, we can see how evolution may have gone on without assuming new species have been formed through a process of competition. Nature's supreme test is survival. She makes new forms to

> bring them to this test through mutation, and does not remodel old forms through a process of individual selection.

MORGAN: THE DAWN OF FRUIT-FLY POPULATION GENETICS

In 1904 Morgan went to Columbia University in New York City, where he was hired as Professor of Experimental Zoology. There he met Edmund Beecher Wilson, who introduced the younger scientist to his research in cell structure and breeding experiments. Nevertheless, it would be virtually half a decade until Morgan changed his Bateson-like stance against the theory that chromosomes were the bearers of the "stuff" of heredity, fully embraced Mendelism, and warmed somewhat to Darwin's emphasis on natural selection playing a role in evolution.

In his presidential dinner address at the 1909 annual meeting of the American Society of Naturalists, Morgan again criticized Darwin (Morgan, 1910). He espoused the view that rather than a new feature emerging from an unknown cause and then surviving because natural selection favored it, the persistence of a novel feature was also a matter of chance as to whether it was compatible in the surroundings in which the organism bearing it existed. Consequently, despite the title of his best-known work, Darwin had not made a case at all for the origin of species. Rather, he had merely described the role of natural selection in the molding of adaptations of plants and animals. In fact, by incorrectly conflating processes of adaptation with those of evolution, Darwin had blurred the very real distinction between the two, which, Morgan argued, must be maintained if evolutionary biology is going to make any progress.

Morgan also reiterated his view that because a new feature emerged by chance, it was devoid of "purposeful adaptation," which could only be identified in retrospect anyway. And he continued his attack on struggle for existence:

> Is the battle always to the brave— for the brave is sometimes stupid—or the race to the swift, rather than to the most cunning? . . . [For] [a]n individual advantage in one particular need not count much in survival when the life of the individual depends on so many things—advantages in one direction may be accompanied by failures in others, chance cancels chance. (Morgan, 1910)

Morgan concluded his presidential address: "The time is past when it will be any longer be possible to speculate light-heartedly about the possibilities of evolution, for an army of able and acute investigators is carefully weighing by experimental tests the evidence on which all theories of evolution and adaptation must rest. . . . To them belongs the future." And the very next year Morgan acted on his admonition and turned to experimental genetics. The organism he chose to study was the common fruit fly of the genus *Drosophila*, which had already gained favor among geneticists because fruit flies were easy to obtain and breed, it took only about twelve days for members of a new generation to hatch, mature, mate, and lay eggs, and generational populations were enormous. The experimenters could also easily control the conditions of the experiment.

There were two other factors that made fruit flies attractive experimental organisms. First, there were lots of fruit-fly "variants." Of the approximately 15 species of fruit fly that had thus far been identified, more than 400 races or subspecies of the most common species, *Drosophila melanogaster*, were known, as were over 125 races of the next most common fruit fly species, *D. ampelophila*. As such, even though a specific feature—such as eye color, wing length, or wing shape—might distinguish one race of fruit fly from another, individuals of each race could nonetheless interbreed. The second reason was that fruit fly chromosomes are not only large, but also few in number (ranging between three and six pairs, with four pairs being the most common, of which one pair represents the sex chromosomes). Because of this fruit fly chromosomes are fairly easy to study under the microscope and there aren't that many to compare between individuals (which is significant if one is comparing thousands of individuals within and between experimental generations).

It's hard to believe that a creature as small, annoying, and at the same time humorous as a fruit fly could have had such an impact on evolutionary thinking as it did. Yet this is the case. Although breeding experiments with and observations of spontaneous mutation in the early twentieth century revealed that large-scale changes—such as the disappearance of eyes or wings, or the development of extra body segments, wings, legs, and even a doubled thorax—were just as benign reproductively to their bearers as small-scale changes—such as different eye color, slightly longer or shorter wings, more or fewer body bristles—the leader of this group of scientists, Thomas Hunt Morgan, chose to reject the former as being evolutionarily relevant and base an entire theory of gradual evolutionary change on the latter. And it is because of this overt bias and conscious choice about what is and what is not viable in nature that fruit-fly population genetics rather than developmental genetics came to inform the evolutionary synthesis of the 1940s and subsequent generations of Darwinians.

The unfortunate affect of the synthesis' dogmatic preaching of Darwinism was the elimination of intellectual curiosity and alternative thinking that had characterized evolutionary biology from the mid-nineteenth century through the turn of the century. Indeed, when at times after the synthesis ideas that were not strictly Darwinian, much less non-Darwinian altogether, were proposed, their authors were subjected to scathing ad hominem attacks and even branded "anti-evolutionary." Fortunately, however, the intellectual tide is turning and the centrality of developmental thinking in evolutionary biology is becoming re-established. The result is the sidelining of the pesky fruit fly and a recognition of the significance for evolutionary biology of the non-Darwinian ideas of earlier scholars.

By 1925, Morgan and his collaborators Alfred Henry Sturtevant, Hermann Joseph Muller, and Calvin Blackman Bridges had not only established what quickly became the most well-known fruit fly experimental laboratory in the country, but completed such an astonishing number of experiments that they were able to publish a volume on this work, *The Mechanism of Mendelian Heredity* (1926). One reason for their rapid success was the sudden and inexplicable appearance within their pure lines or races of different mutants. That is, in comparison with the normal morphologies of typical fruit flies found in the wild—the so-called normal—types the abruptly appearing mutants differed in, for example, aspects of body color and size, wing shape and size (including absence), eye color and size (also including absence), antennae size and shape, number of abdominal segments, and distribution of body bristles on thorax and abdominal segments. Although in 1909 the Danish botanist Wilhelm Johannsen had coined the word "gene" to refer to units or factors of heredity, Morgan didn't use it, most likely because the Dane, like Bateson, rejected the chromosome theory of inheritance. (Since Morgan eventually used the word "gene," I shall now use it.)

But there was a major stumbling block that Morgan had to overcome, namely, the question of what exactly a unit or factor of inheritance might be. If, according to Mendelism, a feature was based in a gene, and, according to the chromosome theory of inheritance, these genes were represented by chromosomes, then how could an organism as morphologically complex as a fruit fly be produced by only a handful chromosomes? Clearly this couldn't be the case if there was a one-to-one correspondence between the number of chromosomes an organism had and the number of features it possessed. This was the unanswerable question Walter Stanborough Sutton faced after finally demonstrating in 1901–1902 that the replicas of the chromosomes that each parent contributed to produce a zygote were later involved in that individual's production of gametes and the very enigma that prevented many scientists from embracing the chromosome theory of inheritance.

The answer to this question first came from Morgan and his colleagues' experiments in breeding, crossbreeding, and back-breeding different races of fruit flies as well as wild-type and mutant individuals. One reason this worked is that even though races of the same species had the same number of chromosomes, the chromosomes themselves were of different sizes and shapes in different races. Consequently, when comparing the chromosomes of offspring generations with those of the parental generation, Morgan and his collaborators could see through the microscope which chromosomes came from which parent and how they did or did not pair up in resultant zygotes.

The task of comparison was even simpler when parents did not have the same number of chromosomes—which was the case with the emergence of a mutant strain they called "diminished-bristles." In contrast to the wild type, the "diminished-bristles" strain had one instead of a pair of fourth chromosomes. Upon crossbreeding wild and mutant, some offspring would have an uneven number of chromosomes but others an even number. The importance of this observation became clear with subsequent breeding experiments.

When an individual from the diminished-bristles' strain was mated with one from the mutant "eyeless" strain, offspring with only a single fourth chromosome were eyeless. But when an individual from the "eyeless" strain was crossed with a normal, wild-type individual, their offspring developed eyes because the condition in the wild type was the dominant state. The explanation seemed straightforward: The recessive state for "eyeless" must reside on the fourth chromosome, so that when "eyeless" (with four pairs of chromosomes) and "diminished-bristles" (with three pairs of chromosomes and only one fourth chromosome) mutants were crossed, the recessive "eyeless" state could be expressed as if it was in the dominant state. Once aware of this, Morgan and colleagues proceeded with similar breeding experiments, which resulted in the compilation of a map—a "chromosome map"—of where on each chromosome a specific gene for a specific feature lay.

Another way to hypothesize where on a given chromosome a gene resided came from the work of the Belgian cytologist F.A. Janssens, which focused on the behavior of chromosomes during body cell duplication (mitosis) and gamete production (meiosis). In a 1909 publication Janssens illustrated how the "arms" of adjacent chromosomes sometimes twisted around one another, seemingly fusing together (Janssens, 1909). When during mitosis or meiosis chromosomes pulled apart and went to opposite poles of the dividing cell, chromosomes that had fused broke at the point of fusion, resulting in one or more chromosomes that were part "original" and part "other" chromosome. Such "crossing over" leading to rearrangements of parts of chromosomes could explain why even in Mendel's simple pea experiments hybrid crosses only approached but never equaled the three-to-one ratio.

Morgan and his team used the likelihood of crossing over and chromosomal rearrangement in their chromosome mapping. Beginning with the assumption that crossing over could occur anywhere along the length of a pair of chromosomes, they predicted that the farther apart two genes were, the greater the likelihood that these genes would be separated upon a break in the chromosome and its subsequent rearrangement. Conversely, the closer together two genes were on a chromosome, the greater the likelihood they would remain on the same piece of chromosome if breakage and rearrangement occurred. By crossbreeding strains of fruit flies characterized by specific features, Morgan and his colleagues could calculate how far apart or close together the genes for these features were and on which chromosomes they were located by seeing which features appeared together or separately in offspring. Eventually, they mapped the chromosomes of all 400 races of *Drosophila melanogaster* and discovered that characters were inherited in four groups—the same number as pairs of chromosomes. Through similar experiments on other *Drosophila* species, they found the same correspondence between number of linkage groups and pairs of chromosomes.

MORGAN: THE MELDING OF DARWINISM AND MENDELISM

In the early twentieth century, evolutionary thinkers were divided into two camps: Darwinism versus Mendelism. The reason for this dichotomy was that Darwinism was grounded in notions of continuous variation and graded series between individuals as well as between species, while Mendelism, because of the concept of discrete units of heredity, led to viewing variation as discontinuous and differences between individuals and species as real. How then did the saltational mutationist and sometime–anti-Mendelian Morgan come to bring Darwinism and Mendelism together? It is an interesting case of deciding which observations should be considered evolutionarily relevant.

Beginning in 1916 with *A Critique of the Theory of Evolution*, Morgan argued that observations and experiments done in the laboratory were valid for interpreting nature. From witnessing the sudden appearance in his fruit-fly colonies of mutants with large-scale changes (such as absence of eyes or wings, or double thoraxes or wings) and others with small-scale changes (such as extra bristles or different eye color). Morgan summarized: ("some of the changes (are) so slight that they would be overlooked except by an expert, others so great that in the character affected the flies depart far from the original species" [Morgan, 1916]), Morgan concluded that the underlying cause of change in both instances was the same—spontaneous mutation—and that the same process would obtain in nature. He also observed that many if not most mutations "appeared independently several times" and through various breeding experiments demonstrated, as Bateson had earlier predicted, that many if not most mutations arose in the recessive state and were not all injurious when expressed. Having thus originated in the recessive state, the mutation could spread silently through the colony across generations until there were sufficient heterozygotes with the mutation that they would produce offspring that were homozygous for it. Even though many of the mutant conditions were profoundly different from the parental norm, they bred true. These mutants could also be successfully bred with individuals of the normal parental generation with the result that the progeny of these crosses were not always sterile and usually displayed the mutated or normal condition rather than an intermediate one.

Clearly these observations do not inexorably lead to a model of gradual change based on small mutations. Rather, they present a picture whereby the silent spread of a mutation over some number of generations leads to the abrupt appearance in a subsequent generation of some number of individuals with the same novel feature, whether profound or slight. Why then did Morgan either not see or choose to ignore the implications of his observations and experiments? As he stated in 1925, "animals and plants are extremely complex machines that are highly adapted to the conditions of life in which they live . . . [a]ny change, and especially any great change in them, is far more likely to throw them out of balance with their environment than to bring to them an advantage" (Morgan, 1925). Consequently, since something has to change in order for "evolution" to occur, a slight change would be preferable to a large one because it would be less likely to endanger the "fit" of an organism to its circumstances. This in spite of the fact that mutants with profound differences not only survived but bred successfully among themselves as well as with normal, wild-type individuals.

Another contradiction in Morgan's 1916 presentation begins with his pointing out that there is no evidence that a mutation producing change in a feature would be followed by another mutation affecting the same trait—like tossing a coin, "the number of heads obtained has no influence on the number of heads that will appear in the next throw." Yet, two pages later, he declares, "Owing to this property of the germ plasm to duplicate itself in a large number of samples not only is an opportunity furnished to any advantageous variation to become extensively multiplied, but the presence of a large number of individuals of a given sort prejudices the probable future result." Thus, contrary to the reality of a coin toss, an increase in the number of individuals bearing a beneficial trait will increase the likelihood that another mutation will arise that will also affect that trait. As he stated in 1925 (p. 150), "[E]volution once begun in a given direction is in a favorable position to go on in the same direction rather than another, so long as the advance does not overstep the limit where further change is advantageous."

Having denied profound change any evolutionarily relevance and decided that a favorable mutation in one direction predisposes another in that direction, Morgan set the stage for melding Darwinism and Mendelism: While "units" or "factors" of inheritance may be discrete the variants of the features they underlie differ so slightly that they are continuous and thus these "factors" essentially are as well. In short, small mutations produce small changes from which natural selection can pick and choose, paving the way for other mutations to continue the path of change now set in motion.

Although Morgan never embraced the notion of a struggle for existence, he was otherwise Darwinian in promoting gradual change via natural selection. In fact, beginning in 1916, Morgan included in his monographs an illustration from the paleontologist Lull, in which various elephant "ancestors" have been reconstructed in the flesh, with the explanation that this represents an example of directional evolutionary change. How odd, considering his complaint about paleontologists:

> When the biologist thinks of the evolution of animals and plants . . . [h]e thinks of series of animals that have lived in the past . . . whose bones and shells have been preserved in the rocks. . . . He thinks of these animals as having in the past given birth, through an unbroken succession of individuals, to the living inhabitants of the earth today. He thinks that some of the simpler types of the past have in part changed over into the more complex forms of the present time.
>
> He is thinking as the historian thinks, but he runs the risk of thinking that he is explaining evolution when he is only describing it. (Morgan, 1925)

MORGAN'S LEGACY

Before Morgan left Columbia University in 1928, the Russian Theodosius Dobzhansky came on a postdoctoral fellowship to study fruit-fly chromosomes, especially how different-sized chromosomes from different parents could or could not pair up along comparable segments. This interest led him both to the German developmental geneticist Richard Goldschmidt's studies of intersexuality in various moth species and to favoring chromosomal rearrangement as a major factor underlying mutation and the emergence of morphological novelty because this would affect entire series of genes (Goldschmidt, 1934). Although he dwelt at some length on chromosomal rearrangement in the 1937 edition of *Genetics and the Origin of Species*, Dobzhansky admitted that he could not provide a complete theory for the origin of species.

In 1940 and reminiscent of Bateson and de Vries, Goldschmidt argued that the genetic phenomena underlying variation within a species was different than that involved in the origin of species. Single gene mutations or micromutations could lead to variation or "microevolution." But the origin of species or macroevolution must be rapid and due to a profound, systemic mutation (macromutation), which chromosomal rearrangement seemed to provide (Goldschmidt, 1940). Harking back to the nineteenth-century usage of "sport of nature" Goldschmidt called the bearer of a systemic mutation a "hopeful monster." Unfortunately, the image of a single individual—and a monster to boot—was used against him. In 1941 Dobzhansky published the second edition of *Genetics and the Origin of Species*, in which he attacked Goldschmidt and dug his heels firmly into Darwinian gradualism (Dobzhansky, 1941). The next year in *Systematics and the Origin of Species* (1942) Ernst Mayr followed suit. Even the paleontologist George Gaylord Simpson, who admitted in his 1944 *Tempo and Mode in Evolution* that paleontology contributed little to discussions of micro- versus macroevolution, lambasted Goldschmidt for rejecting gradual evolution, in spite of the fact that Simpson's "quantum theory" advocated periods of rapid, although smoothly transitional, change (Simpson, 1944). Sadly, the legacy of melding fruit-fly population genetics with Darwinism into the evolutionary synthesis had the stultifying effect of squelching alternative thinking in evolutionary biology. Fortunately, the recent emergence of the field of evolutionary developmental biology ("evo-devo") has begun to return evolutionary thinking to a mindset in which it is "OK" to question received wisdom and to entertain alternative ideas, and through this it is becoming clear that many of the concepts that the Victorian saltationists, Bateson, de Vries, the early Morgan, and Goldschmidt had offered as alternatives to Darwinism are the more viable.

FURTHER READING

Bateson, W., and Saunders, E. R. "Reports to the Evolution Committee of the Royal Society," in Report 1, Experiments Undertaken by W. Bateson, F. R. S., and Miss E. R. Saunders (London: Harrison & Sons, 1902).

Darwin, C. *The Variation of Animals and Plants under Domestication* (London: John Murray, 1868).

Goldschmidt, R. B. *The Material Basis of Evolution* (New Haven, CT: Yale University Press, 1940).

Mendel, G. *Experiments in Plant Hybridisation* (Cambridge, MA: Harvard University Press, 1965).

Morgan, T. H. *Evolution and Adaptation* (New York: MacMillan, 1903).

Schwartz, J. H. *Sudden Origins: Fossils, Genes, and the Emergence of Species* (New York: Wiley & Sons, 1999).

Neo-Darwinism

Adam S. Wilkins

> [T]he beautiful train-ride and the wonderful weather in Genoa have restored my equilibrium. I told myself all the time that if on this trip something really should happen to me it would be in the midst of a beautiful experience and in the midst of the most important phase in my profession, that is, in free research and that under such circumstances death would be a glorious end to my life and all the incidental accompanying circumstances would be quite unimportant. I enjoy every day on which I can work and make use of the experiences of the past days.
>
> Ernst Mayr, age 23, 5 February 1928, writing to his mother from Italy on the way to his first bird-collecting expedition in New Guinea

INTRODUCTION

"Neo-Darwinism" is the shorthand term for the contemporary theory of evolution, a theory that emerged between 1920 and 1950. Any historical account of the genesis of a theory is, inevitably, in part an account of the *dramatis personae* involved in its origins and perhaps, for Neo-Darwinism, none is more central than the great twentieth century biologist Ernst W. Mayr (1920–2005).

In principle, an evaluation of Mayr's role in the creation of modern evolutionary biology should be a simple task. The emergence of the evolutionary synthesis involved a dozen or so seminal figures, and many subsidiary ones, but most biologists would probably name Mayr as the emblematic, even iconic, figure of twentieth-century evolutionary biology.

Ernst Mayr. University of Konstanz. Meyer, A. (2005), "On the Importance of Being Ernst Mayr." PLoS Biol 3(5): e152 doi:10.1371/journal.pbio.0030152.

He was, indeed, sometimes referred to as the "Darwin of the twentieth century." The centrality of his role is attributable both to the multiplicity of ways in which he participated in the creation of modern evolutionary biology and the effectiveness with which he played these parts. These roles included those of systematist, evolutionary theorist, catalyst for cross-disciplinary discussions, organizer of conferences, journal editor, publicist within biology for the new ideas, and, in the final part of his life and career, as a public educator on evolution through a series of books for the general reader. The evolutionary synthesis was one of the great achievements of twentieth-century biology and Ernst Mayr became its most energetic participant and promulgator.

Yet, a close look at the emergence of twentieth-century evolutionary biology and Mayr's participation in it reveals some interesting ambiguities and questions. Some of these concern the relative degrees of importance of Mayr's different roles. Were his intellectual contributions as crucial, for instance, as his educational and organizational ones? The more fundamental questions, however, deal with the nature of the evolutionary synthesis itself, in particular with the task of identifying its central ideas and most significant features. Unlike many theories, the formulation of modern evolutionary theory was a gradual one, occurring over a period of about thirty years, involving quasi-discrete stages, each of some duration. Furthermore, it was a creation with many fathers. In this respect, too, it contrasts with the genesis of many other theories, for instance, Isaac Newton's laws of motion or Ernest Rutherford's theory of the atom, or Einstein's theories of relativity, or Francis Crick's central dogma or Peter Mitchell's chemiosmotic theory. The prolonged period of development of twentieth-century evolutionary theory, in combination with its multiple parental sources, ensures the existence of differing interpretations about the relative significance of its component parts. Furthermore, there is bound to be a strong "observer effect": Each evaluation is strongly influenced by the disciplinary background of the person attempting the assessment. Geneticists, for instance, will, for the most part, tend to view these historical matters differently from paleontologists or systematists.

In turn, such differences in opinion about the different component ideas inevitably affect one's view of the relative importance of the different participating individuals. Many systematists and naturalists, for instance, will tend to assign a higher degree of importance to Mayr than they will to Fisher or Haldane or Wright while, for population geneticists, it will be the reverse. Not least, any evaluation of the role(s) of Mayr in the evolutionary synthesis has to acknowledge, as Mayr himself did without hesitation, that some critically important steps—many might argue, *the* critical steps—had taken place well before he appeared on the scene as a key actor. Unlike classical Darwinian evolution, which unambiguously begins with Darwin—whatever his debts to his intellectual precursors—and the publication of *The Origin of Species* (full title: *On the Origin of Species by Means of Natural Selection or the Preservation of Favoured Races in the Struggle for Life*), the modern synthesis clearly did *not* begin with Ernst Mayr. It was an emergent intellectual creation, already half-formed, at least, at the time he became an active participant in 1942 with the publication of his seminal book, *Systematics and the Origin of Species from the Viewpoint of a Zoologist.*

A further complicating factor in evaluating just what the evolutionary synthesis consisted of, and Ernst Mayr's role in it, is the fact that he himself became the principal historian of the synthesis, hence of his own role in it. Though he always meticulously and fairly documented the contributions of others, it was inevitable, given his own role in recording the history that he would emerge in these accounts as a key player in the critical events of the 1940s and the 1950s in which the mature synthesis emerged. Other major contributors, in particular the palaeontologist George Gaylord Simpson, are known to have felt that they were not always given their full due in the resulting histories.

This chapter will, therefore, attempt to provide an independent perspective of what the evolutionary synthesis consisted of and Ernst Mayr's importance in relation to it. No one can pretend to full objectivity, balance, and accuracy about historical matters and I do not claim such. I had the privilege of getting to know Ernst Mayr in the last two-and-a-half years of his life, principally through correspondence. I greatly admired him and counted him as a friend, despite the more than forty-year age difference between us. Hence my own account may be more Mayrian than other comparable accounts would be or, perhaps, than mine would be had I not known him personally. Nevertheless, I come to these events at one temporal remove, with my own history and a different disciplinary orientation. I hope that a new perspective from someone other than Ernst Mayr, one that attempts a weighing of the different factors, complexities, and not least, ambiguities, may therefore have some value as a contribution to discussion of these events, events that shaped twentieth-century biology and which continue to exert an important influence on twenty-first-century biology. A different, but complementary, view can be found in the longer treatment by V. B. Smocovitis (1996).

A brief preview of the organization of this chapter might be helpful. I will begin with a brief note on terminology—what term best describes the theory of evolution that had emerged by the mid-1950s? If one is going to describe the history of a complex idea, one needs a term to designate that idea. The interesting fact is that none of the names given to this body of theory is fully satisfactory, which highlights the ambiguity of defining just what modern evolutionary theory consists of, a point to which we will return at the end of this chapter.

I will then give a capsule history of what one might term the "prehistory" of the synthesis. This period stretched more than sixty years, from 1860 to the early 1920s. It has been the subject of a large body of scholarship but a précis of some of the key features might be useful. I will then describe the growing concrescence and synthesis of ideas between 1930 and the late 1940s, during which the theory first assumed a distinct shape and identity. The penultimate section will then describe Mayr's entrance on to the scene, in 1941–1942, and discuss his multiple contributions—both conceptual and organizational—to the growth of evolutionary biology. The chapter will end with an attempted re-evaluation of just what the evolutionary synthesis was at mid-twentieth century and the significance of Ernst Mayr's contribution to it.

"NEODARWINISM," "THE MODERN SYNTHESIS," "THE EVOLUTIONARY SYNTHESIS," OR JUST PLAIN "DARWINISM?"—WHAT *SHOULD* WE CALL THE CORE THEORY OF EVOLUTIONARY BIOLOGY?

It is customary, at least among biologists (if not the general public), to distinguish Darwinian evolution from contemporary evolutionary theory, often designated as "Neo-Darwinism." The reason is simple. Darwin's theory was incomplete in ways that the current theory is not. Darwin himself fully appreciated and acknowledged the lacunae in the evidence and the corresponding uncertainties in interpretation. His resulting tentativeness is obvious in any reading of *The Origin*, which can be seen as a long meditative summation of Darwin's view of the evidence for evolution—and the gaps in that evidence—in the light of his proposed solution for the driving force (natural selection). Darwin referred to the book as "one long argument," a phrase that implicitly acknowledges the uncertainty of his position. Indeed, Darwin's theory of evolution by natural selection may be seen as a complex syllogism of propositions and deductions (Huxley, 1942; Mayr, 1991). Such a chain of theses, in which the solidity of the final conclusion is dependent on the validity of all the component ideas and of their linkages, is bound to be a somewhat shaky conceptual edifice.

The provision of Darwin's argument was, of course, directly related to the incompleteness of the data sets he worked with. Whether one is concentrating on the comparative zoological data, the geology, the fossil evidence, or the biogeographic findings, the information available to Darwin was far less than what his successors had, even forty or fifty years later. Yet, the resolution of the validity of Darwin's theory would clearly involve more than simply finding and compiling more facts, similar in character to those that had been obtained. Rather, certain fundamental questions had to be answered and, as Darwin knew, those answers would have major bearing on the fate of his theory. None was more fundamental than that of the nature of the "variations" that provided the raw material for the operation of Darwinian natural selection and the mechanism of inheritance of those variations. These issues were briefly addressed in *The Origin* but essentially saved for his later book, *The Variation of Plants and Animals under Domestication*, first published in 1868.

In contrast to Darwin's theory, modern evolutionary theory is built on a firm understanding of the nature of genetic variations and their inheritance, foundations that were laid in the first two decades of the twentieth century, beginning with the rediscovery of the "laws" of Mendelian inheritance in 1900 and the proselytizing efforts of William Bateson (1861–1926) on their behalf. Indeed, it was the fusion of this understanding with a sophisticated set of treatments of the behavior of those variations, over time, in populations, in response to selective influences, that comprised the first stage of the twentieth-century synthesis (1922–1932). That fusion of Darwinian thinking with genetics and population thinking is the essence of "neo-Darwinian evolution." Thus, if "Darwinism" is taken to refer to evolution as Darwin understood it, then it is safe to say that today no biologist is a Darwinian. Instead, we are all neo-Darwinians.

Yet, "neo-Darwinism" is itself a somewhat ambiguous term because it is associated with at least four different meanings and implications. It was first used in a different context entirely. August Weissman's vindication, both experimentally and through argument, of "hard" genetic inheritance against the inheritance of acquired traits, led George Romanes (1848–1894), a naturalist friend and supporter of Darwin, to state that "Darwinian natural selection + hard heritance" constituted a new theory, which he termed "neo-Darwinism." Although few today use "neo-Darwinism" in this sense, Ernst Mayr, in his last book (Mayr, 2004), insisted that Romanes's sense was the correct usage. At least for a time, however, "neo-Darwinism" connoted the synthesis of Mendelian genetics with population thinking and natural selection, the conceptual development pioneered by Ronald A. Fisher (1890–1962), J.B.S. Haldane (1892–1964), and Sewall Wright (1890–1988). Yet, this usage is far from uniformly accepted. Mayr himself came to prefer the term "the Fisherian synthesis" (Mayr, 1999, 2004) for this phase of conceptual unification. This term itself, however, has not yet caught on and with its slighting, however unintentional, of Haldane and Wright, might not. Nevertheless, when many genetically-minded evolutionists refer to "neo-Darwinism," they are referring to this first phase of the synthesis, Mayr's "Fisherian synthesis."

The third usage of "neo-Darwinism" makes the term a synonym for the whole body of contemporary evolutionary theory, namely the Fisherian synthesis plus everything added to that conceptual edifice in the 1930s and 1940s. This is the expanded sense of the term and is probably the one that most biologists have in mind when they refer to "neo-Darwinian evolution."

Yet, there is a final and distinct usage. In this form, it designates the gene-based selection theory popularized by Richard Dawkins (1976) and adhered to by many others. In this version of evolution, the key object of selection is not the individual organism, as Darwin and most of the mid-twentieth-century evolutionary biologists thought, but rather the gene. This is a considerably more narrow definition than the broad-sense version of neo-Darwinism and only attention to the context, in any specific instance, will alert the reader or listener to which sense is being employed.

What about calling the main theory today "the modern synthesis" or simply "the evolutionary synthesis?" (Hyxley, 1942). The particular problem with the slogan "the modern synthesis" is the word "modern." Ironically, the concept of "modernity" has itself acquired a feeling of being slightly dated, indeed associated with the last part of the nineteenth century and the first half of the twentieth century. Furthermore, what was indubitably new and "modern" in the early 1940s is now more than sixty years old. Evolutionary biology has moved on in many ways since the publication of Huxley's book, even though the basic tenets of present-day evolutionary theory were in place by then.

As a descriptive term, therefore, the "evolutionary synthesis" is, perhaps, the best. Nevertheless, it is hardly perfect, either. It amounts to an assertion that there really was a true synthesis and connotes both an inclusiveness and a finality that, from the perspective of the first decade of the twenty-first century, look hugely overstated. Given the contentiousness today of all sorts of issues—the levels of selection debate, the prevalence (or not) of truly "neutral" mutations, continuing debates about the variability of rates of evolution and their significance, the important ways that developmental evolution has to be integrated into evolutionary biology as a whole, the precise role of selection versus "developmental constraints" and other factors in shaping morphological evolution, the various controversies surrounding speciation mechanisms, and many more—can one truly assert that there has been a true "synthesis" of views? Even in the late 1940s, the claim might have seemed overblown to many paleontologists and developmental biologists who were dubious that the problem of evolution had been, in principle, truly solved. Can one really speak of "the" synthesis if a fair number of biologists are immediately provoked upon hearing the expression to start discussing what it left out and what it got wrong? Matters of scientific truth are, of course, not settled by vote but to speak of "the" synthesis implies a general level of agreement that does not exist today and which, indeed, may never have existed. Biologists with a morphological bent—morphologists, paleontologists, and developmental biologists—always felt that the synthesis never did full justice to their concerns.

Another possibility is "synthetic Darwinism" favored by some German evolutionary biologists. This has a certain strong appeal but the term has not achieved wide currency outside of Germany and is not likely to, at this point. There is yet another possibile solution and it was suggested by Mayr in his last book. He proposed that the simplest course was just to call the contemporary theory "Darwinism" because in its essentials—gradual evolution, "hard" inheritance of "random" variations, the overwhelming importance of natural selection—the current theory has many of the same fundamentals as Darwin's. Mayr further argued that to use the term in this sense risked no confusion with Darwin's Darwinism. There is something to be said for this proposed Gordian knot-cutting sort of resolution but it is unlikely to be widely adopted. Mayr himself had earlier enumerated nine distinguishable meanings associated with "Darwinism," many associated with distinct periods during the nearly century and a half that have elapsed since publication of *The Origin*. In light of this fact, it would seem that the term is doomed to perpetual ambiguity and really cannot serve as a clear designation of current ideas about evolution.

Everything considered, I would suggest that "the evolutionary synthesis" is probably the least ambiguous of the umbrella terms to denote the theory of evolution as it had been formulated by the mid-twentieth century and the one that seems to be accepted most widely. Despite its problematic aspects, it is, therefore, the one that will be used in this account though we will come back to its implicit claims of completeness at the end. The matter that we will now turn to is the nature of the long gap between the formulation of Darwinian evolution, as set out in *Origin* in 1859, and first steps in the development of the evolutionary synthesis in the 1920s.

THE "PRE-SYNTHESIS" PERIOD: FROM *ORIGIN OF SPECIES* TO THE "FISHERIAN SYNTHESIS"

Many histories, both by biologists and historians, treat the arrival of Darwinian evolution as an immediate and irreversible transformative event in biology and society. This view is seen both in treatments that celebrate Darwinism and those that are hostile to it. In effect, Darwinism in many accounts is depicted as a wave that instantly swept all before it. Yet, the historical evidence shows that this picture is not accurate.

In reality, Darwinian evolution, in the period 1859 to the 1920s, is best seen either as an aborted revolution or a half-completed one. The respect in which it was an unambiguous success, and thus can claim to be at least a half-completed revolution, was that it convinced biologists of the reality of evolution, accounting for the immense diversity of forms of plants and animals. Though sentiment among scientists for evolution had been growing, the evidence marshaled by Darwin in *The Origin* and by Alfred R. Wallace and others was simply too strong to dismiss. In this sense, Darwinism as "evolutionism" was a clear success, at least for the majority of biologists. (That it has not triumphed even in this regard among the general Western publics, especially in the United States, hardly needs mention.)

Darwin's goal, however, was considerably more ambitious than simply to document evolution as the definitive shaper of living things: He felt that he had found the motor of evolution in the process that he termed "natural selection." His gathering of evidence in support of the reality of evolution was intended to provide the necessary prelude to his explanation of how evolution works: his theory of natural selection. Yet, this central plank of Charles Darwin's theory, the role of natural selection in shaping evolutionary trajectories, was regarded with skepticism right from the beginning and had been rejected by most biologists by the 1880s. Even Thomas Henry Huxley, Darwin's presumed great champion ("Darwin's bulldog")—who famously said after reading *Origin*, "How stupid of me not to have thought of that!" in fact—soon had second thoughts. He was never convinced by the idea of natural selection, at least as based on the selection of small-effect variations. The majority of biologists of the late nineteenth century rejected the explanation even more completely, believing that evolution must be driven by other causes or forces. Virtually all those who rejected natural selection, and they were the great majority, favored a "transformationalist" or "developmentalist" mode of evolution, one that operated along quasi-predetermined tracks. Such explanations discard the idea of "random variations" as the substrate for evolutionary change. In doing so, they simultaneously eliminate the "population thinking" (Mayr, 1963, 1982) that was so central to Darwin's own ideas. It seems highly likely that by the time of Darwin's death, he himself felt that he had failed in his most important self-appointed task, namely convincing other scientists of the importance of natural selection as the driving force of evolution. In this respect, Darwinism was an aborted revolution.

Nor was the dismissal of natural selection a short-lived phase. Indeed, Darwinism as a theory of change failed to find widespread acceptance for more than sixty years. During this period, Darwinian natural selection was undeniably in "eclipse" (Bowler, 1983). Yet the word "eclipse" does not signify cessation of existence. An entity that is eclipsed, as for instance the sun by the earth during a total solar eclipse, may well be invisible during such an event but it has not ceased to exist nor, necessarily, to exert influence. The gravitational influence of the sun on the earth, for instance, remains just as strong during a solar eclipse, however completely the moon blocks the view of the sun. A biological example of "eclipse" illustrates the point. When a single bacteriophage viral particle infects a susceptible bacterial cell, there may be a period, perhaps thirty to forty minutes, when it seems, from observation of the cell under a microscope that nothing has happened. Yet, at the end of that time, the cell suddenly bursts, releasing anywhere from dozens to hundreds of new bacteriophage particles (depending upon the species of virus and the conditions). The invisible events during the eclipse period are the crucial ones of viral genome replication and viral particle assembly.

The eclipse of Darwinism from 1860 to the 1920s had a similar character. On the surface, belief in natural selection seemed to have nearly vanished from the scene. To posit it as the main driver of evolution, as Darwin had, seemed deeply implausible, indeed ridiculous, to most biologists. Furthermore, as Mendelian genetics bloomed, from 1900 onward, it seemed less and less likely to many geneticists that the kind of small-effect "variations" Darwin had posited as the stuff that natural selection worked upon could possibly be the source material of evolution. After all, the only hereditary factors that Mendelians could study were variations that created large, visible phenotypic effects. It appeared to the early Mendelians that the long-sought mechanism of inheritance, the "missing link" in Darwin's theorizing, had put an end to Darwinian ideas about the process of evolution. Of course, Mendelians did not monopolize discussion about heredity. There were the biometricians. This group did champion the power of small additive effects. The biometricians' case, however, was hobbled by their rejection of Mendelism as a *general* explanation of inheritance even as the evidence for Mendelian inheritance grew progressively stronger.

Nevertheless, despite the general rejection of Darwin's ideas about the selection of small-effect variations and gradual evolution, they never really vanished, in part because they could not be disproved. Indeed, the Darwinian explanation had the singular trait of being a fully consistent materialist mechanism of evolution, in contrast to the developmental-transformationist modes supported by the non-Darwinians. Darwin's ideas may have suffered eclipse but they were still present, in the background, as it were. Furthermore, the evidence that certain traits were the product of additive effects of individual Mendelian factors gave new life to the possibility that mutations of small effect not only existed but, conceivably, might be a substrate for natural selection, if combined in the right way. These discoveries, in effect, put biometrical genetics as an ideology out of business though some of their analytical methods have lived on and continued to be useful. Furthermore, none of the alternatives to natural selection ever received independent, empirical confirmation. Darwinian natural selection may have been regarded for decades as an unsatisfactory explanation but it remained the default idea that would simply not go away. It was that reality, along with the demonstration of the additivity of small-effect Mendelian genes, that paved the way for the work of Fisher, Haldane, and Wright and the beginnings of the evolutionary synthesis.

THE EVOLUTIONARY SYNTHESIS: A ROLLING CONCEPTUAL CASCADE

The evolutionary synthesis is customarily, and reasonably, seen as a process occurring in stages over a period of roughly three decades, each stage marked by one or (usually) more key books or long publications. In the first part (roughly 1922–1932), Mendelian genetics, cytology (chromosome studies), and population thinking were synthesized imaginatively and combined with the idea of Darwinian natural selection to produce the first phase of "neo-Darwinian evolution" (in its post-Romanes sense). This was unquestionably a true synthesis, in which the whole was far more than the sum of the parts. The second stage (approximately 1937–1942) involved the addition of insights, data, and methods of systematics to the developing brew. The key figures here were Theodosius Dobzhansky (1900–1975) and Ernst Mayr. The third stage was the accretion of paleontology to the developing structure of ideas and can be traced directly to the influence of the seminal book *Tempo and Mode of Evolution*, by George Gaylord Simpson (1902–1984), in 1944. The final addition was that of the evolution of plants, with G. Ledyard Stebbins' book, *Variation in Plants* (Stebbins, 1950), but by then the battle of ideas for the new science of evolutionary biology had already been largely won.

Yet, the early stages were hardly independent of one another. The first stage was absolutely crucial: Without the fusion of transmission genetics, population thinking, and the appropriate mathematics and statistics with the idea of natural selection, it is safe to say that none of the rest could have happened. Following the laying of those foundations, Dobzhansky deliberately built on the work of Fisher, Haldane, and Wright, while both adding his own special cytogenetic insights and simultaneously presenting the mathematically hard-to-digest material of the genetic triumvirate in palatable form for the majority of biologists. Ernst

Mayr, in turn, inspired by Dobzhansky, took up where Dobzhansky left off, again adding his own distinctive insights to the mix. Simpson's groundbreaking book in paleontology was also prompted by the key advances in population genetics to attempt a synthesis of paleontology and genetics. Like Mayr, he was especially inspired by Dobzhansky's book rather than the original publications of Fisher, Haldane, and Wright. And all of this conceptual advance, which was largely based on animal biology, served as the foundations for Stebbins's treatment, which brought plant biology into the evolutionary synthesis. What was crucial in this cascade of scientific insights and achievements was that all the participants consciously and deliberately transcended their disciplinary training. Each must have sensed *the possibilities* of a large synthesis, well before the synthesis assumed mature form in the late 1940s, and acted accordingly. The following is a more detailed account of this sequence of stages and critical contributions.

NEO-DARWINISM: BRINGING MENDELIAN GENETICS TOGETHER WITH POPULATION THINKING AND NATURAL SELECTION

The three key figures here were R.A. Fisher, J.B.S. Haldane, and S. Wright. In an age such as ours when the term "genius" has been greatly devalued by overuse and should therefore be used cautiously, it is safe to say that all three men fully deserve the appellation. While all three today are remembered as theorists, it is worth remembering that they all had had field and/or experimental experience, Fisher and Haldane particularly with plant breeding and Wright with animal breeding. Fisher had also done mouse breeding at his home, as an avocation, and had had fairly extensive field experience with his colleague E.B. Ford (1901–1988), a distinguished figure himself and a contributor to the synthesis. In effect, the theorizing of the three key founders of population genetics had been informed by hands-on experience with living things.

Despite the links and communication between these three men, and the fact that they made their key contributions to the synthesis within a few years of each other, Fisher was undoubtedly the most crucial figure of the three, at least for the genesis of the synthesis. He made the earliest contributions, in 1922, and certainly influenced Haldane and Wright. Had Haldane and Wright never been born, Fisher's influence in paving the way for both a re-legitimization of Darwinism, along with its great strengthening, might have been just as powerful. It seems appropriate, therefore, to call this first phase of the synthesis the "Fisherian synthesis," as per Mayr (Mayr, 1999, 2004). Nevertheless, it should be mentioned that Wright's work stands apart from that of Fisher and Haldane in certain respects. The key difference involves the question of "structured populations" (in Wright's formulation), in which chance factors affect the genotypes of demes, as opposed to large panmictic populations as the field of operation of natural selection (Fisher's view). This remained a significant point of dispute and source of estrangement between Fisher and Wright until the former's death in 1962. Although, in this respect, most population geneticists today are Fisherians, most systematists with an evolutionary bent are sympathetic to Wright's ideas. It is not clear that the last word on this debate has been said.

To return to the history, however: The crucial first paper by Fisher was titled "On the Dominance Ratio" (Fisher, 1922). In it he introduced both stochastic considerations to gene survival, the interplay of selection (at differing selective strengths) and these stochastic factors, and the analytical methods (the "chain binomial" method and diffusion approximations) that were crucial to the analysis. Fisher's work culminated in his book *The Genetical Theory of Natural Selection* (1930), which summarized the previous eight years of work and remains his landmark publication for the evolutionary synthesis. Haldane began somewhat similar analyses in 1924 though he did not seriously grapple with stochasticity in his models until 1928. His work is summarized in *The Causes of Evolution*, first published in 1932, a book that is more accessible than Fisher's and that, in consequence, may have had more impact on more biologists but which is considered a lesser work than Fisher's. Wright began work on evolutionary theory in 1925 but did not publish anything in this area until 1929. His seminal contribution was a long paper in *Genetics*, "Evolution in Mendelian Populations." A detailed description of the key events, papers, and interactions can be found in Edwards (2001).

BRINGING SYSTEMATICS INTO THE MIX AND SPREADING THE GOSPEL: DOBZHANSKY AND MAYR

Although Fisher, Haldane, and Wright had all had practical experience with plant or animal breeding, their experience had not included extensive study of populations and the natural extent of variability within wild populations, of either plants or animals, though Fisher was exposed to this area through his collaborations with E.B. Ford. To naturalists and systematists, therefore, their work, which largely dealt with—to the naturalist's cast of mind—highly abstract and hypothetical situations, might have seemed to lack much reality. Indeed, even experimentally based transmission genetics, upon which the population geneticists built, seemed largely divorced from the reality that the zoologists and botanists studied in the field.

This latter point was remarked upon by the distinguished Columbia University geneticist L.C. Dunn in his preface to Dobzhansky's book in 1937. Dunn wrote:

> The requirements of this search [for conceptual rigor] drove genetics into the laboratory, along an apparently narrow alley hedged in by culture bottles of *Drosophila* and other insects, by the breeding cages of captive rodents, and by maize and snapdragons and other plants. Biologists not native to this alley thought sometimes that those who trod along it could not or would not look over the hedge; they admitted that the alley was paved with honest intentions but at its end they thought they could see a red light and a sign "The Gene: Dead End." (Dunn, 1937)

Furthermore, the additional layers of abstraction and mathematics that Fisher, Haldane, and Wright laid over experimental transmission genetics to create the new science of population genetics would almost certainly have been mystifying even to most experimental geneticists. In the early 1930s, there might have been no more than a few hundred biologists who truly grasped the mathematical reasoning of the three founders of population genetics and probably, in reality, no more than a few dozen. Even today, Fisher's book remains a dense and difficult read. Fisher himself was aware of this probable lack of impact. Being informed in 1931 by Oxford University Press of somewhat better sales of *The Genetical Theory of Natural Selection* than had been expected, he remarked, "It was so long before I heard from them that I had quite made up my mind that it was one of those which everybody praised and nobody read, and would have no influence on biological opinion" (quoted in Edwards, 2000). In this, he was undoubtedly too pessimistic: It seems more likely that like Einstein's ideas of relativity into the 1920s, there was a general sense of the drift of the argument and its importance even though only a relatively small number of individuals followed the mathematical reasoning in detail.

Nevertheless, despite the perception by some biologists that a breakthrough of sorts had been made, the budding evolutionary synthesis, initiated by two Englishmen and an American, might itself well have become a "dead end." That it did not happen, however, is because bridges between the new conceptual edifice and the realities familiar to most biologists were soon constructed. The beginnings of this exercise in conceptual bridge-building had its origins in what might seem an unexpected place: Russia. The largely forgotten hero of this phase was the Moscow-based insect biologist, Sergei Chetverikov (1880–1959), whose publications on evolution spanned the period 1906–1926. Originally a naturalist, specializing in butterflies, he branched out to the study of genetics in natural populations of *Drosophila*. His first critical contribution was the demonstration that naturally occurring mutations, those that arise and are found in natural populations, range in selective value from adaptive (of various strengths) to neutral to slightly deleterious to highly deleterious. His accompanying insight was that most mutations, when they arise, are hidden from selection, in heterozygous form, and thus have an initial chance to spread, or at least be maintained in populations. His second critical contribution was the demonstration, by breeding captured wild *Drosophila*, that natural populations in fact harbor many visible recessive mutations. When homozygous, such mutations undoubtedly lower fitness but in heterozygous form, they can be maintained and passed on with high efficiency.

Chetverikov's work, indeed, foreshadowed the molecular work of Richard Lewontin and Harry Henry Harris (both in 1966) more than four decades later, work that showed that populations possess huge amounts of genetic variation. But, by then, the earlier classical work of Chetverikov and his school had been largely forgotten, with the result that the molecular findings of the 1960s were treated as a revelation by most geneticists and evolutionary biologists.

Chetverikov's work and that of his students also convinced him of the reality and pervasiveness of natural selection, even when the positive selective value of an allele was small relative to the most common allelic form. As Mayr has emphasized, Chetverikov arrived at such conclusions from his field and laboratory work long before the theoretical work of Fisher and Haldane established the same points. Furthermore, Chetverikov obtained much evidence for the ubiquity and significance of genetic interactions in populations as a substrate for natural selection. In Fisher's work, such epistasis was treated primarily as a factor slowing down the operation of natural selection while Wright would later treat it qualitatively as a positive source of change but never incorporated this thinking substantially in his more mathematical and theoretical work.

Chetverikov's contributions were not only direct but indirect, in the students he helped train, some of whom went on to major careers. The most eminent of these was, undoubtedly, N.V. Timofeeff-Ressovsky, who later became the mentor of Max Delbruck and thus, through this star pupil, one of the intellectual grandfathers of the nascent science of molecular biology in the 1940s and early 1950s. Yet, possibly the most significant of Chetverikov's influences was on someone who was not one of his students but that of a colleague in Leningrad. This student was the Ukrainian-born Theodosius Dobzhansky (1900–1975), whose career and thinking can be seen as a continuation and extension of the paths laid down by Chetverikov. Like Chetverikov, his work was a constant dialogue between field work, involving captures of flies from wild populations, and experimental analysis in the laboratory. Dobzhansky left Russia in 1927 to pursue his career in the United States and, after a brief spell at Columbia University, joined T.H. Morgan's group at CalTech.

A decade after his arrival in the States, in 1937, Dobzhansky published the first edition of his most important book, *Genetics and the Origin of Species*. In it, he showed, in nontechnical language, how the insights of Fisher, Haldane, and Wright were consistent with the approaches of Chetverikov's school and the perspective of the systematists-naturalists more generally. Though not a mathematician or a theoretical population geneticist himself, Dobzhansky was the son of a mathematics teacher and comfortable with, if not proficient in, the approaches taken by the population geneticists. He thus had a special capacity to act as a bridge-builder between that group and the naturalist-systematists. In terms of its impact on biologists generally, the evolutionary synthesis probably truly began with Dobzhansky's *Genetics and the Origin of Species* although much of the crucial foundations were laid down in the Fisherian synthesis. If the "eclipse" of Darwinian evolution had continued for the great majority of biologists well beyond the Fisherian synthesis into the 1930s, that period had definitively come to an end with the attention that Dobzhansky's book received.

Yet, there is a paradoxical feature about *Genetics and the Origin of Species*. The title suggests a primary emphasis on the process of speciation. Yet, just as Darwin's *Origin of Species* does not really examine speciation overtly—speciation is treated there implicitly as an inevitable byproduct of genetic divergence, promoted by natural selection—Dobzhansky's book also does not deal as extensively with the actual process of speciation as it does with the general genetic basis of evolutionary change. Ernst Mayr would later claim that only 4.5 pages of Dobzhansky's 321 pages in the 1937 edition dealt overtly with speciation. This is clearly incorrect since Dobzhansky devoted one whole chapter to isolating mechanisms and a second chapter to hybrid sterility, phenomena that are essential elements in the speciation process; we will return in a moment to the probable basis of Mayr's meaning. Nevertheless, there is clearly less about speciation mechanisms in *Genetics and the Origin of Species* than one might expect from its title. Indeed, the terms "speciation" and "species formation" are not in the index.

This gap provided the opening for Ernst Mayr's entrance to evolutionary biology as a major figure. By the early 1940s, Mayr was a noted ornithologist and systematist, serving as deputy curator of ornithology at the American Museum of Natural History (AMNH), where he had worked since 1931. Having received both an M.D. and a Ph.D. precociously at twenty-two, he had made the beginnings of his reputation through two major bird-collecting expeditions, the first in New Guinea, the second in the Solomon Islands. That experience had led to his appointment to the AMNH to catalogue and organize the newly acquired Whitney-Rothschild collection of birds.

Though all of this early work was clearly well within the domain of systematics, not evolution, Mayr's interest in speciation can probably be traced to his collecting expeditions. A key part of his assignment had been to determine whether there were natural populations of several seemingly unique birds of paradise found in several European zoological collections. Mayr, despite extensive success in observing and collecting tropical birds including birds of paradise, did not find any individuals in the tropical wilds that matched the particular zoo specimens in question. The unavoidable conclusion was that these individuals had been produced by matings in the zoos between birds that would ordinarily never mate. A thoughtful biologist reaching this conclusion would be bound to wonder what accounts for the scarcity of such matings in nature, if they can occur among captive birds.

To explore and understand such things, however, would require some understanding in genetics, which initially Mayr did not have. He arrived at the museum still a believer in Lamarckian inheritance. But he sought out expertise. His first, though preliminary, tutelage in modern genetics came from another colleague at the AMNH, James Chapin. Nevertheless, Dobzhansky was undoubtedly the major influence in Mayr's education in genetics. Furthermore, Dobzhansky, as a fellow continental European and one who came, no less, from a similar background in systematics, was undoubtedly a more natural and comfortable tutor in genetics for Mayr than the traditional American lab-trained geneticists. Dobzhansky, throughout most of the 1930s was at CalTech but he and Mayr first met at a conference in 1935 and following a symposium organized by Dobzhansky in December 1939, at which Mayr was a speaker (on speciation), and Dobzhansky's move to Columbia University in 1939, he and Mayr had increasingly more contact.

In a series of lectures on evolution, for the prestigious Jesup lecture series at Columbia University, the principle organizer, L.C. Dunn recruited two scientists to talk on speciation processes in 1941. The first was Edgar Anderson, a maize geneticist whose assignment was to talk on speciation in plants. The second was Ernst Mayr who was to give an equivalent set of lectures on species formation in animals. Speakers in the series were expected to then produce a book. This was the origin of Dobzhansky's 1937 book, in fact. Both Mayr and Anderson gave their commissioned set of four lectures each, Mayr delivering his in March 1941. The two men were expected to co-author a single book. When it came to producing the book, however, a problem developed. Anderson, who suffered from periodic depression, reported that he was unable to get on with his writing. Mayr, therefore, expanded his contribution to make a complete book, which became *Systematics and the Origin of*

Species from the Viewpoint of a Zoologist, which was published in 1942. The title was clearly a tribute to Dobzhansky's book, while simultaneously indicating how it would differ from the latter, just as the title of Dobzhansky's book paid direct homage to Darwin's. Mayr's somewhat cumbersome title, with its subsidiary clause, is also an implicit recognition of the limits of the treatment and of Anderson's default.

Although Mayr always maintained that the purpose of his book was to fill the perceived gap in Dobzhansky's book about actual speciation processes, another reason may also have been at play. In his book, Dobzhansky recognized but somewhat played down the importance of geographic separation in speciation, in his crucial chapter on isolating mechanisms. Indeed, he only gave four-and-a-half pages to this aspect of the problem. For Mayr, however, initial geographical separation was the *essential* condition for the beginnings of speciation. In Mayr's view, and throughout most of his life, speciation was overwhelming "allopatric" in character (involving geographical separation as the key initiating factor). His reference to "only 4½ pages" in Dobzhansky's book devoted to speciation is undoubtedly a reference to those particular pages in *Genetics and the Origin of Species* that are focused on geographical isolation as a factor in speciation. For Mayr, this was the crucial factor. (Toward the end of his life, however, he conceded that speciation without prior geographic separation, namely "sympatric speciation," was important in certain animal groups and, perhaps, even more widely [Mayr, 1999, 2002].)

His book can thus be seen as much as an argument for the signal importance of allopatric speciation as it is a conscious attempt both to fuse systematics with the budding synthesis and to focus attention on the key role of speciation in evolutionary processes. Conceptually, its main contribution was the "biological species concept" (BSC) but that was hardly original to Mayr, who freely acknowledged his nineteenth-century antecedents in this idea. The principal significance of the book was, undoubtedly, that it put speciation mechanisms as a central and essential, though hitherto somewhat neglected, aspect of evolutionary mechanisms. It did this, in part, because Mayr was able to draw on a much wider array of organisms, especially among vertebrates, for his material, while Dobzhansky, limited to organisms where something about the genetics was known, had perforce concentrated on *Drosophila* and plants.

PALEONTOLOGY AND THE EVOLUTIONARY SYNTHESIS

If there was one field that had remained stubbornly resistant to the idea that Darwinian natural selection had been the primary shaper of the evolution of animal and plant forms, it was paleontology. Though biochemistry and physiology still remained largely outside the growing penumbra of evolutionary biology, despite some early hints that proteins (specifically enzymes) also evolve, their practitioners did not exhibit the same overt hostility to Darwinian evolution as did the paleontologists. Paleontology remained the last redoubt of the skeptics and disbelievers, who for the most part were devoted to alternative views (Lamarckian evolution, orthogenesis, and saltatory evolution).

The person who changed this was George Gaylord Simpson (1902–1984). Trained initially as a geologist but having a strong interest in biology generally and paleontology more specifically throughout his university and postgraduate years, Simpson earned his Ph.D. at the early age of twenty-four, and almost immediately began his researches on early Cenozoic mammalian fossils, which would remain his central area of paleontological research throughout his career. But, in the 1930s, the winds of change had begun to blow through the halls of paleontology, and perhaps no paleontologist was as receptive to them as the young G.G. Simpson. He had read Fisher and Haldane soon after their books appeared (but Wright only later), and was already groping toward some synthesis of paleontology and genetics, but it was Dobzhansky's book that galvanized him. As he recollected later:

> My own [early] thinking along theoretical lines was nevertheless mostly along lines of historiography and organismal adaptation, in fossil and recent organisms, until the first edition of Dobzhansky's *Genetics and the Origin of Species* (1937). That book profoundly changed my whole outlook and started me thinking more definitely along the lines of an explanatory (causal) synthesis and less exclusively along the lines more nearly traditional in paleontology. (quoted in Mayr, 1980b)

In 1938, Simpson began writing the book that would become *Tempo and Mode in Evolution*, his first and undoubtedly his most important contribution to the growing evolutionary synthesis. It would be interrupted by Simpson's war service, in 1942–1943, and was not published until 1944. But its effect was immediate, especially on those paleontologists who had intuitively accepted Darwinian natural selection as a main driver of evolution, even if these were not yet the majority. In his book, Simpson showed how the population genetic thinking of Fisher and Haldane and Wright was, contrary to the seeming evidence of the fossils and the prejudices and assumptions of generations of paleontologists, fully consistent with the evolutionary patterns that could be reconstructed from fossil evidence. Simpson, in this book, for the first time also carefully set forth the evidence for strikingly different tempos of evolutionary change, and the suggestion that these implied different "modes." Yet that conclusion, he emphasized, did not imply non-Darwinian mechanisms. It simply called for additional thought and work to understand the different bases or mechanisms of those modes. Simpson was a true

Darwinian, in his beliefs in incremental evolution over time and the importance of natural selection, and the effect of his influence was to make Darwinian natural selection a plausible explanation of the patterns of evolutionary change, even as reflected in that seeming most un-Darwinian data set, the fossil record. Indeed, all the other mechanisms favored by paleontologists were in retreat from the time of publication of *Tempo and Mode in Evolution* but what was missing was a cogent Darwinian explanation and that is what Simpson supplied in *Tempo and Mode*. Though he had built on the earlier work of the Fisherian synthesis and Dobzhansky's expansion of the latter, his intellectual achievement in bringing paleontology into the fold is probably nearly on the same order as that of Fisher, Haldane, and Wright.

ERNST MAYR: FROM ORNITHOLOGIST TO EVOLUTIONARY "BRIDGE BUILDER" TO "COMMUNITY ARCHITECT"

It will be clear that the evolutionary synthesis was itself an evolutionary sequence, in the conceptual realm. Each stage had an influence on and served as a foundation for the next, not least because the participating individuals were aware, often acutely, of the other's contributions. The idea of the power of natural selection had been validated—in effect, Darwin had been vindicated—and fields that had been thought of as quite distinct were beginning to find connections among themselves. Yet, by the early 1940s, there was still no recognized field of "evolutionary biology" and many biologists still remained outside the arena of the new thinking, with many still disdainful of it as a proper subject.

Something more was needed. And that something was increased cross-communication and venues (both physical places and journals) for discussions of problems of mutual interest—or areas of conflict. The birth of a scientific discipline requires more than an intellectually coherent and consistent set of ideas, focused on a set of related problems. It needs structure and organization. And the creation of those features requires the work of individuals. For evolutionary biology, there was a small group of such individuals, conscious intellectual "bridge builders," individuals devoted to the cross-fertilization of ideas, relevant to evolution, between the different subject areas of biology. The term itself was, in later years, often invoked by Mayr, who named half a dozen individuals as the key participants in this exercise. These were Dobzhansky, Simpson, Huxley, Bernard Rensch (a mentor of Mayr's), Stebbins, and, of course, Mayr himself.

Of these, Mayr might, in the late 1930s, have seemed the least likely to play a role in the building of a consensus about the workings of evolution. As mentioned earlier, he was originally a talented bird watcher, then a proper ornithologist and systematist, but he had been slow to come to genetics, still believing in Lamarckian inheritance, when he arrived at the AMNH, three decades after "hard" (Mendelian) inheritance had been verified. Yet, he had some outstanding personal qualities that would certainly overcome any initial handicaps. These included great energy, enthusiasm, a capacity of joy in his work, endless curiosity, and a large ability to absorb new information. Not least, his analytical capabilities, though non-mathematical, were strong. He learned how to ask the right questions and follow them up. Some of these personal characteristics—the energy, the enthusiasm, the delight in biology—are apparent in the quote given at the start of this chapter as a young man starting out on his first great adventure, contemplating the possibility of death in the jungles of New Guinea but feeling that it was worth the risk and that he had already had a lucky and fulfilling life. Ernst Mayr was an optimist who loved life. His later specific optimistic belief, in the late 1930s and early 1940s, that a true evolutionary synthesis could be created was a manifestation of his basic temperament. It would be Ernst Mayr who would be the most active "bridge builder" for the whole enterprise of creating the science of evolutionary biology, doing so with great energy and intensity over a span of seven to eight years, roughly from 1942, when *Systematics and the Origin of Species* appeared, to 1949–1950, when he stepped down as chief editor of the journal *Evolution*. In all this activity, he went from being one of the bridge-builders to the crucial "community architect" (Cain, 1994).

We have looked at his first book and its significance for the nascent synthesis. His burgeoning activities as a builder of the structures that promoted the development of the synthesis will now be considered.

THE COMMITTEE ON COMMON PROBLEMS OF GENETICS, PALEONTOLOGY, AND SYSTEMATICS (1942–1947)

An often overlooked chapter in the history of the synthesis was the formation of a special body, by the National Research Council, in Washington, D.C., to promote dialogue about the growing consensus on evolution. This was "The Committee on Common Problems in Genetics and Paleontology," which soon added systematics to its brief, to become "The Committee on Common Problems in Genetics, Paleontology, and Systematics." The original committee, founded in 1938, was initially seen as a talking shop of some sort but appears to have been fairly inactive during its first years of existence. Its most significant venture was the circulation of questions and answers, via written correspondence, among its members, who comprised a group of about thirty scientists across the

United States, in the form of four mimeographed bulletins, in 1944. The procedure was that members were to write to each other when they had questions about the other's specialty, and send a copy to Ernst Mayr. The reply would also go both to the initiating member and to Mayr, who, when a sufficient number of letters and responses had accumulated, would compile them, and send them out as a bulletin. The first of these bulletins is dated 15 May 1944, the last 13 November 1944. Not all of the thirty or so members participated but the correspondents included geneticists, systematists, animal paleontologists, botanists, and paleobotanists.

The most striking feature of these exchanges is, perhaps, simply that they were taking place. Geneticists, for instance, were, at last, talking to paleontologists, and vice versa. This kind of interchange had hardly existed a mere four to five years before. Perhaps the most significant prior attempt at generating dialogue between paleontologists and geneticists had been a special conference dedicated to that aim, held in Tubingen, Germany, in 1929. Yet, as had happened before in less structured encounters between members of these two disciplines, the two groups largely tended to talk past one another. The Tubingen effort was apparently judged by most or all of the participants as a failure. In contrast, in the letters that constituted the exchanges of the committee, the questions were serious and probing and the answers were always frank and often detailed and full. Many of the exchanges dealt with differing rates of evolution—Simpson's great theme in his book—and the significance of those differences. An interesting feature is that it is apparent that the paleontologists in the group were all fairly well-committed Darwinians. They would have been ready to receive the treatment of evolution and paleontology that would come forth in *Tempo and Mode of Evolution*. The committee seems to have gone back to its fairly inactive state by early 1945 but it had helped lay the groundwork for two key developments, the first being the establishment of the first society devoted to evolution, which would become known as the Society for the Study of Evolution (the SSE) and the founding of the journal *Evolution*, the first journal devoted to evolutionary biology. A discipline has begun to have an identity when it starts acquiring societies and journals devoted to its propagation. For evolutionary biology, the immediate postwar years, 1946–1947, were crucial in this respect. The society was founded at a meeting in St. Louis, in June 1946. The attendance was small (fifty-nine) but the meeting was of signal importance for the birth of the new discipline. And the first issue of *Evolution* appeared in January 1947. Perhaps an equally noteworthy event occurred that month: the holding of a meeting at Princeton University, which provided the first general recognition, ratification in effect, of the fact that there was now a comprehensive and believable modern theory of evolution. The excitement of the meeting is still evident today, more than sixty years later, in the book that was produced from the conference proceedings.

The histories of both the society and the journal are intriguing but need not be given in detail here. The stories can be found in Cain (1994) and Mayr (1997b). A crucial fact, however, is that Ernst Mayr was a key participant in both ventures, providing much of the vision and the organizational energy. Along with Dobzhansky and three others, he was a key organizer of the new society, originally called the Evolution Society, whose goal was officially stated as "the promotion of the study of organic evolution and the integration of the various fields of biology" (Mayr, 1997b). And he was the chief editor of the new journal, from 1947 through 1949. In that latter capacity, he was a very active editor, trying to shape and balance the contents and promote useful dialogue. Those editorial efforts were essential. The early issues tended to be dominated by the *Drosophila* geneticists, which produced complaints from members of other disciplines that their subjects were not being given due attention. Mayr found himself explaining, more than once, that he was doing what he could but that ultimately it was up to members of these disciplines to submit their material to *Evolution*. And, indeed, better balance was eventually achieved. By the time he stepped down, he undoubtedly felt that he had done his major work in bringing about the evolutionary synthesis.

CONSOLIDATION

Of course, Mayr did not stop there. He lived another fifty-five years and was active until the end of his life, publishing his last book six months before his death. In the 1950s, he continued to produce important ideas on speciation and in 1963, he produced his major book on the subject, *Animal Species in Evolution*. And, in his long life, he continued to publish on virtually every subject in evolution, from hominid evolution, to the evolution of the eye, to critiques of the Three Kingdom hypothesis of Carl Woese, to the value of the search for extraterrestrial intelligence (SETI). As new subjects in evolutionary biology developed, such as evolutionary developmental biology in the 1990s, he welcomed them and followed them with interest. In addition, from the 1970s onward, he devoted more and more of his time to the related subjects of the history of biology and the philosophy of science as it pertains to biology. His magisterial *The Growth of Biological Thought* (1982) remains his landmark work in these areas but the fact that his last book, *What Makes Biology Unique?* (2004), was focused on these questions illustrates the importance he attached them. All these efforts were not distinct from his evolutionary interests but, as he saw them, an extension of them. It was, after all, his long immersion in the issues of evolutionary biology that convinced him that most of the previous thinking about the philosophy of science had been unfortunately skewed toward the peculiarities of physics. He was determined to show the nature of the subject matter in the biological sciences required a different kind of philosophy of science, one that was not inferior to the physics-based kind but simply different, with its own requirements.

All of this work he undertook out of interest for his subject and love of biology but also, one suspects, in the interest of consolidating his substantial intellectual legacy. Ernst Mayr in the last decades of his life regarded himself first and foremost as an evolutionary biologist—not just as an ornithologist or systematist or a historian of science—and that is the way he wanted to be remembered.

THE EVOLUTIONARY SYNTHESIS: HOW COMPLETE WAS IT? HOW IMPORTANT WAS IT?

In this coda, one can only indicate the interest of the questions rather than provide a definitive accounting. With respect to the completeness and definitiveness of "the evolutionary synthesis" of the 1940s, its gaps are only too apparent today. And it could hardly be otherwise: In 1949, the year that Ernst Mayr stepped down as editor of *Evolution*, the world was still more than three years away from the announcement of the Watson-Crick model of DNA structure. In effect, the synthesis was a conceptual accounting of evolution that lacked any foundations in detailed knowledge of genes or genomes. The advent of the revolution of molecular biology has changed the scope and face of evolutionary biology, if not its underlying precepts. In particular, the new molecular approaches ushered in the birth of whole new fields that were hardly dreamt of at mid-twentieth century, such as molecular evolution, comparative genomics, and evolutionary developmental biology. Not least, phylogenetics, the study of the actual history of life, was hardly touched upon by the synthesis—the latter was almost entirely a discussion about processes and mechanisms, not about the actual, detailed history of life on earth. The first modern phylogenetic trees were produced in the late 1960s, but it is probably only in the last ten to fifteen years that the science of phylogenetics has really taken off. From this perspective, "the" evolutionary synthesis can be seen more as "a" synthesis than as any sort of final or complete pulling together of the various strands. It was a great consolidation, which cleared away cobwebs of great misunderstanding, and, in effect, a staging post for further progress in evolutionary biology rather than as the final summit of understanding, the picture that was often presented in the 1959 centenary celebrations. Of course, we can see how woefully incomplete it was in terms of present-day knowledge. Furthermore, many of its key assumptions, indeed basic postulates, have been subject to continuing evaluation and reevaluation. These include: the evolutionary significance of "small" versus "large" effect mutations; the nature of variability of rates of evolution (Simpson's "modes"); the importance and extent of truly neutral evolution; the significance of sympatric versus allopatric speciation; the ways in which developmental evolution has to be treated and how it can be incorporated into and expand evolutionary biology; the importance of epigenetic inheritance mechanisms; the nature of the exact foundations for the evolution of growing complexity; and many, many more. Yet, whatever its apparent gaps as visible today, the synthesis of the 1940sand 1950s remains a striking intellectual concrescence and coming together of diverse strands of thought. It also continues to be the key reference point for all subsequent thinking about evolution, including, not least, the highly critical commentaries on it that so often dominate discussions today.

Evolutionary biology remains a live, vibrant, growing science and it is hard to imagine that its founders would have envisioned or wanted anything else for their "child." And, among those founders, Ernst Mayr retains a special place, as both a father and a midwife of mid-twentieth century evolutionary biology. It is, of course, possible that had he never been born, that place would have been taken by one or more others. But the proper subject of history is what happened, and who was involved, not the might-have-beens of counter-factual history. As a conceptual bridge-builder, Mayr was one of a half-dozen or so key contributors. He was undoubtedly not a creative genius on the order of Fisher or Haldane, nor was his contribution as original as Simpson's. Yet, in his multiplicity of roles, he deserves a special, honored place. Like Dobzhansky and a small group of others, he could envision in the late 1930s and early 1940s that a synthesis *was* possible—no small feat in itself. Then, more than anyone else, he worked to make sure that others, well outside the initial small fold of critical thinkers, would understood that too—and to make their contributions.

ACKNOWLEDGMENTS

I am very grateful to Ms. Alison Pirie and Ms. Dana Fischer of the Museum of Comparative Zoology, Harvard University, who kindly provided access to materials at the Ernst Mayr Library, in particular the letter quoted at the beginning of this article and the four bulletins of the "Committee for Common Problems in Genetics, Paleontology and Systematics." I would also like to thank Francisco Ayala, James F. Crow,

Anthony W.F. Edwards, Evelyn Fox Keller, Jurgen Haffer, and Eva Jablonka for their careful reading and helpful comments on an earlier draft of this chapter. Nevertheless, responsibility for the final interpretations, as well as any remaining errors, is entirely mine.

FURTHER READING

Haffer, J. "Mayr, Ernst Walter," *Dictionary of Biography*, in press.
Mayr, E., Provine, W. B. (eds.). *The Evolutionary Synthesis: Perspectives on the Unification of Biology* (Cambridge, MA: Harvard University Press, 1980).
Smocovitis, V. B. *Unifying Science: The Evolutionary Synthesis and Evolutionary Biology* (Princeton: Princeton University Press, 1996).

Cladistics

Jeffrey H. Schwartz

INTRODUCTION

Cladistics, or cladism, is an approach to reconstructing the evolutionary (= phylogenetic) relationships of organisms that was formalized in the mid-twentieth century by the German entomologist Willi Hennig (1966). It contrasts markedly with other approaches to phylogenetic reconstruction especially in the way in which demonstrated similarity between organisms is interpreted. But in addition to being an approach to determining phylogenetic relationships, Hennig's cladism was also a method for generating taxonomies or classifications. For, as Henning envisioned the endeavors of systematists (those who seek to determine patterns of evolutionary relationship), the ultimate goal is to translate one's hypothesis of relatedness into a classification. Consequently, classifications were hypothetical schemas that had to be continually revised as new hypotheses of relatedness were generated and/or old ones rejected. There are thus two elements to Hennig's cladistics: an approach to generating theories of relatedness, and an approach to generating classifications. The question is: Are they necessarily interrelated? Or, can we dissociate the two pursuits? Before answering these questions directly, it would be useful first to visit the relevant history of evolutionary biology in order to place cladism in it.

TAXONOMY, THE NAMING AND CLASSIFYING OF LIFE

Trying to make sense out of nature has probably been a part of the human experience since the emergence of our species. Witness the pictographs

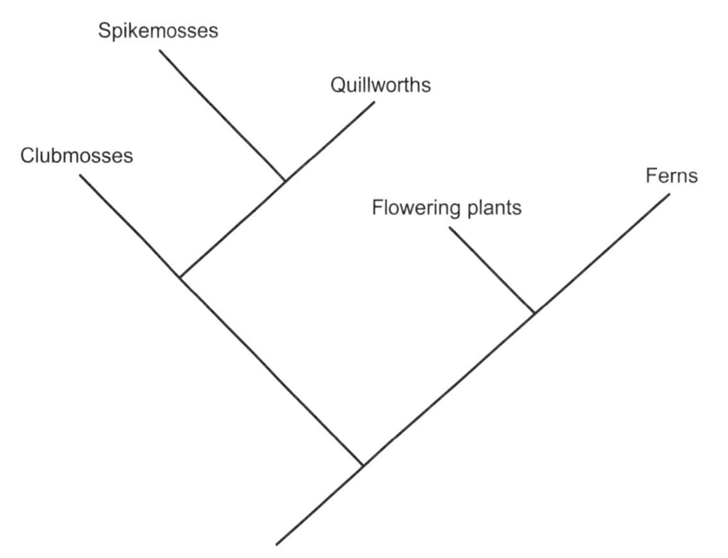

A typical cladogram. All the plants on the chart are related at the base, and then as their characteristics become more different they spread out onto separate arms. This shows the course of their evolutionary divergence.

and paintings on bone and cave walls, some of which now date to almost 80,000 years ago. Witness also the Venus figurines of the Upper Paleolithic and the more recent standing stones, such as Stonehenge, in the British Isles and northern Europe of some 7,000 years ago. Assyrian tablets provide lists of animals grouped according to various criteria, and the Old Testament's Leviticus catalogues animals that, because of certain features of the anatomy, are fit to be eaten, and those that are not. Greeks scientists also sought to delineate order in life, and Aristotle's view of it dominated intellectual thought for nearly two millennia thereafter. In one way or another, humans have tried to find "sense" in what sometimes seems like an "insensible" world (Schwartz, 1987).

In the second century C.E., the Greco-Roman tradition of doing science because of curiosity and an excitement about discovery was replaced by the Christian dogma of inspiration and nature and natural phenomena were interpreted through the restricted lens of biblical literalism (Schwartz, 1999). But even as the subsequent intellectual Dark Ages began to give way to the Renaissance, naturalists of the sixteenth century still produced classifications that reflected as best as possible the story of creation, which was represented as a Great Chain of Being. Taxonomists were united in the endeavor to arrange organisms in an ascending scale of "becoming human" (often with the most primitive organisms ranked just above the sludge at the bottom of a pond). But beyond that common goal, the classifications these scholars generated were typically idiosyncratic, without common structure or methodology. Furthermore, the sources that constituted the "links" of any taxonomist's classification could run the gamut from being as tangible as the kinds of teeth an animal had, to imagining the "essence" of a crustacean-phase in the skin folds of a malformed human stillbirth. In an attempt to bring some rigor into classification, the German taxonomist (often referred to as the Pliny of his time) Konrad Gesner (1551–1558) introduced the taxonomic rank of the genus. Although the recognition of species, or even anything vaguely resembling a concept of a species, would not become part of taxonomy for more than 100 years, Gesner's "genus" served to unite organisms on the basis of features they had in common.

In the late seventeenth century, John Ray introduced the species rank and added another piece of clarity to the art of classification. While earlier, contemporaneous, and even subsequent taxonomists often based their classifications on single specimens or vague definitions of organisms, Ray's "species" was meant to convey something specific about the basic "unit" of nature. And indeed, the word "species" is related to "spice," which means a specific kind.

Because of Gesner and Ray, Karl von Linné, better known as Linnaeus, had available to him the basis of the schema of binomial nomenclature that in 1735 he would apply consistently in his first major classification of plants and animals, the *Systema Naturae*. Above all else, Linnaeus's system demanded that a taxonomist provide testable definitions of what became the fundamental biological unit of nature: the species. Whether the taxonomist focused on dental, skeletal, or soft-tissue anatomy, he was obliged to inform fellow taxonomists of the criteria that formed the basis of the newly proposed species. To add further information that would allow another taxonomist to assess whether a specimen he was studying represented a previously named species or a new one, Linnaeus regularized the association of a species with the next highest rank that subsumed and thus helped to define it: the genus. Consequently, the two names—genus and species—constituted the basis of Linnaeus's system of binomial nomenclature, which still forms the core of classifications, even though the number of taxonomic ranks that have been created since the eighteenth century has multiplied many times over.

Although Linnaeus's approach to classification and his attention to repeatable detail were novel for the time, and dictated his consideration of the groupings of all taxonomic ranks, his approach to ordering these groups was guided by the Great Chain of Being. Thus while his eye for detail led to the delineation of major groups, such as mammals, on the basis of tangible morphology, his sequential ordering of them was less firmly situated in morphology. This was, however, in sharp contrast to his ordering of species within groups. Although still molded by the scala naturae of the Great Chain of Being, Linnaeus's ordering of genera (the plural of genus) within higher taxonomic groups, and of species (spelled the same in the singular and plural) within genera, followed a more or less linear progression in which overall similarity was the primary criterion of comparison. Indeed, it was Linnaeus's emphasis on overall similarity that provoked him in 1735 to classify humans with other "anthropomorphous" ("man-shaped") primates in the Order Anthropomorpha (which he changed to Primates in the 1758 tenth edition of the *Systema*). As such, even though other taxonomists saw his act as heretical, Linnaeus was not disavowing that humans were created last and occupied the top of the Great Chain. He was merely following the logic of his classificatory system. Thus while other taxonomists echoed his predecessor John Ray's recognition of great similarity between humans and other primates but classified humans separately, Linnaeus used this demonstration of overall similarity to classify all primates together. In increasing numbers, subsequent taxonomists generally followed Linnaeus's example of emphasizing similarity in classifying or grouping organisms together, although sometimes an organism's distinct dissimilarity from other organisms led to its being classified by itself.

FOSSILS FIRST, EVOLUTION LAST

There was, however, a limit to how much of the Great Chain of Being a taxonomist could illustrate by studying living organisms alone, even with pathological specimens included. Fortunately for the history of evolutionary biology, Nicolaus Steno demonstrated in 1667 that the chemical and structural configuration of rock-hard objects that looked like clamshells, sharks' teeth, or mammal bones were indeed the fossilized remains of once-living organisms and markedly different from the minerals, rocks, and other objects that also came from the earth. Thus paleontology was born and with it came the potential for finding the "missing links" in the Great Chain. Although still interpreted in the context of the creation story, geological strata and the fossils they contained were seen as portraying a chronological history.

When, beginning in the mid-nineteenth century, the Great Chain was transformed into an evolutionary chain, extinct species were linked to younger ones, and these to living organisms on the basis of overall similarity. Paleontology was seen as a historical science, with its practitioners hoping to close the geological gaps between extinct species. Thus, even though the "unit" of evolution was a biological entity—the species—the picture of evolution was supposed to emerge through study of a nonbiological source: geology. Consequently, the act of reconstructing the evolutionary relationships of extinct organisms and linking any of them to extant organisms relied first on the proper ordering of stratigraphic sequences, and then on aligning fossils through time on the basis of how similar they were to one another. In this context, the process of linking taxa (= any taxonomic group, from species to genera and even higher; sing. taxon) forces one to make judgments of ancestry and descent, which perforce will change as fossils representing new and different taxa are discovered.

WHAT DO YOU DO WITHOUT FOSSILS, OR WITH AN INCOMPLETE RECORD OF LIFE?

Although twentieth century paleontology proceeded apace in its endeavor to find the missing links between stratigraphically disconnected species and to fill in the gaps between them, not all systematic enterprises could be addressed through the study of the fossilized remains of once-living organisms. As frustrating as the discontinuities between fossil taxa were, paleontologists still clung to the hope that intermediate taxa might some day be found. Not so for the study of insects, or spiders (arachnids), or even birds. Rock-hard fossils can only form when minerals derived from the soil in which the remains of the dead become buried replace the calcium carbonate in the shells of invertebrates such as crustaceans and mollusks, the calcium phosphate in the bones of vertebrates, or the hydoxyapatite in vertebrate teeth. This process is facilitated by water, but fast-moving water tends to break up bones and shatter jaws, which then lose the teeth anchored in them. Consequently, the likelihood of an organism becoming fossilized is increased if it lives and dies, or at least dies, in a slow-moving aquatic environment, and if in life at least some of its anatomy was calcified. Not so with soft-bodied organisms, such as worms, or insects, whose bodies have little in their anatomy that can become mineralized.

What to do in the absence of a fossil record? One could still suggest, from a sample under study, that the two organisms that were most similar overall were most closely related, that the next most similar organism was related to these two, and then the third most similar, and so on, until all taxa become clustered in a pattern from most to least similar in overall morphology. This seems like a logical way in which to generate schemes of relatedness. For, after all, I am more similar to my parents than to a cousin, or to a cousin four times removed, or to a member of my community. Thus, although this approach was present in many of the classifications that strove to portray the Great Chain of Being, it remained intact even with the recognition of evolution. In fact, in the early latter half of the twentieth century, it was modeled mathematically by statisticians such as Sokal and Sneath (1963), who called the approach of using overall similarity "phenetics."

But is overall similarity necessarily a reflection of closeness of relatedness? No, according to Hennig (1966). Being interested in the phylogenetic relationships of beetles, Hennig was confronted with a dilemma. First and foremost, the subjects of his study—with fragile wings, a lifestyle that was often not near water, and with only a very thin exoskeleton available for mineralization—were unlikely candidates for fossilization. Yet, he reasoned, even without a record of extinct relatives, surely extant taxa still had some sort of evolutionary relationship to each other. The same was true if one were studying a group that was known only from the fossil record, without any living representatives. All known organisms had to have some pattern of relationship to one another.

How, then, to approach the sorting out of a hypothetical pattern of relationship?

Without a fossil record, how could one endeavor to determine the phylogenetic relationships of organisms? For, is it not through the chronological arranging of taxa that one can recreate the lines of descent and the paths through which lineages gave rise to daughter lineages?

Hennig was nonplussed by this concern, for the simple reason that no matter how incomplete (or complete) the fossil record of any "group" of taxa, it remained the case that two of them will always be more closely related to each than to any other known taxon. And this is the crux of his brilliance: recognizing that what matters most is not a complete or even incomplete fossil record, or a completely documented record of extant taxa, but that those taxa that are known will have some evolutionary relationship to each other, regardless of whether they are represented only in the fossil record, in the present day, or in some combination of fossil and extant taxa.

But how can one represent any scheme of phylogenetic relationship among taxa—even if one acknowledges that the picture of all life that had ever existed will never be complete—if not by discovering relative degrees of overall similarity?

At base is the question of the evolutionary significance of similarity. Is its demonstration really a reflection of phylogenetic propinquity? For instance, does it contribute to understanding the relatedness of, say, humans and the large-bodied apes (orangutans, gorillas, and chimpanzees) if we demonstrate their shared possession of a solid braincase composed of bone, five fingers and toes, seven neck (cervical) vertebrae, upper and lower molar teeth with rather straight sides and relatively low-cusped chewing surfaces, an eye socket (orbit) entirely encased from behind by bone, a bony tube extending from the tympanic ring at the margin of the bony ear region, a thoracic cavity that is compressed from front-to-back (dorsoventrally)? Although these features do indeed characterize humans and all large-bodied apes, all are not unique to a group that includes only humans and large-bodied apes.

If we look broadly among animals, we find that most vertebrates (sharks and their kin being the exceptions to the rule) have solid braincases formed of bone, but no invertebrates do (even those with "heads," such as the cephalopod octopi, squid, and cuttlefish). Among vertebrates, a subset of them is distinguished by its possession of five digits at the end of each limb. Within this group of four-legged vertebrates, which we may refer to as the tetrapods, some are distinguished by having three distinct types of vertebrae, with seven of a particular configuration occurring in the neck or cervical region, between the base of the skull and the first of the set of vertebrae associated with ribs. Within this group of tetrapods with seven cervical vertebrae, which we may refer to as the mammals, some are distinguished by having lower cusps on their upper and lower molar teeth, which are also straighter on their sides. Among this dentally distinct group, which we may refer to as primates, some are distinguished by the development of outgrowths from various bones of the skull that encase the eyeball from behind, creating a true bony eye socket or orbit. Within this group, which we may refer to as anthropoids, some are distinguished by the development of a tubular bony outgrowth from the lateral margin of their bony ear region (which houses the inner ear bones and semicircular canals). Within this group, which we may refer to as catarrhines, some are distinguished by the shape of their rib cage or thoracic cavity. In contrast to most other mammals, in which the thoracic cavity is deep dorsoventrally (from the vertebral column to the breast bone or sternum) and narrow from side to side, the thoracic cavity of these catarrhine primates is compressed dorsoventrally and wide from side to side. And within this group of primates, which we may refer to as the hominoids (= humans and small- as well as large-bodied apes), some are distinguished not only from all other primates but mammals in general by having a relatively thick bony palate. And this subset of hominoid primates, the large-bodied hominoids, includes humans and the large-bodied apes.

Clearly, while it may be true that humans and large-bodied apes do indeed possess all of the features listed above, it is also the case that most of these features are not unique to large-bodied hominoids among vertebrates, tetrapods, mammals, or primates. And even among primates, as far as this example goes, large-bodied hominoids are distinguished by only one feature: a thick bony palate. As such, we might again ask the question: Is the demonstration of overall similarity between any two or three taxa phylogenetically relevant? The answer is obviously "no." So why, then, has the criterion of overall similarity enjoyed such popularity among systematists, whether dealing with extinct forms, living forms, or some combination of both?

The answer to this question may seem elusive. The example given above, of my being more similar to my relatives (albeit in some descending hierarchical relationship from my parents through my far-flung cousins) than to neighbors, may seem to suggest that overall similarity is a viable criterion. But, if we think about it, we realize that the same "deconstruction" of relative degrees of similarity that can be applied to delineating increasingly smaller and more restricted subsets of vertebrates, culminating in the large-bodied hominoids, actually also applies to our view of the relatedness of ourselves to "others." For the features that I share uniquely with my parents are few indeed compared to the vast number that contribute to my being a functioning and viable organism.

Linnaeus partly understood the relevance and need to discriminate between features that merely describe the totality of an organism and those that are distinctive of it (the species) and yet others that are possessed by only a handful of organisms (species within a genus). But he also sought to order taxa within these groupings on the basis of overall similarity. Oddly, this schizophrenic dichotomy between defining features and general features continued to determine how evolutionary minded systematists, whether interested in extinct or living organisms, approached the question of the relatedness of taxa. A possible entrée into the intellectual enterprise that would distinguish between features that are specific to a taxon, to a few taxa, and to increasingly inclusive sets of taxa can be found in the work of the British primatologist Sir Wilfred Le Gros Clark.

In his 1949 monograph on the relationships of fossil and living primates, Le Gros Clark anticipated Hennig by a year when he articulated the necessity of distinguishing between features that are merely descriptive of taxa and those that are distinctive of taxa. The former traits Le Gros Clark actually referred to as "primitive" features. In conformity with accepted scientific jargon, Le Gros Clark referred to the more restrictively possessed features as "advanced" or "specialized" characteristics. And it was the sharing of advanced, not primitive features that Le Gros Clark saw as being important for delineating close evolutionary relationships between taxa. Although this distinction is not trivial and represented a step forward in systematic theory, Le Gros Clark did not actually use this approach in his ensuing discussion of primate relationships. Rather, he fell back on the notion of "mosaic evolution," which had become popular after the evolutionary synthesis, and his own construct, the "total morphological complex."

"Mosaic evolution" embodied a Darwinian perspective with its emphasis on organisms being in states of continual change (Simpson, 1953). But since it was obvious that not all aspects of an organism's biology would change at the same rate, the "mosaic" constituted the entire package, in which different features were believed to change or to have changed at different rates. The "total morphological complex" was in part similar to "mosaic evolution" in noting distinctions between different features, but lumped together primitive and advanced features to produce a package that, overall, was believed to be unique to any given organism.

Thus it was not until 1950, with the publication in German of Hennig's first outline of his approach to systematics and classification, *Grundzüge einer Theorie der phylogenetischen Systematick* (Fundamentals of a

Theory of Phylogenetic Systematics), that not only would the importance of distinguishing between primitive and advance characters be fully articulated, but also that this distinction would be put into practice. The vast majority of English-speaking evolutionary biologists would not, however, be introduced to Hennig's ideas until the publication in 1966 of an expanded discussion of the subject titled *Phylogenetic Systematics*.

CLADISM AND CLADISTIC TERMINOLOGY

The root of the term "cladism" is "clade," which is synonymous with the term "monophyletic group," which refers to a group of organisms or taxa that are united evolutionarily by virtue of their sharing a common ancestor. A large clade can contain any number of nested sets of smaller clades (somewhat analogous to the nesting of a series of Russian dolls, one within another). Consequently what is fundamental to a cladistic approach to determining phylogenetic relationships is the generation of theories of monophyly that lead to a hypothesis of nested sets of monophyletic groups or clades.

This might sound like an unnecessary emphasis. For, after all, isn't the only evolutionarily meaningful group that in which its members are related by virtue of their descent from a common ancestor? Naturally. But the important question is how one goes about generating a theory of monophyly.

By hypothesizing a theory of relatedness on the basis of unspecified overall similarity or on the basis of concepts of mosaic evolution or total morphological complexes, one is treating the myriad features that comprise an organism as if they were of equal phylogenetic significance. Since, however, one can never compare every feature an organism possesses, one is not actually assessing overall similarity, but merely an array of similar features that one has managed or decided to catalog and compare. Obviously, if most of the similarities between two or more taxa are not unique to them, then a claim of the relatedness of these taxa that is based on "lots" of shared similarities is based on little of phylogenetic merit. No less critical is the fact that one or more taxa may be less similar to other taxa because each of them has some number of features that are not shared with any other taxa, but, rather, that are unique to each. The situation becomes more complicated if fossils are part of the analysis and biochronology becomes an arbiter of comparison or if one includes geographic locale as a criterion of significance.

At issue is not the demonstration of shared similarity, but the sorting out of this shared similarity into its different levels of meaning or evolutionary significance. Is a feature that is shared by taxa A and B also seen across a broad array of taxa? If it is, it is a shared primitive feature or, to use Hennig's terminology, a symplesiomorphy ("sym" refers to shared, and "pleisomorphy" to a primitive feature). Consequently, while a comparison of taxa A and B may correctly reveal shared similarity, the fact that the feature is not restricted to only these two taxa means that it cannot reflect anything about their specific evolutionary relationship to each other. But if, in the context of a broad comparison across an array of diverse taxa, one can demonstrate that taxa A and B share a feature that is not seen in another taxa, then one does have something potentially phylogenetically significant with which to hypothesize relatedness. Such a uniquely shared feature is what Le Gros Clark and others would call an advanced feature, but in Hennig's terminology it is a shared derived feature or synapomorphy.

The theory is this: Because a feature is shared uniquely by a subset of taxa, and not ubiquitously by the broad array of taxa, a potential synapomorphy is likely to have originated more recently than the ubiquitous, primitive feature. Another way to think of this is in terms of ancestors. For instance, the more recent an ancestor, the fewer potential descendants it will have. The more ancient an ancestor, the more potential descendants it will have. Because of this likely situation, the more broadly shared features will reflect inheritance from a more distant ancestor, while the more restrictively shared features will reflect inheritance from a more recent ancestor. And that is why even if a comparison reveals true similarity it is imperative that one carry the analysis further, by comparing one's taxa of interest to a much larger sampling of taxa. For it is only in the context of the broadest comparison possible that one can make any assessment of primitiveness (pleisomorphy) versus derivedness (apomorphy). The ancestor's own uniquely derived features (which Hennig called autapomorphies, in which "aut" refers to self) would be inherited by its descendants and thus would make the shared similarities derived from this ancestor their synapomorphies.

But, you may ask, wouldn't at least one or perhaps more of the features present in a distant ancestor be autapomorphic for it? Is it only ancestors of very recent origin that can be hypothesized to have possessed uniquely derived features? The answer is yes. And this gets us to the next point: A theory of symplesiomorphy is a theory of synapomorphy, just at the "wrong" level in the analysis. In other words, if we were asking the question about the potential phylogenetic significance of possessing four limbs, it would be inappropriate to look at our own order, Primates, or even the group Mammalia for the answer. But if we broadened our sample to include all vertebrates, we would see that a subset was distinguished by four-leggedness, which would be their synapomorphy, because four-leggedness had been the autapomorphy of the last common ancestor of all tetrapods. For mammals and, within Mammalia, for primates, four-leggedness would be a shared primitive retention, a symplesiomorphy. Synapomorphy at one level (for a group); symplesiomorphy at another (within the group). Once synapomorphy is hypothesized for a clade, that feature or those features cannot be invoked to delineate evolutionary relationships within that clade. Thus for the tetrapod example, if I wanted to hypothesize clades within this clade, I would have to return to my broad-based comparison to see whether more restrictively shared features emerged that would potentially delineate tetrapod subclades.

One last thing, though: the matter of autapomorphy—that is, when a taxon has derived features that are unique for it. In a phenetic analysis using overall similarity, if a taxon, say A, has a lot of uniquely derived features but its closest relative, taxon B, does not but retains a lot of primitive features, it could very well end up that taxon B will be united with another taxon, C, if taxon C also retains a lot of the same primitive features as taxon B. So while the real situation is that taxa A and B are the closest relatives (sister taxa), taxa B and C will be grouped together because they appear more similar overall. In turn, and incorrectly, taxon A may very well be deemed "primitive" relative to taxa B and C because taxon A is "more different."

Sounds like a lot of work, doesn't it? It is. But it is necessary for the very reason that at the beginning of any investigation into evolutionary relationships, one doesn't know if or how many features will emerge in support of a set of nested theories of monophyly. But in order to do this kind of work, one has to use as broad a comparative sample as possible so that one can discern for the taxa of interest whether the features they share are symplesiomorphies or synapomorphies. In this context, overall similarity just doesn't cut the mustard.

CLADISTIC ANALYSIS AND THE HYPOTHETICO-DEDUCTIVE APPROACH TO SCIENCE

You might have been wondering why I've been using words such as "potential" and "hypothetical." The reason is simple. In evolutionary studies, and particularly when reconstructing phylogenetic relationships, one can never know or prove anything for certain. As such, while there is only one phylogenetic history of life, it is unlikely that we will ever discover all the players or know that we have actually demonstrated the true phylogenetic relationships of those taxa we do know. But we can be fairly certain when we're wrong. And the way in which we know we're wrong is when a hypothesis that we generate on the basis of one set of data is contradicted by additional information we bring to the analysis.

This approach to science is the hypothetico-deductive approach, or more simply put, hypothesis testing. It differs markedly in practice and theory from an inductive approach. In the latter, a mass of information is collected and a general theory or hypothesis derived from it. As more information is accumulated and some bits don't fit the theory, they are explained away as exceptions to the rule. Sociobiologists and evolutionary psychologists are typically inductively oriented in their presentation of Darwinism-predicated theories. But to a deductively minded scientist, even a single bit of information that doesn't fit (that is, corroborate) the hypothesis, falsifies it. That's it. Out it goes. Theoretically one can generate a theory or hypothesis from little data and then continually test it as additional data becomes available. The longer the theory or hypothesis resists falsification—the longer it remains uncontradicted by additional data—the more "robust" or highly corroborated the theory is said to be. This doesn't mean that the theory is true. Only that is has resisted falsification when tested with more data.

How the hypothetico-deductive approach relates to cladism is that every statement that one generates is considered a theory or hypothesis. As such, claims of demonstrated similarity are actually theories of similarity. In more technical jargon, if similarity is evolutionarily meaningful—that is, if taxa A and B have a feature in common because they inherited it from an ancestor they shared—then the shared similarities are said to be homologous. Because homologous features reflect common ancestry, only homologous features can be used to generate theories of relatedness. Shared primitive features (symplesiomorphies) and shared derived features (synapomorphies) are shared homologous features. Consequently, some shared homologous features will reflect more ancient ancestors while others will reflect more recent ancestors. It is the goal of the systematist, whether a cladist or not, to try to demonstrate "homology" (Hennig's "homoiology") between features.

It is, however, possible that what seem to be similar features in different taxa are actually "falsely" similar. For example, taxa A and B may have some features that appear similar that were not inherited from a common ancestor. Rather, these features had arisen independently in each taxon. Features that appear similar but became so through different evolutionary paths are referred to as "parallelisms" or in Hennigian terminology "homoplasies." But how can one distinguish homology from homoplasy?

That's where the hypothetico-deductive approach, or hypothesis testing, comes in. We begin with demonstrating shared similarity between two taxa, A and B. Our broad comparison indicates that only taxa A and B share this feature. We conclude that the feature is homologous in both taxa. Because a statement of homology is actually a *theory* of homology, it is testable. And the way in which it can be tested is with another theory of homology. If this other theory of homology doesn't lead to a different theory of relatedness of the taxa under study, then we may conclude that the second theory of homology corroborated the first.

But since we're cladists, we have to test this more robust theory with yet another theory of homology. If our original theory of relationship still withstands contradiction (falsification), it becomes even more robust. And so the process goes, until or unless one eventually stumbles upon a shared similarity or set of shared similarities that lends itself to the generation of a different theory of relationship between the taxa under study.

When, however, one delineates a shared similarity that leads to a different theory of relationship than another shared similarity does, a question arises: Which shared similarity is the evolutionarily "real" one, the homology, and which is the evolutionarily "false" one, the parallelism or homoplasy? It is only when faced with this dilemma that one can then make statements about parallelism because only one theory of homology (which is based on shared similarity) can be correct—which means that the other similarity has to be false, that is, the result of homoplasy. Because of the nature of phylogenetic reconstruction, one cannot demonstrate parallelism. One can only hypothesize it after embracing one of the alternative comparisons as representing the evolutionarily "true" similarity, that is, as representing homology.

This is perhaps a bit much to wrap your brain around on the first go. But at some point one has to stop tiptoeing around the guts of a subject and dive in. Nevertheless, if you can grasp the concept—that theories of parallelism can only be generated after embracing an alternative set of shared similarities as being *the* evolutionary significant, homologous features—then you are farther along than many of my colleagues in evolutionary biology are. All too often one sees in the literature an attack on someone's theory of who's related to whom in the form of: "Your phylogeny doesn't work because the similarities on which you based it are really parallelisms!" What the attacker isn't realizing is that he already has in mind a preferred theory of who's related to whom that is based on a different set of shared similarities. But let's leave this point for now and proceed with the theory of cladistic analysis.

You've amassed a lot of comparative data on a vast array of taxa and sorted out the different levels of shared unique features, the synapomorphies. In some cases you've discovered that among, say, three taxa—A, B, and C—different sets of shared similarities can group A with B, A with C, and B with C. How do you choose among these alternative schemes of relatedness? The simplest and most straightforward way in which to proceed is to test one theory of homology and synapomorphy with other theories of homology and synapomorphy. In the end, the theory of relatedness that is most highly corroborated (for the time being at least) will be the preferred theory of relatedness, while the similarities underlying the less well-corroborated theories will be interpreted as parallelisms.

We represent a theory of relatedness in a diagram called a "cladogram," so-named because it is a diagram based on the hypothesizing of clades (Eldredge and Cracraft, 1980). A cladogram may look like a tree, with trunks dividing into smaller and smaller clusters of branches and then twigs, but it is not an evolutionary tree. A phylogenetic tree incorporates all sorts of information, most of which is untestable, such as ancestor-descendent relationships, times of divergence of branches, and even places in which and circumstances under which evolutionary events occurred. You can recognize when someone is defending a particular evolutionary tree if they use words such as "what, when, how, why, and where." Many armchair systematists think they're generating a theory of relationship when they're actually only defending a particular theory of relationship, such as when mounting an argument as to how such and such an evolutionary event could have happened. Functional morphologists are particularly fond of such explanations, which Stephen Jay Gould called "just so" stories or scenarios. I can, though, in part empathize with the appeal of spending one's time concocting scenarios about, say, how and why human ancestors became bipedal or reduced the size of the canine teeth, or how and why birds acquired feathers and flight. It's lots more fun than doing the grudge work of generating theories of relatedness. But enough complaining. Back to the task at hand.

The backbone of a phylogenetic tree is a cladogram, which is a diagrammatic representation of a theory of relationship. There are, however, no taxa or ancestors specified at its branching points or points of divergence. Why? Because a statement of ancestry and descent is untestable. Indeed, how good can my claim be of having found *the* ancestor of some group if during your next paleontological field season you find a slightly older or better-looking specimen that you declare is the real ancestor of the group? The fact of the matter is that although paleontologists make claims of having discovered ancestors, they can't prove it. A claim of ancestry and descent cannot be falsified because it is not based on data.

In generating theories of relatedness you can only use the taxa you know, whether they are all extant, all fossil, or a combination of living and extinct organisms. Although paleontologists may traditionally use the age of a fossil to determine whether its features are primitive or derived relative to taxa older and younger than it, cladists don't do that. First of all, there is no a priori reason why something that is extinct is necessarily more primitive than something that is alive (I doubt that any paleontologist would consider *T. rex* a primitive reptile). And second, since we will never sample everything that was ever alive or that is alive today, we would be making false assumptions about primitiveness and derivedness were we to interpret features of a given fossil one way or the other because of its antiquity. Whether fossil or extant, all taxa are treated equally in a cladistic analysis, with relatives states of primitiveness or derivedness determined on the basis of broad outgroup comparisons.

The taxa under study are presented in a cladogram at the end of each terminal branch, with nothing specified at the branching points. The arrangement or pattern of relatedness of the known taxa derives from the analysis of levels of synapomorphy, so that the largest clade will include everything in the group of interest (but not necessarily the array of other taxa that constituted the outgroup comparison) with all of its members united on the basis of at least one feature unique to them all (not present in the comparative group, which in cladistic terminology is referred to as the "outgroup"). Clades within the hypothesized clade will be delineated in similar fashion: each clade by at least one shared feature that all of its members uniquely possess. It might turn out that the large clade can be subdivided into two subclades, each united by its own synapomorphy or synapomorphies. Or it might turn out that only one taxon, perhaps only a genus and its species, is the relative of all other members of the larger clade, the latter being united by a synapomorphy not shared with the former. And on the process goes, with ever-decreasing numbers of taxa being united by more restrictively and more uniquely shared features, until you get to the fine relationships between individual taxa, perhaps between genera or even between species.

What you now have generated is a theory of relationship, in which the data you used to inform your theories of homology and synapomorphy are fully explained to your colleagues so that they can, if they choose, use your data as well as new data to test your theory of relationship. As will no doubt have happened during your analysis, you will have embraced one set of shared similarities as being the homologies and rejected another as being the homoplasies. You will be obliged to make this clear, to offer an explanation as to why the "other" similarities are the homoplasies, the false similarities, so that your readers can also test your theories of homoplasy, either with your own or with different data. There's no hiding behind your mother's apron strings in a well-done cladistic analysis. If you stick your neck out with a particular theory of who's related to whom (and why you rejected alternative theories), you have to take the consequences.

For most cladists, the end result of a cladistic analysis is the cladogram: a stripped-to-the-bone diagram of a theory of closeness of relatedness. No specified ancestors or descendants, only patterns of relationship. Back in the 1970s and into the 1980s, the scientific literature was filled with debate about the details of cladistic analysis. In addition to an emphasis on the hypothetico-deductive approach and hypothesis testing, another point worth mentioning is the concept of the three-taxon statement, which lies at the core of a cladistic analysis.

Imagine three taxa, A, B, and C. Two of them are more closely related to each than either is to the third. Whether A, B, and C are all extant taxa, or all fossil, or some of each, and regardless of how many other taxa might be or have been more closely related to any one of them (but we don't know these taxa), it is still the case that among taxa A, B, and C, two of them are more closely related to each other than either is to the third. Because of this truism, one can generate a theory of relationship based on as few as three taxa, perhaps that A and B are more closely related to each other than either is to C. Another way you can present this relationship is to say that taxa A and B shared a common ancestor not shared with C. These two formulations embody the three-taxon statement. It's the simplest statement of relationship one can make, and it also points out that one can pursue a cladistic analysis in more than one way. Not just from the bottom up—by hypothesizing the all-inclusive clade and then its subclades—but from the top down, from the relationships between the smallest taxa to increasingly larger and more inclusive clades. I find that I go back and forth between the two approaches—top down and bottom up—as I bring in more data, expand my comparison to include more taxa, and encounter more contradictory sets of shared similarities.

CLADISM AND CLASSIFICATION

Although most present-day cladists continue in the tradition of the rigorous approach to phylogenetic reconstruction that Hennig outlined, few if any would subscribe to his demand that the ultimate goal of the systematist is to translate his cladogram into a classification. And it was because of his insistence on making the classification the bearer of a theory of relatedness that Hennig's ideas about phylogenetic reconstruction were initially misconstrued and even maligned by many of the leading evolutionary biologists of the 1960s and 1970s.

Since one never knows beforehand what pattern of relationship will emerge from one's analysis, all sorts of possibilities lie ahead. It could be that relationships within a clade will present themselves as nicely symmetrical sister subclades, each of which is also symmetrically subdivided into increasingly smaller subsclades. If this were the case, according to Henning's rules of classification, each subclade of equal status will be classified at the same taxonomic rank. For example, if a large clade is subdivided into two subclades and you decide to classify the former at the taxonomic rank of order, then you might classify each subclade at the rank of suborder. If each subclade were subdivided equally, you might classify each of those groups as an infraorder. If each infraorder were to be subdivided equally, you might classify each of these groups as a superfamily. And so on.

Problems, however, begin to intrude when the subdivisions are asymmetrical. For instance, let's say you are analyzing relationships between taxa at the genus level and that you hypothesize that all species under scrutiny constitute a monophyletic group. Great. You classify this group at the level of an order. But as you proceed with the analysis you find that rather than hypothesizing within this order nested sets of sister subclades that you can classify at equivalent ranks, you end up with a cladogram in which, in succession, each genus is the single relative of a clade that contains all other genera. How do you classify that?

According to Hennig, you would have to give equal taxonomic status to each sister group, even if that sister group consisted of one large clade and one lonely genus. Consequently, if you make the clade a suborder, and the clades within it infraorders, and work your way down the taxonomic hierarchy through superfamily, family, subfamily, and finally to genus, you have to classify that single genus in as many ranks: first in a suborder, then an infraorder, a superfamily, a family, and finally a subfamily. And that's just the one genus. Imagine having to do that for every subsequent set of sister groups consisting of a single genus plus its sister clade. The end result is an enormous and unwieldy classification that no one would be able to memorize. But that's not all, because if this classification represents the pattern of relationship of your cladogram, and a cladogram is a theory waiting to be falsified, if it's falsified tomorrow and another substituted for it, you have to memorize the new taxonomic concoction.

Most of the old-guard systematics, such as Ernst Mayr, couldn't stand for that. So in opposition to Hennig's "phylogenetic systematics" they countered with "evolutionary systematics," in which classifications would continue to be generated in the tried and true manner of hoping that one was classifying monophyletic groups and then sorting out smaller taxonomic units according to whatever criteria you thought were important. Not yet ready to throw out the classificatory part of Hennig's brainchild, but also sympathetic to the unnecessary generation of a series of taxonomic ranks that would in the end subsume only a genus and its species, two fish paleontologists, Colin Patterson and Don Rosen (1977), proposed a classificatory device they called the "plesion."

As you will recognize "plesion" resembles "pleisomorphy," and each embodies the notion of "primitive." In this case, the plesion is a single taxon that is the (primitive) sister taxon of a clade larger than it. In a classification the rank plesion is used as the equivalent of the rank of its sister clade. So if I conclude that a single genus is the sister taxon of a clade that I would rank at the level of a suborder, then plesion is put into the classification at the same level. However, rather than going down through all the ranks that a suborder normally subsumes, I would go straight to the genus. For example, the large clade would be classified within the order thus:

Order
Suborder
Infraorder
Superfamily
Family
Subfamily
Genus
Superfamily
Family
Subfamily
Genus

But the plesion would be included in the classification only as:

Order
Suborder
Infraorder
Superfamily
Family
Subfamily
Genus
Superfamily
Family
Subfamily
Genus
Plesion
Genus

I actually used this approach in two major reviews of primate relationship I undertook many years ago. But in the end it was more an intellectual exercise than anything else. These days I tend to present a cladogram as the visual representation of a particular theory of who's related to whom and skip the classification. People get crazy when you start messing around with things they've spent long hours memorizing, which includes classifications. So if you want colleagues to give due thought to the results of your cladistic labors, it's best not to distract them by forcing them to digest a new classification on top of everything else.

ANOTHER PERSPECTIVE ON MORPHOLOGY

Up until now I have concentrated on the consideration of phylogenetic relationships from a historical as well as a practical perspective, grounding the discussion in morphology, which, of course, has been the traditional focus of systematists and taxonomists. And until the early and especially the later twentieth century, when studies in immunological cross-reactivity were first explored and then later became widespread, morphology was basically all that a systematist could work with. For most comparative anatomists, the source of comparison was the adult organism. As such, features of an adult human would be compared with features of an adult ape, or features of an adult dolphin would be compared with features of an adult fish. The scala naturae and the Great Chain of Being were predicated on this perspective.

But just as some pursuers of the details of the Great Chain of Being reveled in Steno's demonstration of the reality of fossils as once-living creatures because these newly recognized forms of life could provide evidence of some of the missing links in the chain, so, too, did other comparative anatomists turn to an organism's development for added proof of the Great Chain: that somewhere in the record of an organism's development may lie clues to the creation of life on earth. In this context, even congenitally malformed and miscarried fetuses were interpreted as bearing witness to a divine being's acts of creation.

But there were two ways in which one could interpret the developmental "stages" through which organisms passed. One view was that each "stage" represented a miniature version of an adult, in which case, for example, the supposed gill-slit stage of vertebrate development represented a small but fully formed adult fish. The sequence of development therefore represented the history of creation of the adult individuals of increasingly more complex organisms. The other way in which developmental stages were interpreted grew out of the embryological studies of the German anatomist Karl Ernst von Baer, who discovered the ovum in humans and thus brought the understanding of inheritance from the vagaries of the ovary to female germ cells themselves.

In 1828 von Baer articulated four "laws" of development or ontogeny, of which two were the most important. First, the ontogenetic stages of an organism do not represent the adult stages of organisms created before them. Instead, the ontogenetic stages that organisms have in common represent only their shared embryonic or fetal stages of development. Thus, rather than representing a fish stage, a so-called gill-slit stage represents only a stage of development that organisms may share that are specific to the organism's larger taxonomic group or species. This then led to von Baer's second law of importance, which is that the features that are specific to an organism by virtue of its membership in a group that includes its species were acquired as the organism "deviated" from the common embryonic state as it followed its own group's and species' path of development. For example, a fish may appear to share a gill-slit stage with other vertebrates, but it only becomes a fish after it departs from this common state and develops the features of being a fish. The significance of organisms sharing a greater number of ontogenetic states is that it reflects their greater similarity, which, of course, is an attribute that in one way or another taxonomists incorporated into the ordering of their classifications.

To my way of thinking, the evidence in support of von Baer's ideas should have had an immediate impact on the study of development, imbuing it with a more central role in systematic studies. They did not violate a belief in a divine creator, and they correctly embodied the essence of development. But von Baer's ideas did not do so, in part perhaps because the discipline of comparative anatomy had traditionally focused on studying the adult. Another part of the answer may lie in the history of evolutionary thought, when paleontology and embryology were supposedly expunged of their roots in the Great Chain of Being and incorporated into the mindset of evolution.

ERNST HAECKEL, CHARLES DARWIN, AND DEVELOPMENT

Many historians of evolutionary ideas might immediately think of the German evolutionist and comparative morphologist Ernst Haeckel and his "biogenetic law," the essence of which is encapsulated in the phrase, "ontogeny recapitulates phylogeny." By this Haeckel meant that during its ontogeny or development an organism repeats or recapitulates its evolutionary history. Consequently, if one was interested in understanding an organism's evolutionary past, one need only study its developmental stages. But in the context of the biogenetic law these developmental stages were a reflection of the adult stages of ancestors. For example, an apparent gill-slit stage represented not only a fish, but also an adult fish in a presumed fish stage of evolution. But more than that, Haeckel's biogenetic law allowed one to take fossils, which especially with regard to terrestrial animals are more frequently represented by the remains of adults than juveniles, and to insert them into a reconstructed ontogenetic and thus evolutionary sequence. As such, not only could Haeckel view the then-known Neandertal fossils as representing evolutionary stages that preceded the emergence of modern humans, he could also speculate on the existence of not-yet-discovered ancestral species. And for human evolution he did just that when he created the genus and species "Pithecanthropus allalus" in anticipation of the discovery of a fossil representing a stage in human evolution that he was convinced had existed: an "ape-man" lacking the capacity for speech (Haeckel, 1876).

But it was not only by proposing the biogenetic law and reiterating it through his phenomenally influential monographs that Ernst Haeckel assumes historical significance. He was also the first major proponent of Darwin's ideas, which had been both severely criticized in reviews of *Origin*, including one by Thomas Henry Huxley, and marginalized by the dominant Victorian comparative anatomists, such as Huxley and

St. George Mivart, who believed that the emergence of the organismal novelties that produced evolution could come about only through an abrupt reorganization of an organism's development, not by the gradual accretion of minute change. The changes that were the "stuff" of evolution thus occurred not slowly over long periods of time, but in the course of a generation, by way of developmental leaps or saltations.

In the context of historians of science emphasizing as one example of Darwin's holistic grasp of biology his appreciation of development, I had for years been puzzled by Haeckel's attraction to Darwin's evolutionary ideas, because twentieth-century systematists and historians had worked so hard to remove any vestiges of the biogenetic law and to expunge from the image of Darwin any hint of unacceptable thinking. But, finally, in re-reading various editions of *Origin*, including and especially the sixth and last (1872), which was published in response to Mivart's 1871 saltationist treatise titled not entirely by coincidence *On the Genesis of Species*, the source of the connection between Darwin and Haeckel became clear. The only way in which one can conceive of Darwinian evolution—the gradual accumulation from one generation to the next and over long periods of time of minute changes that were favored by natural selection—is by its enactment on mature or at least fairly mature postnatal individuals, not embryos and fetuses. Indeed, the winnowing of favorable versus unfavorable attributes can only occur once an individual is born with its array of attributes. And Darwin's theory of inheritance, which he called "pangenesis," attests to this emphasis.

Although he had begun to articulate the basics of pangenesis in his notebooks, it was only in 1868 in his two-volume treatise, *Variations of Plants and Animals Under Domestication*, that Darwin formally proposed his theory of inheritance. In brief, Darwin imagined that throughout an individual's life all of its anatomical parts were sloughing off particles, which he called gemmules, which came to reside in its reproductive organs. As such, gemmules recorded the life history of an individual, including not only aspects of its natural biology but also the effects on it of use and disuse, which reflected Darwin's belief in the so-called Lamarckian idea of an organism's desires being capable of provoking or engendering its own change. Upon mating, the gemmules of parents would blend and the result would be passed on to their offspring. Since gemmules were like a tape-recorder of one's life history, Darwin could explain the emergence of features, such as secondary sexual characteristics, that appeared later in life. Through the incorporation into gemmules of the effects of use or disuse and their subsequent blending in offspring, Darwin believed that he provided natural selection with sources that would expand the pool of variation from which selection could choose the more over the less favorable. Thus while one might argue that the theory of pangenesis does embrace consideration of the embryo and fetus, because an organism must proceed through prenatal development, and even postnatal growth and development, prior to attaining adulthood, the very essence of Darwin's theories of natural (as well as sexual) selection makes it obvious that the focal individual is the adult (or at least the near-adult). No wonder when Haeckel visited Darwin at Down House not long after the publication of the first edition of *Origin* did each find intellectual resonance in the other.

With regard to our discussion of cladism, the ordering of individuals or of fossil specimens in an unbroken continuum that is predicted by Darwin's gradualism is not consistent with the persistent gaps in the fossil record and it is certainly not cladistic in approach. Indeed, as we have discussed, statements of ancestry and descent are not even testable.

A DIFFERENT APPROACH TO DEVELOPMENT THAT IS CLADISTIC

Although it is neither with Ernst Haeckel nor with Charles Darwin that we can situate a meaningful understanding of development and relate it to testable hypotheses of relatedness, we can turn to Thomas Henry Huxley for insight into this matter. For it was in one of three essays that were collectively published in 1863 under the title *Man's Place in Nature* that we find what I believe is the essence of the proper way in which to think about ontogeny and phylogeny.

In this particular essay, "On the Relation of Man To Lower Animals," Huxley (1863) employed a von Baerian approach to the phylogenetic interpretation of development in order to argue not only that humans should be grouped with vertebrates, but that within vertebrates humans should be grouped with mammals, and within mammals with primates, and within primates in a group that also included the "tailless" apes; (in the nineteenth century, what we refer to as monkeys were called tailed apes and true apes, as they are, tailless apes). To my knowledge, this was the first time that any systematist or taxonomist pursued this argument since 1735, when Linnaeus first classified humans with other primates.

In this essay, which was fairly anatomically sophisticated for the time and especially for an essay meant to be read by lay folks as well as by professionals, Huxley used von Baer's principles in the following manner. First he compared the embryos and fetuses of a monkey, dog, and chicken and demonstrated that for some amount of time they are all similar. He then pointed out that at a later point in time the fetuses of the monkey and dog remain similar, but that in their commonality they have come to differ from the chick. And at an even later point in time, the fetuses of monkey and dog become less like one another as they acquire the characteristics, respectively, of primate versus carnivore, and then within each group, of monkey versus dog. Huxley then demonstrated that one can make the same comparisons for a human as for a monkey fetus and, further, that the human and monkey share an even longer period of similar development. If, Huxley asked of his audience, we can embrace the suggestion that the common course of development demonstrates something special between the monkey and the dog and chicken, and then more so between the monkey and the dog, then surely we must accept the same set of nested relationships for humans and, further, that the longer period of shared development of human and monkey reflects their closer relationship.

Although somewhat simple by today's standards, Huxley's application of development to the endeavor of hypothesizing relationships among taxa is clearly within the framework of the hypothetico-deductive approach and eminently amenable to cladistic analysis. For example, the features of development that are shared by a wide array of taxa can be hypothesized as derived for the largest clade, in whose members their presence then represents primitive retention. Likewise, the more restrictively shared developmental state(s) would reflect synapomorphy for a clade within the larger group, and so on, until we end up with aspects of development that only potential sister taxa share. Of course, by common state(s) of development I do not mean all common state(s) of development. Rather, from a cladistic perspective, I am referring to shared derived state(s) of development, with the hypotheses of derivedness emerging from the broad comparison that characterizes a cladistic analysis.

Nevertheless, as logical and scientific as was Huxley's use of von Baer's understanding of ontogeny to address questions of phylogenetic relatedness, the dominance of Haeckel's biogenetic law contributed in the early twentieth century to a general dismissal of comparative embryology and developmental studies as being applicable and particularly relevant to research in systematics. Although there were some studies during the early 1970s that strove to reestablish a sensible approach to considering ontogeny and phylogeny (my own studies, in which I cladistically evaluated data on tooth development for purposes of determining phylogenetic relationships among primates being among them), it was not until the 1980s that broader discussion of the systematic implications ontogenetic analyses took place. Unfortunately, this discussion was not followed by an increased application of this approach to systematic questions. As a result the adult individual has remained the focus of comparison, especially by mammalian systematists, with paleoanthropologists being the most resistant to incorporating developmental data into their studies.

NOT ALL PURPORTED CLADISTIC ANALYSES ARE CLADISTIC

As the appeal of the hypothetic-deductive approach underlying cladistic analysis spread, abuses and misuses of Hennig's idea also emerged. One of the most egregious was the co-opting of cladistic terminology by traditionally minded systematists who used anything but a cladistic approach to reconstructing phylogenetic relatedness. Another abuse was claiming that a cladistic analysis had been done when, in fact, the results of a non-cladistic analysis had been translated directly into a classification (thus giving lip service to Hennig's ultimate representation of cladistically determined relationships). These abuses are most frequently seen these days in molecular analyses.

Although at the very beginning of the twentieth century the British bacteriologist H. Nuttall was among the first to suggest that phylogenetic relationships might be revealed by investigating the immunological reactions of blood sera of different taxa, it was not until 1962 that this approach was formalized by Emile Zuckerkandl and Linus Pauling. In that article, which focused on the biochemistry of the blood protein hemoglobin, they tested immunological reactions between human, gorilla, horse, chicken, and fish hemoglobins and found that human and gorilla were more similar than either was to the horse, that these three were more similar than either was to the chicken, and that these four were more similar than either was to the fish. Because this arrangement paralleled the traditional, morphologically based theory of relatedness among these taxa, Zuckerkandl and Pauling proposed that their demonstrations of similarity could be explained by a (Darwinian) model in which it was posited that molecules are constantly changing in continually evolving lineages. As such, greater difference between taxa reflected a more distant separation of their lineages—with the earlier-divergent taxon accumulating its own particular molecular changes—while greater similarity between taxa reflected a more recent divergence from a common lineage. When protein and then DNA sequencing became feasible, this assumption was applied to the interpretation of these data.

From a cladistic perspective, one should immediately question the interpretation of overall similarity as reflecting phylogenetic propinquity because one does not know how much of this similarity represents primitive retention. Furthermore, because the assessment is in terms of overall similarity, rather than a comparison of identified characters, one does not know whether the same parts of molecules or amino acid or nucleotide sequences are yielding the signals of similarity across all taxa under study. Nevertheless, in spite of the fact that the demonstration of overall similarity, whether of molecules or of hard- and soft-tissue anatomy, does not constitute a cladistic analysis, beginning in the 1980s, molecular systematists have increasingly proclaimed that their demonstrations of similarity are demonstrations of shared derivedness. Witness the publications of Morris Goodman, Maryellen Ruvolo, and Todd Disotell.

The basis for their assertion lies, however, in the belief that molecules are constantly changing and thus that changes that accrue in a lineage prior to the divergence of descendent lineages constitute shared derived changes. At best, this is tautological. At worst, this elevates to the level of truth an assumption of continual change that has been uncritically imposed upon a static demonstration of similarity. If, however, this were the case, we would expect to see in every generation of multicellular animal species the appearance of novel forms as well as the disappearance of taxa because while only about 2 percent of the metazoan genome codes for proteins, the rest—basically introns and promoter regions—does not. The potential for introns to be relevant to embryonic development aside (although this has interesting implications), we are faced with the realization that vast expanses of the metazoan genome comprise promoter regions, whose functions, grossly, include extracting information from coding regions and maintaining the integrity of an organism's development. Consequently, if molecules were continually changing, we would expect promoter regions as well as introns to be affected more frequently than coding regions. And since the stability of promoter regions is so critical to the development and survival of an organism, we should also expect to see as common phenomena the emergence of novel forms as well as ongoing extinction. But we do not. As Darwin, Huxley, and others well knew, and we also know: Like tends to beget like.

Yet, in spite of the uncritical interpretation of similarity and the untenable assumption underlying its interpretation, the general public has come to think of molecular "data" as indisputable proof of relatedness and molecular systematists as doing science properly. The abiological aspects of the molecular assumption notwithstanding, until molecular systematists begin questioning the significance of overall similarity, they are not doing systematics, and certainly not cladistic analyses.

CODA

Just as we will never know the true pattern of the relationships of all life, we will never know the perfect way in which to pursue phylogenetic reconstruction. But the hypothetico-deductive approach embodied in cladism at least affords us a way to make this pursuit as scientifically rigorous as possible. And that is all a scientist can ask for.

Most of us have heard of the saying, "blood's thicker than water." What this refers to is that there is something in our blood that binds the more closely related among us to each other rather than to a stranger or someone to whom we are not closely related. In turn this simple statement lends itself in evolutionary biology to an approach to reconstructing evolutionary relationships that relies on the demonstration of degrees of overall similarity: The more similar two species are, the more closely related they are supposed to be. Not so, at least not necessarily. And this is where the German entomologist Willi Hennig comes in. For beginning in the 1950s it was he who revolutionized systematic biology and the methodological approach to determining who's related to whom. This approach is generally called "cladism," which derives from the word "clade," which in turn is synonymous with the phrase "monophyletic group," which is a group of organisms related by virtue of sharing a common ancestor.

The essence of a cladistic approach is that once comparisons between species have been made and shared similarities delineated the hard work actually begins. This entails a hierarchical sorting out of shared features from those that are the most commonly shared to those that are the least commonly shared. The logic is that the totality of an organism results from the combination of features that emerged in a series of ancestors, from the most ancient to the most recent. As such, a demonstration of overall similarity is not necessarily a demonstration of closeness of relatedness. Only the demonstration of relatively uniquely shared features can do that. So even if it is true that humans and chimpanzees share 98 percent of their DNA this doesn't mean they're closely related. It may take a while to convince the molecular anthropologists that they've got it wrong, but maybe you can help out.

FURTHER READING

Clark, W. E. L. G. *History of the Primates: An Introduction To the Study of Fossil Man* (London: British Museum (Natural History), 1949).

Darwin, C. *The Variation of Animals and Plants Under Domestication* (London: John Murray, 1868).

Darwin, C. *On the Origin of Species by Means of Natural Selection, or the Preservation of Favored Races in the Struggle for Life* (London: John Murray, 1872).

Gaulin, S. J. C. "A Jarman-Bell Model of Primate Feeding Niches," *Human Ecology* 7(1979): 1–20.

Goodman, M., Porter, C. A., Czelusniak, J., Page, S. L., Schneider, H., Shoshani, J., Gunnell, G., and Groves, C. P., "Toward a Phylogenetic Classification of Primates Based On DNA Evidence Complemented By Fossil Evidence," *Molecular Phylogenetics and Evolution* 9(1998): 585–598.

Gould, S. J. *Ontogeny and Phylogeny* (Cambridge, MA: Belknap Press, 1977).

Haeckel, E. *Generelle Morphologie der Organismen* (Berlin: Georg Reimer, 1866).

Haeckel, E. *The Evolution of Man: A Popular Exposition of the Principal Points of Human Ontogeny and Phylogeny* (New York: H.L. Fowle, 1876).

Huxley, T. H. "Review of 'The Origin of Species,'" *The Westminster Review* (1860).

Huxley, T. H. *Man's Place in Nature* (New York: D. Appleton, 1863).

Linnaeus, C. *Systema naturae per regna tria naturae, secundum classes, ordines, genera, species cum characteribus, differentiis, synonymis, locis* (Stockholm: Laurentii Salvii, 1735).
Linnaeus, C. *Systema naturae per regna tria naturae, secundum classes, ordines, genera, species cum characteribus, differentiis, synonymis, locis*, 2nd ed. (Stockholm: Laurentii Salvii, 1758).
Lovejoy, A. O. *The Great Chain of Being* (Cambridge, MA: Harvard University Press, 1942).
Mivart, S. G. *On the Genesis of Species* (London: John Murray, 1871).
Nuttall, G. H. F. *Blood Immunity and Blood Relationship* (Cambridge: Cambridge University Press, 1904).
Patterson, C. "Verifiability In Systematics," *Systematic Zoology* 27(1978): 218–222.
Patterson, C., and Rosen, D. E., "Review of Icthyodectiform and Other Mesozoic Teleost Fishes and the Theory and Practice of Classifying Fossils," *Bulletin of the American Museum of Natural History* 158(1977): 81–172.
Ray, J. *Synopsis Methodica Animalium Quadrupedum et Serpentini Generis* (London: Samuel Smith, 1693).
Ruvolo, M. "Molecular Phylogeny of the Hominoids: Inferences From Multiple Independent DNA Sequence Data Sets," *Molecular Biology and Evolution* 14(1997): 248–265.
Schwartz, J. H. "Primate Systematics and a Classification of the Order," in D.R. Swindler and J. Erwin (eds.), *Comparative Primate Biology* (New York: Alan R Liss, 1986), 1–41.
Schwartz, J. H. *The Red Ape: Orang-utans and Human Origins* (Boston: Houghton Mifflin, 1987).
Schwartz, J. H. *Sudden Origins: Fossils, Genes, and the Emergence of Species* (New York: John Wiley, 1999).
Schwartz, J. H. "Darwinism versus Evo-Devo: A Late 19th C. Debate," in S. Mueller-Wille and H.-J. Reinberger (eds.), *A Cultural History of Heredity III: 19*th *and Early 20*th *Centuries* (Berlin: Max Planck Institute, 2005), 67–84.
Schwartz, J. H. (2006). "'Race' and the Odd History of Paleoanthropology," *The Anatomical Record (Part B: The New Anatomist)* 289B: 225–240.
Schwartz, J. H., and Maresca, B. (2006). "Do Molecular Clocks Run At All? A Critique of Molecular Systematics," *Biological Theory* 1: 1–15.
Schwartz, J. H., Tattersall, I., and Eldredge, N., (1978). "Phylogeny and Classification of the Primates Revisited," *Yearbook of Physical Anthropology* 21(1978): 95–133.
Simpson, G. G. *The Major Features of Evolution* (New York: Simon and Schuster, 1953).
Sokal, R. R., and Sneath, P. H. A. *Numerical Taxonomy* (San Francisco: W. H. Freeman, 1963).
Von Baer, K. E. *Über Entwicklungsgeschicte der Thiere: Beobachtung und Reflexion* (Königsberg: Bornträger, 1828).
Wiley, E. O. "Karl R. Popper, Systematics, and Classification: A Reply To Walter Bock and Other Evolutionary Taxonomists," *Systematic Zoology* 24(1975): 233–243.

Winter, J. G. *The Prodromus of Nicolaus Steno's Dissertation Concerning a Solid Body Enclosed by Process of Nature Within a Solid* (New York: Hafner, 1968).

Zuckerkandl, E., and Pauling, L., "Molecular Disease, Evolution, and Genic Heterogeneity," in M. Kasha and B. Pullman (eds.), *Horizons in Biochemistry* (New York: Academic Press, 1962), 189–225.

Louis Leakey

Anne Katrine Gjerløff

INTRODUCTION

Paleoanthropology—the science of prehistoric humans and their evolution—is a science of enormous public interest. Thousands of books, magazine articles, TV documentaries, and fictions present the search for our ancestors as the perfect mix of a treasure hunt, a noble scientific endeavour, and as a quest for the ill-heard truth, threatened by ignorance, animosity, and debate. No one has contributed more to this public interest and to the heterogeneous view of the science than Louis Leakey.

Louis Seymour Bazett Leakey (1903–1972) was a paleoanthropologist and the man behind several of the most important fossil discoveries in the twentieth century, but he was also a human and it is human to err. Leakey's scientific legacy consists of both triumphs and errors, but his importance for paleoanthropology is not only based on the discoveries he made, the species he described, or the scenarios he envisioned. Leakey had something to do with almost every aspect of human prehistory, as this chapter will try to show, from African Stone Age cultures, to experimental archaeology, to primate studies, and last, but not least, to the question of getting all the results out to the public to make people aware that there is such a thing as human prehistory and that the study of it is important and need support.

Louis Leakey was an expert in string figures, an unfaithful husband, a visionary, a forensic and handwriting expert, a dedicated scientist, a spy, a lover of Africa, a Kikuyu elder, a father; he was toothless and corpulent but also the "sexiest man" (Morell, 1995, p. 243); he was the man who pronounced *Proconsul*, *Zinjanthropus*, and *Homo habilis* to the world and a man who tried to skin rabbits with his fingernails; he created modern primate studies, but believed that man had an early non-anthropoid forefather; he was funny, curious, and intelligent, but absentminded and sometimes sloppy with his work. Louis Leakey is the man who stands as the symbol not only of the hunt for early man but for any prolonged and hard struggle for scientific breakthrough and approval.

Apart from Louis Leakey's scientific career, his greatest importance can actually be acknowledged as that of an icon; an icon of the idea of human evolution and the struggle to unearth the past to understand the present. This chapter on the life and work of Louis Leakey shall try to deal with both his greatest glorious moments, his many smaller projects, his scientific triumphs and his biggest blunders, his work as a pioneer of paleoanthropology, and his legacy as a godfather of the field.

A BIOGRAPHICAL RESUME

The story of Louis Leakey's life has been told many times before. It is perhaps an indication of his personality that Louis himself wrote his first memoir—*White African*—in his early thirties. In the book he recounts his childhood and youth, and the sequel *By the Evidence* was published posthumously. After his death his assistant Sonia Cole wrote the biography *Leakey's Luck*, and several obituaries have biographical sketches. The most authoritative biography though is Virginia Morell's *Ancestral Passions*; a group portrait of the Leakey family, and written with the collaboration of same. Even though the Leakey family history in Morell's book seems fully exposed, some skeletons are probably still left in the closets.

It is difficult to add anything new to the knowledge of a figure about whom so much has been written, and hence that is not my ambition. Rather I will try to give an impression of the person Louis Leakey was, but especially linger at those topics where he contributed most or was most at odds with paleoanthropological science. The most fascinating thing about Louis Leakey is that even though he is often credited as one of the greatest paleoanthropologists and fossil discoverers of the twentieth century, only few of his celebrated fossils were actually found by himself and his theories of human evolution were only shared by very few other researchers: "although the outstanding importance of his fossil discoveries is universally acclaimed, Leakey's own interpretation of them is not" (publisher's introduction, Leakey, 1974, p. 7). A colleague put it this way: "In his lifetime [Leakey] accomplished and promoted more work that resulted in the accumulation of information of human origins than anyone had ever done before. At the same time he did more to hinder our understanding of the course of human evolution than anyone of his time" (Loring Brace, 1977, p. 171). If this is true, what was it about Louis Leakey that earned him his reputation? What were his contributions to paleoanthropology? Some of the answers to this must be found in his personal background.

Louis Leakey holding pieces of 600,000 year-old human skull. (Library of Congress, Prints & Photographs Division, LC-USZ62-132443.)

Louis Leakey was born 7 August 1903 at a small missionary station in the Kikuyu district of Kenya, west of Nairobi, where his parents had just a few years before arrived as missionary representatives for the Church Missionary Society. Apart from stays in England in 1904–1906 and 1910–1913, Louis Leakey was raised with his siblings and local Kikuyu children as his only company, and he became completely bilingual in English and Kikuyu—a fact he was proud of and benefited from on several occasions. There is no doubt that the childhood that embraced two cultures was important in shaping the person Louis Leakey would later be. He achieved immense insight in aspects of African natural history and folklore, and at the same time experienced formal British customs and doctrines of Christianity, and in both these converging spheres Louis Leakey was given more responsibility and evolved more self independence than is usual for children in more traditional and protected environments. The double role—as British missionary son and a born African—also created a deep love and interest for everything African: history, nature, language, and culture.

Louis Leakey's childhood was spent—apart from lessons from parents and governesses—hunting, playing, and studying every aspect of the nature that surrounded him. As a young teenager he even built his own hut on the mission's grounds where he moved himself and his selection of naturalia and other stuff. Among this were also archeological artifacts for which Louis Leakey had developed an interest after an aunt had sent him a book on Stone Age Britain. After being encouraged by the curator at the small museum in Nairobi "Louis Leakey was addicted to prehistory" (Cole, 1975, p. 37).

Thus it was not a normal, well-behaved, or typically educated boy that arrived in England in 1919 when the family returned from the mission and Louis Leakey began a formal school education. "Louis Leakey felt completely adult and the boys at Weymonth seemed utterly childish" (Cole, 1975, p. 38). After all accounts Louis Leakey genuinely hated the strict rules and the fact that his achievements and knowledge did not matter or were regarded as unnecessary or aberrant.

By bending some rules and with a certain inventiveness Louis Leakey was accepted at Cambridge in 1922. A loophole was found by Louis Leakey in the formal requirement of two modern languages. He spoke French and suggested Kikuyu as the second—his fluency was testified to by a thumbprint-signed document from the senior chief Koinage. The master at St. Johns was horrified but had to accept the language skill as valid. At Cambridge the following lessons consisted of Louis Leakey teaching his teacher Kikuyu. His main interest and favorite field of study was anthropology though, and as a part of this prehistoric archaeology.

His studies were interrupted by a rugby accident in 1923 where he was kicked in the head and suffered severe headaches and epileptic seizures, which he would suffer periodically for the rest of his life. The doctor prescribed a year away from Cambridge. This was a chance to get back to Africa and fieldwork and Louis Leakey applied for a position at the Natural History Museum which was about to send an expedition to Africa to excavate dinosaur fossils in Tanganyika. Louis Leakey was hired as organizer of the expedition—he was then twenty years old.

Louis was back at Cambridge in 1925 and with financial help of the family and scholarships he passed his finals in 1926 with very good grades. After this he was rewarded a six-year research fellowship, and seemed destined for a traditional and glorious career in academia.

The dinosaur expedition was the first of several expeditions Louis Leakey organized and participated in during these years and he brought back impressive fossil collections of prehistoric African fauna. The three expeditions he arranged during his fellowship gave him an enormous experience in fieldwork and knowledge of African prehistory, and his reputation was building up both in scientific circles and in the public that learned about his work from public lectures and articles in magazines.

In 1928 Louis Leakey married Frida Avern whom he had met in Africa and who shared his interest in prehistory. But his busy life in Cambridge left her and their daughter Pricilla (born in April 1931) alone in their cottage. In 1933 Louis Leakey met the young archaeological artist Mary Nicol and hired her to illustrate his book *Adam's Ancestors*. Louis Leakey and Mary became lovers and in early 1934 shortly after the birth of their son Colin, Frida's and Louis's marriage collapsed. Louis Leakey did not see his children until several years later.

The divorce combined with scientific controversies which will be explored later made Louis Leakey an unpopular man in Cambridge and effectively ruined his good reputation and his academic career. In 1936 he and Mary got married and went back to Africa where they began what would eventually be more than forty years of excavations and research. Among their achievements—described in some detail later—are discoveries of Miocene primates, excavations in Olduvai gorge, and the discovery of the hominids *Zinjanthropus boisei* and *Homo habilis*. Also Louis Leakey made East Africa the center of attention for the international paleoanthropological and archaeological community both by personal contacts and by arranging the first of many pan-African congresses on prehistory in 1947.

Until the 1960s the excavations were made under very tight budgets and Louis Leakey survived on grants; worked for the government authorities as an interpreter, handwriting expert, and intelligence agent during World War II; and in the 1950s was an informant and interpreter during the Mau Mau uprising (see sidebar Scientist and Spy). In 1945 Louis Leakey got a permanent position as curator at the Coryndon Museum in Nairobi which secured a regular income. Mary and Louis Leakey had three sons, and also tragically lost an infant daughter. The sons—Jonathan (1940), Richard (1944), and Philip (1949) participated in the archaeological work which took place as a family enterprise with also employed local workers, some of whom eventually became very skilled fossil hunters.

Scientist and Spy

During World War II when Louis Leakey had no permanent position, he was on several occasions employed by the British African Intelligence Department. Among his services were providing weapons to Ethiopian guerrillas and organizing an information-gathering network among the Kikuyu.

Louis Leakey also served as an interpreter in interrogations during the war, and also, more controversially, during the Mau Mau uprising in the early 1950s. Louis Leakey had deep insight into Kikuyu matters and felt that his warnings about secret activities in the late 1940s had been overlooked and could have prevented much bloodshed. As it was, Louis Leakey participated in the intelligence work and because of his fluency in Kikuyu served as an interpreter on several occasions, and among them at the trial of Jomo Kenyatta, who in 1964 became president of Kenya, but who in 1952 was accused of being one of the leaders of the Mau Mau, which had that same year caused a national state of emergency by a series of murders. Kenyatta was sentenced to seven years of labor. Louis Leakey was torn between the loyalty to Britain and to the Kikuyu and later supported the independent state of Kenya and Kenyatta's presidency. But during the emergency he was on the British Kenyan side, which is also apparent in his book *Defeating the Mau Mau*. His sympathies resulted in death threats toward himself and his family, a fate that was later shared by his son Richard Leakey in his fight against poachers and exploitation of protected wildlife. In 1993 Richard lost both legs in an airplane accident that might have been caused by sabotage related to his political activities.

In peacetime Louis Leakey also liked to use his ability to solve riddles, and his autobiography lists many examples of how he joined police investigations to solve murders and thefts. He also assisted the government during his whole life as a handwriting expert. This ability was probably evolved during his years of studying medieval handwriting as a part of his studies in Cambridge. With his flair for drama and sensation this was not only a source for income, but also a kind of intellectual game for Louis Leakey. He found satisfaction both in coveralls and with a hand axe; cloak and dagger.

After sensational finds in Olduvai gorge in the years around 1960 grants were no longer hard to get, but at the same time Louis Leakey became more and more absorbed in administrative work, public lectures and writings, and fundraising for still more projects, especially in the field of primatology. Mary more or less took over the responsibility for the excavations in Olduvai, with impressive results described in a large monograph series.

Their separate lives and Louis Leakey's other project, of which Mary did not always approve, gradually estranged the couple. Louis's interest in young women from whom he craved admiration and affection didn't make matters better. Also his health was deteriorating; overweight, bad hips, and a couple of strokes took their toll, but didn't slow down his pace or level of activity, which included guest professorships, public lectures, TV and radio broadcasts, and the responsibility to an ever-increasing number of projects and admiring fans.

Many things point in the direction that Louis Leakey lost some of his critical sense toward the end of his life, but he lost none of his enthusiasm. Perhaps it is just so that Louis Leakey who had always been ahead of things, was for once left behind. Louis Leakey died 1 October 1972 leaving behind an administrative and economical mess, but also a glorious scientific career and an enormous number of fans and protégés among whom many would make important work of their own. At the time of his death Louis Leakey was a public figure, his name synonymous with paleoanthropological work and the hunt for fossils of early man. Through the work of Mary and Richard Leakey (and Richard's own family) the Leakey name itself stands for a dynasty of dedicated researchers who have discovered many of the most important fossils in the twentieth century.

The struggle—both physical, administrative, and economical—of the 1930s to the 1950s, as well as the scientific triumphs in the 1960s have been described in detail in a number of biographies and this chapter shall only focus on a few selected themes of the life and work of Louis Leakey.

His life was remarkable in several ways, but here I will point only to a few central facts. Leakey was in many aspects an "in-between" person. He was comfortable both as an elder in the Kikuyu tribe and as a Cambridge don. He was the son of a preacher but a defender of Darwinism. He was born British in colonial Kenya, but was an active citizen in the independent state of Kenya as well. He was an honorary doctor at several universities, but only got a formal education from the age of sixteen, wrote tons of scientific papers but was equally comfortable as a public lecturer and a TV star. Leakey was himself always a mediator between separate spheres, and moved between separate cultures and norms, which both caused him trouble and brought him fame, but never fortune.

FIRST FOSSILS, FIRST FAILURES

In Louis Leakey's early career it seemed like he was destined for fame. He won acclaim for his participation in a number of expeditions that brought home to England fossils of extinct fauna and great geological knowledge. But Louis Leakey's own scientific interest was the human prehistory, not just paleontology, and it was this interest that also brought him to the first of many scientific controversies. Louis Leakey had his own interpretation of human evolution that was shared by many when he was in the beginning of his career in the 1920s, but which he did not discard during the 1950s and 1960s when new discoveries and dating methods had all but changed paleoanthropological theories for most other scientists. To put it simply: Louis Leakey firmly believed that *Homo sapiens* had a distinct line of evolution that had not been shared by other ape-like species. That means that the line eventually leading to modern man had split out and followed its own course very early in evolution. Thus *Homo sapiens* did not share immediate ancestor with, for example, the great apes, but had sprung from an even earlier and more primitive or unspecialized branch on the evolutionary tree. This was a common view in the beginning of the twentieth century when no other fossils were known besides *Pithecanthropus* (*Homo erectus*), Neandertals, and *Homo sapiens* itself. Compared to modern man *Pithecanthropus* and the Neandertals looked very ape-like and it was not illogical to think that they and modern man had separate, but convergent, evolutions. Contributing to this view was also a general dislike toward an evolutionary connection to the African great apes, and many paleoanthropologists looked toward even older Asian form to find the ape or even money that bore resemblance to the earliest common ancestor of man and apes.

In addition to this the Piltdown fossil (which was later to be exposed as a hoax) also pointed in the direction that modern man had developed quite early. Piltdown's skull closely resembled a modern skull (in fact it was a modern skull) but was found in geological layers that dated it as more ancient than *Pithecanthropus* and Neanderthal Man, and this was again a "proof" of their status as side branches and not as human ancestors.

The thought that all known fossils were extinct side branches and that the ancestor of *Homo sapiens* was yet to be found and would prove more *Homo*-like than *Pithecanthropus* is called the Presapiens Theory. The theory was accepted by many paleoanthropologists in the first half of the twentieth century and the result was primarily that the Neandertals were perceived rather as cousins than as ancestors to modern man. This is a view that is also prevalent today, although for other reasons. The presapiens theory was also one of the causes why the fraudulent Piltdown Man was so readily accepted by science. Placed in the Pliocene and with a very modern look, Piltdown seemed the perfect candidate for a presapiens ancestor, but many paleoanthropologists (among them Louis Leakey) put Piltdown on a side branch as well. Thus no real fossils could take the place as the ancestor of *Homo sapiens*—all were excluded from this honor, and this practice has caused historians of science to call the tendency the "Shadow-man paradigm" (Hammond, 1988). The presumed ancestor was but a shadow, and Louis Leakey's hope was to make the shadow solid.

As it is apparent on the evolutionary tree from Louis Leakey's 1934 book *Adam's Ancestors* almost all of the known fossils came from the Pleistocene period. Dating at that time was relative, and no absolute dating could be obtained before the adequate scientific methods were developed in the 1950s and 1960s. Therefore when dates were estimated the periods were notoriously too short. No one would have guessed that the *Pithecanthropus*—now *Homo erectus* lived more that 1 million years ago.

As the illustration shows Louis Leakey put the root for modern humans (which he called the neoanthropoids) down in the Miocene period, and the common ancestor of all human-like creatures even back in the Oligocene. The (much debated) recent estimate is that the common ancestor of humans and apes lived about 5 to 7 million years ago, while the Oligocene is today believed to begin 34 million years ago. These numbers give us some idea about the enormous differences between the belief of Louis Leakey and some of his fellow scientist and the best supported view today. Though it is now assumed that modern man appeared not much earlier than 150,000 years ago, in the youngest period called Holocene (Louis Leakey's "recent"), Louis Leakey wanted to place modern man in the beginning of the Pleistocene and his direct ancestors as early as the middle Miocene now believed to be about 14 to 20 million years old.

Most of Louis Leakey's career had the hunt for early *Homo* as its main focus, and it was also the source for many of his scientific controversies. These began with the case of Olduvai and Kanam man. Olduvai gorge (formerly called Olduway) is a part of the Great Rift Valley that (roughly speaking) separates East Africa from West Africa. The gorge was discovered by European science in 1911 when a German butterfly expert literally fell into the gorge while hunting for a rare specimen. The gorge has a unique geological sequence and is a goldmine for fossil collectors. In 1913 the gorge was studied by the German professor Hans Reck who was then a member of the geological survey of German East Africa. Reck found—among many other faunal fossils—a complete and well-preserved modern-looking human skeleton. The peculiar thing was that he found the skeleton in bed II of the gorge. The geological layers of Olduvai are numbered beds I–V, with bed I being the oldest and V the most recent. The ability to record to which bed fossils belong is a precondition to date them relatively. Bed II meant lower Pleistocene (based on the associated fauna) but this provoked critique from Reck's colleagues, since a fully modern man contemporary with *Pithecanthropus* didn't make sense to most, but a lot of sense to those who believed in an early evolution of modern humans. The common explanation was that the skeleton was a recent burial that had intruded into bed II, and that Reck had overlooked the evidence for this. Due to World War I Reck was not able to go back to Olduvai to check his claims, but when Louis Leakey organized an expedition there in 1931 he invited Reck who had been a friend since the 1920s when Louis Leakey had visited him several times to study the Olduvai man. It seems that Louis Leakey at first agreed with his colleagues that the skeleton was a recent burial, but the visit to Olduvai and the wish for an early modern *Homo* soon changed his mind; in reports to *Nature* Louis Leakey and Reck claimed that the skeleton was indeed lower Pleistocene and thus the earliest known fossil of anatomical modern man. Leakey acknowledged that the skeleton was buried, but he believed it was buried directly into bed II, not from the later bed IV or V, intruding into bed II. The claim received much critique, since many found it improbable that a fully modern skeleton could be of lower Pleistocene age. But Louis Leakey believed that possibility and this is one of the cases where history witnessed that science is a human enterprise and thus is prone to be affected by human wishes, beliefs, and prejudices: "Here, on the very first visit to the site [Olduvai] that was to play such a central role in Leakey's professional life, a pattern was set. He wanted to believe in ancient *Homo*, and so suspended the degree of critical judgement he might otherwise have applied to the evidence" (Lewin, 1997, p. 130f).

At the same expedition Louis Leakey had found two other fossils that to him supported his interpretations of Olduvai man. In early 1932 Louis Leakey's team had found some pieces of skull in situ (which indicated Pleistocene date) at a site called Kanjera. The brow ridges of the skull were nowhere near as protruding as in *Pithecanthropus* of the same age, which made Louis Leakey declare it the true ancestor of modern man. This discovery supported the possible early date of Olduvai man: "The age of the Olduvai skeleton might still be in doubt, but here, it seemed was certain evidence of a modern type of man in early Pleistocene deposits" (Cole, 1975, p. 90).

After discoveries were made at Kanjera, the team went on to a nearby site (Kanam), which judged by the fauna was earlier than Kanjera. Louis Leakey's assistant discovered some human-like teeth in a lump of earth, and Louis Leakey who wasn't present at the discovery would the same evening find an almost-complete jaw in the lump of dirt. This fossil—the Kanam jaw—was first mentioned in *Nature* in May 1932. With these discoveries Louis Leakey had apparently succeeded in finding the Shadow Man. He now had three examples of relatively anatomically modern fossils—Olduvai, Kanam, and Kanjera—from the first part of Pleistocene, contemporary with the fossils of Peking Man and Java Man, whom Louis Leakey now felt was securely put on a side branch. But alas—the next couple of years would turn all this over yet again.

First the Olduvai skeleton, the dating of which Louis Leakey defended for some time—was to be re-dated again. Louis Leakey had studied the Olduvai man in Germany again in 1933 after the expedition, and had to his surprise discovered that it did not look at all like the Kanjera skull fragments. This made necessary another geological investigation. The geologist Wayland had had already in the 1931–1932 season visited Olduvai and had suggested a solution to the geological riddle, and by the second investigation he was proved right, and Louis Leakey had to admit he was wrong. In the 18 March 1933 issue of *Nature* Louis Leakey, Reck, and some colleagues—including the geologist Boswell who we will hear more about soon—concluded that "it seems highly probable that the skeleton was intrusive into Bed II" (Leakey et al., 1933, p. 398). The problem at Olduvai was that the skeleton had probably been buried sometime after the formation of bed I, but erosion had later removed all trace of beds III and IV, which had let Louis Leakey's team to believe that the burial had taken place after the formation of bed II. Olduvai man could now not be argued to be any older than upper Pleistocene and its modern look was then hardly surprising. This was the first instance where Louis Leakey's reputation suffered a blow, but one mistake did not matter much. After all one of the characteristics of science (compared to blind faith) is that science progresses by mistakes being discovered and corrected, and Louis Leakey had acknowledged his.

He also still had two other impressive proofs of his theory. Louis Leakey championed Kanam and Kanjera as the earliest human fossils and in 1933 he even proposed a specific name for the remains: *Homo kanamensis*. In his 1934 book *Adam's Ancestors* he proclaimed: "In 1933 I was fortunate enough to be able to lay before the scientific world the evidence which showed that we had found at Kanam in East Africa the oldest fragment of a real ancestor yet discovered, a real *Homo*, who was the approximate contemporary of the various side branches of the human stem represented by Piltdown man, Pekin man and the Java ape-men" (Leakey, 1934, p. 2).

But the Kanam and Kanjera fossils were also under attack, and this time the critique implied that Louis Leakey was guilty of some degree of scientific misconduct. At a meeting in 1933 where Louis Leakey had proposed the name *Homo kanamensis* he had—to silence his critics—invited the skeptical and respected geologist Professor Percy G.H. Boswell to Africa to clear matters up. The result was that Boswell arrived at the Kanjera site in early 1935. To Louis Leakey's dismay he was unable to relocate where the skull fragments had been found, but additional discoveries of more skull fragments indicated that he at least had been close to the correct original position. Boswell though was not impressed and his mood didn't improve at Kanam where the markings Louis Leakey had left at the find spot were gone. The metal pins that marked the location had been moved and used by local fishermen. But Louis Leakey felt he was safe, since he had a photo of the site, which could then be relocated. The problem was that his own photos had been ruined because of a hole in the camera, but a visitor at the excavation had taken a picture of the location, and this photo had been used to illustrate the site both at the 1933 meeting and in an exhibition of the fossils and was to be printed in Louis Leakey's forthcoming book *The Stone Age Races of Kenya*. Confident, the team and Boswell went looking for the location on the photo, which proved to show an entirely different location than where the field notes had recorded the jaw had been found. Now Boswell was neither impressed nor amused, and Louis Leakey hastened to send a telegram to stop the printing of the photo as proof in his book.

It had proved impossible to locate the exact finding spots of either Kanam or Kanjera, and thus the geological reevaluation of the fossils' age could not be carried out. Boswell returned to England to let the scientific community know about the sloppy science of Louis Leakey: "It is disappointing, after the failure to establish any considerable geological age for Oldoway man (of *Homo sapiens* type) that uncertain conditions of discovery should also force me to place Kanam and Kanjera man in a 'suspense account' "(Boswell, 1935, p. 371).

Returning from Africa Louis Leakey desperately tried to reply and restore his reputation and the case for *Homo kanamensis*, but the battle was lost. Louis Leakey's biographer added that "it took this affair to teach him that in science even a few inches are vital" (Cole, 1975, p. 100). In *The Stone Age Races of Kenya* published in 1935 the controversies are not apparent—the book was already in print, and only an errata slip about the Kanam photos witnessed the trouble Louis Leakey was in. The mistakes Boswell had pointed out were bad enough, but the combination with Louis Leakey's relationship to Mary Nicol that resulted in his divorce from Frida in 1936 effectively ruined both Louis Leakey's scientific and moral reputation. Louis Leakey and Mary had openly lived together since late 1935 and the divorce and remarriage the following year was a scandal that even made the press headlines.

Still, these were very productive years where Louis Leakey wrote many of his better known books. Some of them—such as the autobiography *White African*—were written to secure an income, but some were the result of Louis Leakey's wish to communicate his discoveries to both laymen and experts, for example, *Adam's Ancestors* and *The Stone Age Races of Kenya*. Also Louis Leakey's interest in African politics and history sparked books, among these the controversial *Kenya—Contrasts and Problems*. His largest work from this period though remained unpublished until after his death. Louis Leakey's position at Cambridge had run out and no other jobs were in sight, partly due to the last years' scandals. Mary and Louis were in a financially insecure position and therefore readily accepted when the Rhodes Trust offered Louis Leakey two years of funding to travel to Kenya to research and record traditional Kikuyu culture. The product was a huge three-volume manuscript that did not find a publisher before Leakey's death. It was posthumously published in 1977 supported by the Leakey foundation by request of Mary Leakey (Leakey, 1977). In 1937, the couple arrived in Kenya where they would eventually establish a life, a family and an archaeological career that would make both of them famous.

THE EARLIEST APES

To explore the earliest possible ancestors of both humans and apes, Louis Leakey was naturally very interested when remains of what was then believed to be a very chimpanzee-like primate were found on the shores of Lake Victoria in 1927. The creature was given the name *Proconsul*—Consul being a famous chimp in a British zoo. Additional fossils were found in the following years and when Louis Leakey got a chance to explore the sediments at the lake he readily took it. He pointed his interest mainly to the island Rushinga with very rich fossil deposits, and in the early 1940s he paid short visits to the island whenever his work permitted him. This resulted in several discoveries, among them a perfectly preserved jaw of a *Proconsul*. Later (in 1950) he published an article in *Nature* together with William Le Gros Clark and announced it as a new species *Proconsul nyanzae*. The discovery certainly improved Louis Leakey's scientific reputation, and he enthusiastically suggested that *Proconsul* represented a type akin to the common ancestor of man and apes from the Miocene period.

When circumstances (time and funding) during the 1940s allowed it Louis and Mary visited Rushinga in the following years with many fossils of different fauna as the result. The excavations were made much easier from 1948 when the wealthy Charles Boise began funding the couple's expeditions. The contribution paid off soon after when Mary in September 1948 found a whole well-preserved skull with face and jaws of a *Proconsul africanus*. It is a testimony of the happiness the fossil caused that Philip Leakey was born exactly nine months later. In October 1948 Mary presented the find to the scientific community and the press in London, and the discovery was a sensation. The press very much wanted to report the discovery of a "missing link," but Le Gros Clark found that the primate was more likely the ancestor of the apes than of hominids. Louis though fit the fossil into his perception of a Miocene hominid stem, and thought that *Proconsul* gave many clues about the earliest hominids.

In the 1960s and the early 1970s paleoanthropology followed a theory that much resembled Louis's interpretation of the Miocene primates. It was accepted by many that a group of Miocene primates known as the ramapithecines were hominids; that they belonged to the human and not to the anthropoid evolutionary line. The ramapithecines were primarily known from fossils from the Siwalik Mountains in Pakistan and India, but the leading experts in the fossils, David Pilbeam and Elwyn Simons, also believed that some of the Miocene apes from Africa belonged to this group, and thus wanted to rename them.

This annoyed Louis Leakey who had in 1961 found a species which he named *Kenyapithecus wickeri* at a site called Fort Ternan, and who had redefined the first Rusinga fossils to belong to *Kenyapithecus* as well. Louis Leakey resisted Simons and Pilbeam's suggestion that *Kenyanthropus* was in fact an African *Ramapithecus*, and instead suggested that *Kenyanthropus* and not the ramapithecines could be considered the earliest human ancestor (Leakey 1967; Leakey 1969; Leakey 1970). This debate was one of the last Louis Leakey engaged in after his fossil success in the earlier 1960s which will be dealt with later, and his many articles on the *Ramapithecus* question show both his stubbornness, his sense of irony, and his huge knowledge of the fossil record.

Leakey's view caused Simons to formulate what he called The Leakey Syndrome, and Simons—not entirely incorrectly—accused Louis Leakey of thinking that: "The fossils I find are the important ones and are on the direct line to man, preferably bearing names I have coined, whereas the fossils you find are of lesser importance and are all on side branches of the tree" (Lewin, 1997, p. 132). The idea that a group of Miocene primates were indeed hominids fit like a glove on the Presapiens Theory that Louis Leakey had supported his whole life, with the twist that he did not believe this group to be ancestral to the australopithecines, but to a special line leading to man. Among the defining characteristics of humans was tool use (a prominent one), and Leakey must have loved the fact that Simons and Pilbeam were also inclined to believe that the ramapithecines might have used tools. Louis Leakey adopted this view and applied it on *Kenyanthropus* after discoveries at Fort Ternan in 1967: "The available evidence present here, therefore strongly suggests that the Upper Miocene hominid *Kenyanthropus wickeri* was already making use of stone to break open animal skulls in order to get at the brain and bones to get at the marrow" (Leakey, 1968, p. 530).

This view is very controversial today, especially after *Ramapithecus* in the 1970s turned out to have closer affinities to the Asian great apes than to hominids. At the same time molecular analysis became possible and suggested that the common ancestor of humans and African apes was not older than about 7 million years, rendering it impossible that the 14-million-year-old *Ramapithecus* (now redefined as *Sivapithecus*) had more to do with humans than with apes. Louis Leakey did not live to see this turn of events, and would probably not have liked it either. What he would have liked though was that his son Richard and his wife Meave, in 1981, found fossils of a very old member of *Sivapithecus* in Buluk, Kenya, and that this fossil was, after additional fossil finds in 1985, reclassified to a whole new genus *Afropithecus turkanensis*, one of four new genera found at Buluk. Such diversity was among the ideas that Louis Leakey defended throughout his life. He also firmly believed that one group of Miocene apes had evolved into an early hominid, which was not an australopithecine, and it was this hominid he struggled to dig out of the Kenyan ground.

OLDUVAI AND LEAKEY'S LUCK

Though the Miocene primates were impressive and important fossils and close to Louis Leakey's heart, it was not those who made Louis and Mary world-famous. What made Louis Leakey a paleoanthropological superstar was the skull of what was in public called Nutcracker Man, but which Mary and Louis always referred to as "Dear Boy." It is easy to understand their affection for this fossil which they found in July 1959 after almost thirty years of digging at Olduvai. During the decades they had discovered thousands of faunal fossils, and many from species hitherto unknown—among these extinct creatures like the gigantic elephant-relative *Dinotherium mirabilis*, *Simopithecus jonathani* (a giant baboon, named after Jonathan Leakey who found the first jaw of this creature), and the huge pig-like *Afrochoerus nicoli* (named by Louis after Mary's maiden name Nicol!). They had also, especially through Mary's meticulous excavations and classifications, revolutionized prehistoric archaeology regarding the earliest stone-tool cultures. But besides all these achievements, what Louis Leakey most craved eluded them: remains of early man. This he got with Dear Boy.

The skull was found by Mary, and a thoroughly recorded and accurate excavation by her and Louis Leakey followed—this time the evidence had to be waterproof. Mary report that Louis Leakey's first remark was, "Oh dear. I think it's an australopithecine" (Lewin, 1997, p. 138), and this was no doubt a disappointment to him, since he did not regard any of this group of hominids as being ancestral to man. Australopithecines were at this time only known from South Africa, where several species had been identified, which are now normally split in two groups: the gracile and the robust, where especially the gracile ones have been considered possible ancestors to man. Louis nevertheless found the Australopithecines too specialized to play the part as ancestors of *Homo sapiens*—especially in the light of his belief in a very early evolution of *Homo*-like creatures. To Louis Leakey *Australopithecus* was just another side branch and not what he had really hoped to find as the creator of the abundant stone tools at Olduvai.

But when the skull was excavated and Louis Leakey had examined it closely he changed his mind. Regardless of the skull's close resemblance to the robust Australopithecines, Louis Leakey was convinced that this was not an Australopithecine but an early true man. Louis Leakey believed this was witnessed by several anatomical characters, and by the context of the skull which was found among stone tools and broken animal bones. Dear Boy had made tools and had broken the bones of his prey and could thus be regarded as a human. Even though tool use and tool making can not formally be accepted as a biological definition of humans, many scientists in the 1950s and 1960s believed that this was the defining human character and the phrase Man the Toolmaker was popular.

Louis Leakey tried to curb his critics with the argument that "there is no reason whatsoever, in this case, to believe that the skull represents the victim of a cannibal feast by some hypothetical more advanced type of man. Had we found only fragments of skull, or fragments of jaw, we should not have taken such a positive view of it" (Leakey, 1959, p. 491). The skull definitely did not look like *Homo* and resembled more the robust type of *Australopithecus* (often called *Paranthropus*) which also had a saggital crest and massive jaws. It was hard to believe that such a creature should have evolved into *Homo sapiens* in just the 600,000 years the geological layer was estimated to be. New dating methods saved the day, when the skull was potassium/argon dated to be 1.75 million years old. As the scientist Garnis H. Curtiss wrote to Louis Leakey in the letter reporting the dating results: "One thing is certain—Olduvai man is old, old, old!" (Leakey, 1961, p. 568).

DIY Prehistory

One of the gifts Louis Leakey had which enabled him to communicate how prehistoric people had lived was that of making stone tools. Whether it was to skeptical Kikuyu workmen or admiring tourists Louis Leakey would gladly demonstrate how to make a tool in a matter of minutes, and perhaps even show how to skin an animal with it without problems. Photos of Louis Leakey making tools also often accompanied popular science articles about his discoveries and theories. Louis Leakey liked to test his ideas of prehistoric life. Not often as controlled experiments, but rather as whimsical trial-and-errors that would later make a nice argument or anecdote in his books or lectures. He often told that small-teethed hominids simply had to have had tools, since he himself had once tried to skin a rabbit with his teeth and nails, and that it had not worked. Similarly he tried to strengthen his argument about the existence of prehistoric bolas by making one: "[I] was forced to give up after I nearly killed myself" (Louis Leakey quoted in Morell, 1995, p. 129).

When a fossilized feces was found at Olduvai, Louis Leakey very much wanted to compare it with a fresh one originating from a person who had eaten whole small prey. The anthropologist Iven Devore tells that Louis Leakey tried to persuade him to eat a whole raw rat, with the words "my dear boy, let me make you famous." Allegedly Louis Leakey did not want to do the experiment himself since he had not many teeth left and thus could not chew the animal. Louis's sons also declined the honor (Morell, 1995, p. 203).

Louis Leakey published Dear Boy in *Nature* in August 1959 and declared it to be a new species. He probably caused a bit of giggling among his colleagues with the remark, "I am not in favour of creating too many new generic names among the Hominidae"—Louis Leakey had given every new hominid he had ever found a new name, and this was a habit he continued with the words, "but I believe that it is desirable to place the new find in a separate and distinct genus. I therefore propose to name the new skull *Zinjanthropus boisei*" (Leakey, 1959, p. 491). The name was a tribute to Charles Boise who had supported the Leakeys financially through several years, and Zinj is an old word for

East Africa. The name was thus Boise's East African man. The massive jaw and big teeth caused the nickname Nutcracker Man, but the fossil is often simply known as Zinj.

Louis Leakey had been bold to declare a new species on basis of a single individual, but this was not the only thing that made his paleoanthropological colleagues wonder. The fact was that Zinj to everybody but Louis Leakey himself looked like an *Australopithecus*. Louis Leakey was so sure of the differences that he did not compare the skull with the South African fossils before or after he had completed the publication in which he wrote that he had "recently re-examined all the material" (Leakey, 1959, p. 491). Virginia Morell, biographer of the Leakey family, reports how Louis Leakey and Mary went to South Africa to compare the skull to the Australopithecine fossils en route to the fourth Pan-African Congress of pre-historians in Congo. Louis Leakey had been the driving force behind the first of these congresses and was naturally a guest of honor, and was eager to show off his prize fossil. In Leopoldville other paleoanthropologists were invited to examine the skull and tension grew as everybody thought it looked Australopithecine while Louis Leakey was adamant it was another species close to *Homo*. Raymond Dart who had in 1925 published the first-ever discovery of an *Australopithecus* and who had later found several other examples, among these a female skeleton called Mrs. Ples, broke the tension with the remark: "I can't help wondering what would have happened if Mrs. Ples had met Dear Boy one dark night" (Morell, 1995, p. 193). The implication was clear: Everybody found Zinj to be a robust Australopithecine, everybody but Louis Leakey, who later responded to Dart's joke with the words: "I have no doubts at all she would have run away" (Morell, 1995, p. 194).

The scientific skepticism didn't matter to the press and the public, in whose eyes Louis Leakey became the hero who had found the world's oldest human. The fame was mainly caused by big articles in *National Geographic* that showed pictures of the excavation of Zinj and portrayed Louis Leakey as an adventurer and a scientific hero who had never given up his hope of bringing the earliest man out of the ground for the world to see. Louis Leakey was nothing short of Indiana Jones. The articles had plenty of photos from the field, the excavations and reconstructions of prehistoric people hunting and creating tools. The point that a whole family was engaged in the discoveries also appealed to the readers as did the sensational headlines like "Skilled Family Team at Olduvai Gorge Unlocks Priceless Stone Age Secrets," "Timeworn Skull Solves a Stone Age Mystery," and "Olduvai promises New Secrets" (Leakey, 1960). The popular magazines and the following lecture tour also gave Louis Leakey a chance to air his more outlandish theories, among these that round stones found at Olduvai were remains of prehistoric bolas. This was illustrated by a two-page reconstruction in *National Geographic* of naked prehistoric humans, and a nursing mother who admiringly inspects her mate's bola. The popular science magazines needed sensations, good stories, and scientific heroes—Louis Leakey gave them everything they had ever wished for.

Zinj was also presented to the public on a lecture tour where—as in the articles—Louis Leakey had the opportunity to spellbind his audience by the fairytale-like story of his years of searching before he struck gold. There is no doubt that Louis Leakey had an extreme flair for making prehistory and dry science come to life for an audience and by every account his lectures were impressive shows of knowledge, humor, and anecdotes. Louis Leakey had an immense charisma and personal presence that could not help to affect his audience—men and women alike. Audiences felt a huge sympathy and were deeply impressed by this man who had devoted his life to science, who spoke passionately of Africa, nature, and prehistory, who told them how he had made stone tools to skin prey he had caught with his bare hands, and who spared no personal expense to find knowledge and bring his message about the prehistory of man to the public.

The huge interest that followed the discovery of Zinj made it necessary for Louis Leakey to spend more time touring and lecturing, but it also secured future financial support, which caused great relief to the Leakeys who until then had been in quite dire straits. In spite of the publicity and the fame, the scientific glory of Zinj soon faded and even Louis Leakey himself was forced to recognize that it was nothing but a robust Australopithecine—the first of its kind in East Africa though. No doubt this was not what Louis Leakey had hoped and his reluctance to acknowledge Zinj as an Australopithecine had probably only vanished because of a new and potentially even more revolutionizing discovery.

THE HANDY MAN

Believing that Zinj had been an advanced toolmaker, Louis Leakey had predicted that even earlier toolmakers or tool users than Zinj would be found at Olduvai. He had also claimed that Zinj was not the result of a cannibalistic dinner, eaten by another human being, but when Louis Leakey eventually found that earlier toolmaker, Zinj was easily reduced to leftovers from a meal, devoured by a more sophisticated hominid. The remains of this hominid were found close to the site of Zinj in November 1960 by Mary and Louis Leakey's eldest son Jonathan. The fossils were the jaw and fragments of the skull from a child less than twelve years old. In the following years both parts of a foot, a hand, and ribs were found, and eventually the remains of several individuals of the same type were excavated from Olduvai bed I, at a site called FLK, meaning Frida Leakey's korongo. The sites in Olduvai were traditionally named after people who had worked or visited the gorge, and korongo is a local word for gully.

The child—called Johnny's child—was apparently of a different kind than Zinj. It was more gracile and the skull parts indicated than the brain had been relatively big. This—and that the fossils were contemporary with and older than Zinj—must have been an immense satisfaction to Louis Leakey. An old, large-brained hominid was exactly what he had been searching for—and many times thought he had found, just to be disappointed time and again. Considering this, it is impressive that Louis for once did not make a sensational public declaration of the fossils and neither did he rush to give it a new species name. Many articles, both scientific and popular, mentioned the fossils, but no conclusions or theories were put forth. Perhaps the cases of Kanam and Zinj had taught Louis Leakey a lesson.

Facts are that from 1960 to 1964 the fossils were studied by Louis Leakey and two young paleoanthropologists, Louis Leakey had chosen as allies not to stand alone against the critique that would eventually come. The co-workers were Philip V. Tobias, who had also been asked by the Leakey's to analyze Zinj and who was to become Raymond Dart's successor at University of Witwatersrand in South Africa, and John Napier, an expert on the difficult subject of hominid hands and feet. Tobias and Napier studied the fossils at Leakey's request and they probably had a sobering effect on the claims about the fossils. Tobias's detailed study also delayed the publication since he and Leakey long argued about whether the child was Australopithecine or not. Leakey said no, Tobias said maybe, but in the end he was convinced. When the publication of the fossils was finally ready in 1964 it was scientific, meticulous, and cautious, but nevertheless both sensational and prone to provoke critique.

The article which appeared in *Nature* in April 1964 was titled *A New Species of the Genus Homo from Olduvai Gorge*. The title alone was provoking, since no other new species of *Homo* had been suggested for decades, and only three were generally accepted: *Homo erectus* (the hominid formerly known as *Pithecanthropus*), *Homo neanderthalensis*, and *Homo sapiens*. Not only did the paper propose a new species—*Homo habilis*—it also offered a new definition of the genus *Homo* itself. This was necessary to accommodate the new species in *Homo*, since scientific authorities had traditionally reserved the *Homo* label for species with quite large brains. It had been debated among anthropologists how small a brain a hominid could have and still be considered a possible *Homo*. The result—about 750–800 cc—was referred to as the Cerebral Rubicon. This reference to the river Rubicon which Ceasar had crossed illustrated that the brain size was considered a frontier only *Homo* had been able to reach and conquer. No wonder that critical voices were heard when Leakey et al. suggested to move the frontier and include species with a smaller brain into the human genus with the words:

> We have come to the conclusion that, apart from *Australopithecus* (*Zinjanthropus*), the specimens we are dealing with from Bed I and the lower part of Bed II at Olduvai represent a single species of the genus *Homo* and not an Australopithecine. The species is, moreover, clearly distinct from the previous recognized species of the genus. But if we are to include the new material in the genus *Homo* (rather than set up a distinct genus for it, which we believe to be unwise), it becomes necessary to revise the diagnosis of the genus. (Leakey et al., 1964, p. 7)

The new definition of *Homo* included (non-controversially) erect posture and developed dexterity of hands (e.g., an opposable thumb), several facial and dental characteristics, but also Cerebral Rubicon of 600 cc ranging to 1600 cc. The brain size thus embraced in *Homo* both the new species *Homo habilis* and the *Homo* species with the largest brains, the Neandertals.

Though it was not a part of the formal description and definition of *Homo* in the paper, it was also hinted that the association with stone tools had contributed to the inclusion of the new species in *Homo*. The name *Homo habilis* (suggested by Raymond Dart and meaning "handy man" or "able man") pointed in the direction that the skills of the species was important, and in the end of the paper the cultural context was addressed:

> When the skull of *Australopithecus* (*Zinjanthropus*) *boisei* was found on a living floor at F.L.K. I, no remains of any other type of hominid were known from the early part of the Olduvai sequence. It seemed reasonable, therefore, to assume that this skull represented the makers of the Oldowan culture. The subsequent discovery of remains of *Homo habilis* in association with the Oldowan culture at three other sites has considerably altered the position. While it is possible that *Zinjanthropus* and *Homo habilis* both made stone tools, it is probable that the latter was the more advanced tool maker and that the *Zinjanthropus* skull represents an intruder (or a victim) on a *Homo habilis* living site. (Leakey et al., 1964, p. 9)

It was not formally correct to include culture in a biological classification and this produced much criticism, but even more provoking were the implications for hominid phylogeny. The paper concludes with the sentence: "It thus seems clear that two different branches of the Hominidae were evolving side by side in the Olduvai region during the Upper Villafranchian and the lower part of the Middle Pleistocene" (Leakey et al., 1964, p. 9). Zinj and *Homo habilis* had been contemporaries. The implication was that their common ancestor, and thus the branching off of the *Homo* line, was older than hitherto expected by most other than Louis Leakey himself. The *Homo* line had probably branched off at least in the Pliocene according to Louis Leakey, and as his writings on the Miocene apes show, maybe the branch even started way back in the Miocene. As the case of *Ramapithecus* shows, in the late 1960s and 1970s it was not uncommon to think the human line had started with the Ramapithecines some 14 to 20 million years ago, but when *Homo habilis* went public not many believed in this huge age of the hominid line.

Also the mere idea of creating a new species name was opposed to the common paleoanthropological thinking of the period. It was defined by a tendency to lump species together rather than to split species or declare new ones. The lumping tendency had begun in the early 1950s as a result of the biologist Ernst Mayr's suggestions that all Pithecanthropithecines (Java Man and Peking Man, and some African fossils) should be put together in the species *Homo erectus*, and that the confusingly many South African fossils should all be lumped together in the genus *Australopithecus*. Mayr's suggestion had been followed by most paleoanthropologists (at least in the Western part of the world) and Leakey's tendency to create names had been a irritation to many. That he now even suggested a new *Homo* did not make matters better. Many paleoanthropologists felt that the fossils of *Homo habilis* could comfortably be accommodated in either the gracile Australopithecines or be considered an early *Homo erectus* and thus spare science an unnecessary species name. Even long-time supporters of Louis Leakey such as the English paleoanthropologist William Le Gros Clark criticized Leakey and the new species, and Clark even insisted on writing "H. habilis" and not the correct *H. habilis*, and in this subtle way indicated that he did not considered it a valid species.

Another point that provoked other researchers, but which was absolutely in line with the "Leakey syndrome" was that Louis Leakey saw *Homo habilis* not as an ancestor of *Homo erectus* who again would lead to *Homo sapiens*, rather he regarded *Homo habilis* as a direct ancestor of *Homo sapiens* (possibly with still-unknown intermediate forms) and *Homo erectus* as side branch who had nothing to do with the evolution of *Homo sapiens*: "I believe that *Homo habilis* represents a direct line of *Homo* possibly leading to *Homo sapiens*, and do not see it as an intermediate link between *Australopithecus* and *Homo erectus*" (Lewin, 1997, p. 150).

Thus dismissing many of the accepted ancestors of modern man, Louis Leakey had in *Homo habilis* at last found one of the Shadow Men of his Presapiens Theory. This theory was in the 1960s all but abandoned by paleoanthropology and not many believed Louis Leakey's ideas of the importance of *Homo habilis*. The species would eventually decades later be considered a valid and important species, and many researchers also became skeptical about the Asian *Homo erectus*'s place in the pedigree, but not many thought so in the 1960s, both because of Louis Leakey's earlier mistakes and because the species was interpreted through his belief in an ancient human line.

Why Louis Leakey did not change outlook like the rest of the scientific community seems strange: "What is interesting is the fact that he stayed fast with this tradition—the essence of it at least—while the academic colleagues around him were modifying theirs. This is where the mystery lies" (Lewin, 1997, p. 136).

The mystery can't really be solved, but a couple of explanations can be offered. Perhaps the faith in the Presapiens Theory was just one example of the stubbornness and commitment that affected all of Louis Leakey's work. It has also been suggested that it might be a way in which Louis Leakey could hold on to some kind of semi-religious component in human evolution even though he at an early age turned against the doctrines of his childhood, embraced the evolutionary theory unconditionally, and throughout his life strived to convince others to do the same. Or perhaps the explanation is just that the idea of an ancient human line, represented by fossils that he could find himself, was simply an extremely appealing idea to a person like Louis Leakey who wanted to make a difference and make spectacular discoveries and to vindicate his reputation in light of his earlier failures to do the same. Maybe the stubbornness was Louis Leakey's way of proving himself right, no matter the cost. Louis Leakey had the satisfaction toward the end of his life to experience that *Homo habilis* was gradually accepted, though not the Presapiens Theory—that no longer has any followers among professional paleoanthropologists.

In the 1960s Louis Leakey spent more and more time on fundraising, public lectures and writings, and also on the research on Miocene primates which have been described earlier. Science writer Roger Lewin states that: "The discovery, analysis, announcement and subsequent debate of *Homo habilis* was really the last major event in Leakey's palaeoanthropological career" (Lewin, 1997, p. 150). Parallel with the debates from the late 1950s Louis Leakey made another important long-lasting impact on paleoanthropology, based on two things: his love of nature and his love of women.

THE WOMEN AND THE APES

Even though Louis Leakey believed that humans had followed their own evolutionary line for a long, long time, he nevertheless fully accepted the connection and close biological relationship between man and the great apes. This view, which not all in his generation shared, was probably caused by his conviction that man had originated in Africa and thus resembled the gorilla and the chimpanzee more than the Asian apes and monkeys. Louis Leakey thought that insight into the behavior of modern apes would give clues to the behavior of the common Miocene ancestor back to the earliest humans and the earliest apes, which he believed had not changed so dramatically since then as the human line had. As already mentioned Louis Leakey also saw tool use as an early defining character of humanity, which could perhaps be found even among the Miocene primates. Also, besides his archaeological work, Louis Leakey was very interested in the growing field of animal behavior, an interest that went back to his African childhood and his close relationship to the wildlife and nature in Kenya.

These circumstances were among the reasons why Louis Leakey found research on primate behavior very important, and he became the founder of what we today feel are the normal and correct ways to study and observe wild primates. Before the 1950s primate studies were primarily carried out by psychologists and in laboratories, and before Louis Leakey's initiative only George Schaller's study of the mountain gorillas exemplified observations of wild primates for a longer time period. Louis Leakey, like most other men, enjoyed the company of young good-looking women, who he employed as secretaries, assistants, and scientific helpers, and occasionally also had extramarital affairs with. He believed that women were better suited than men for scientific tasks that demanded patience, thoroughness, and empathy, and observations of wild animals were one of those tasks. In the late 1950s Louis Leakey wanted to launch a study of chimpanzee behavior, and after having employed a young man who failed completely, Louis Leakey began to look for another candidate with the conviction that the best man for the job was a woman. In 1956 Leakey's assistant Rosalie Osborn had conducted five months of study of gorillas in Uganda, but the project was cancelled due to failing financial support.

In 1957 he was contacted by a young woman, Jane Goodall, who wanted to work with animals and asked for his advice. This also illustrates that Louis Leakey's reputation was not only built in the field of archaeology, but that he was considered an authority on diverse subjects such as social anthropology, wildlife, and many other scientific fields. Goodall was a devoted natural history amateur, and on a visit to Kenya friends suggested that she get in contact with Louis Leakey who might help her in her interest of wildlife. Leakey was charmed and impressed and Goodall was employed as his assistant. While he clearly had other motives for the employment, Goodall didn't and eventually he decided that she had every ability needed to succeed in the chimpanzee study. Louis Leakey raised funding from an American businessman, Wilkie, who "felt passionately about the evolution of tools" (Jahme, 2000, p. 45) and in 1960 Goodall and her mother Vanne Goodall (a chaperone was requested by the authorities) went to Gombe to study the chimps. The rest is history: Goodall demonstrated great skill in observing and making the animals feel safe and her records of chimpanzee behavior completely changed both the knowledge of apes and the way animal studies were carried out. Among the most surprising observations were that chimps regularly hunted and ate meat and that they used natural objects for tools, e.g., to fish for termites or crack nuts. Goodall became a scientific superstar in her own right and still serves as an example for young people interested in protecting and observing wildlife.

The success of Goodall's studies created the financial possibility for Louis Leakey to launch other primate projects regarding the other great apes; the gorilla and the orangutan. He also founded the Tigoni Primate Research Centre which he felt strongly for, but which sadly became an economic and administrative burden which especially Mary did not approve of. Perhaps also because many of Louis Leakey's female protégés were employed there for periods.

Following Goodall's example the therapist Dian Fossey approached Louis Leakey on one of his American lecture tours in 1966. Fossey had earlier visited Africa, seen Olduvai and gorilla habitats, and Louis Leakey caught the chance of "getting a girl for the gorillas," and again secured funding from Wilkie. Again the project was a success though with great personal costs for Fossey since civil war and poachers posed a dangerous and in the end a fatal threat. Fossey became famous just as Goodall, and her tragic death in 1981 (when she was murdered by poachers) has made her a mythic figure portrayed in the movie *Gorillas in the Mist* starring Sigourney Weaver.

Louis's third ape-girl was the anthropology student Biruté Galdikas, who began her research in 1971 in Indonesia where it continues today. There is no doubt that the research Louis Leakey initiated had a huge effect on the way people have understood apes and their relationship to man. The knowledge of especially chimpanzees has given possible analogies to the lives of the earliest hominids and the common ancestor of apes and humans. But more than the knowledge the studies have contributed to the field of paleoanthropology, they have set a whole new standard for field studies of wild animals. Goodall, Fossey, and many others have served as role models to the effect that primatology is one of the few natural sciences to have more women researchers than men. "Jane Goodall, Dian Fossey and Biruté Galdikas . . . are known as the trimates, an obvious pun on the word primate. These three are the most famous women associated with his name, but they are certainly not the only women shown favour by Leakey. At least another fifteen women found careers working with apes and monkeys thanks to Leakey's patronage" (Jahme, 2000, p. 40).

This is a legacy by Louis Leakey often forgotten and it is a nice irony that his interest in pursuing young women has resulted in a science with a feminist touch, but his judgement of women's character was probably right:

> In some ways, Louis Leakey's theories have been supported by the evidence. On average men do not stay longer than two years in the field with any one group of primates. . . . On the other hand, time and again there are examples of women sticking out the gruelling existence with their particular group of animals for decades. Leakey's other ape women substantiate this claim. Dian Fossey had spent eighteen years watching mountain gorillas before she was murdered. Louis's third in line, Biruté Galdikas, has studied the orang-utan for twenty-nine years. Women become emotionally connected to the animals they study and do not want to leave them. (Jahme, 2000, p. 50)

Such devotion and patience was akin to the passion Louis Leakey himself felt toward the study of prehistory, and even though he enjoyed being a famous scientist with fans and followers, no doubt the fact that his administrative duties and failing health forced him not to do fieldwork anymore was a loss to him in his last years. Urging young scholars to go into the field perhaps made amends to this loss. In the late 1960s Louis and Mary's marriage had ended in all ways but formally, and Louis Leakey's life was also affected by the ambition of their son Richard, who pursued his own career as a paleoanthropologist and became the director of the National Museum of Kenya, thus being formally his father's boss. The two had equally big egos and confrontations were loud and frequent. Louis Leakey spent more and more time in the States and England, lecturing, making broadcasts, and writing, the last with occasional help from Vanne Goodall, with whom he always lived when in England. Olduvai excavations were now entirely Mary's responsibility and Louis Leakey engaged himself in projects such as the primate center and a disastrous project of finding early man in Calico, California. This caused his already-burdened reputation some damage since many of the artifacts turned out to be products of geological forces, and not man-made at all. Though ever eloquent with words, Louis Leakey's intellect no doubt was threatened by his bad health, and his critical sense became worse than ever. Harsh but probably true are Jane Goodall's words "Louis was gaga toward the end of his life" (Morell, 1995, p. 380).

THE END

Reading the eminent biography of the Leakey family *Ancestral Passions* by Virginia Morell one also senses that Louis Leakey's triumphs all had a bittersweet flavor, and that his last years were full of disappointments and physical pains resulting from strokes and operations.

Therefore it seems appropriate that the last days of Louis's life were happy ones. Richard Leakey had since the late 1960s been excavating sites at Lake Turkana (then Lake Rudolf) and at a site called Kobi Fora his crew had found a fossil skull known by the museum number 1470. Not referred to any species, 1470 seems to prove the existence of early large-brained creatures, and when it was found it was dated geologically to 2.6 million years old—older than Zinj and *Homo habilis* (this dating was subsequently changed to about 1.9 million years). Shortly after the discovery in late September 1972 Richard visited his dad to show him the skull which had every feature that would vindicate Louis's theories about early big-brained man, and in spite of their earlier confrontations Louis Leakey was delighted about his son's fossil. He, Richard, and Mary—who also happened to be in Nairobi that day—spent the evening talking and celebrating the discovery: "Louis was excited, triumphant, sublimely happy" (Mary Leakey quoted in Morell, 1995, p. 400). The happiness was also caused by the fact that Louis Leakey's son Colin (from his marriage with Frida) had just left Uganda with his family and thereby escaped the anti-British regime of Idi Amin (Cole, 1975, p. 404), and Louis Leakey was relieved the family was safely on the way to London.

The day after the happy evening—"it was almost like old times" (Morell, 1995, p. 400)—Louis Leakey flew to England en route to the United States. He never came further than London, where he stayed at Vanne Goodall's flat to write the last of his second autobiography *By the Evidence*. There, early in the morning of 1 October, he suffered another stroke and died in the hospital later that day.

His death made headlines all over the world, and while memorial services were held and obituaries were printed, the Leakey family debated about how to bury the grand old man. In tune with the traditions of Kikuyu he wasn't cremated but buried, but the family debate on the question delayed the decision about a headstone. When Jonathan Leakey a year later returned to the grave with a suitable stone from Olduvai, he found to his surprise that a stone with an inscription had already been placed on the grave without the family knowing. The inscription said:

> Louis Seymour Bazett Leakey
>
> 1903–1972
>
> "Wakaruigi"
>
> "Son of Sparrow Hawk"
>
> You live on

In the minds you inspired

In the projects you pioneered

In the lives you improved and created

In the hearts that loved you

You cannot die.

ILYEA

The riddle of the stone was solved later by Richard Leakey who recognized the ILYEA from letters from the correspondence between Louis Leakey and Rosalie Osborn, who had been Louis Leakey's assistant and first "gorilla girl" in the late 1950s. They had been lovers and the letters meant "I'll Love You Ever Always." As Virginia Morell who reports this story concludes: "Even in death, it seemed, Louis would have his other women" (Morell, 1995, p. 405). In spite of this and probably because of its most appropriate words the stone still stands on the grave. The words quite accurately sum up why Louis Leakey, in spite of his many controversies and unconventional ideas of human ancestry, still stands as an icon for the study of evolution of man. His accepted scientific contributions are only a small part of Louis Leakey's legacy. While the spotlight on African prehistory, the magnificent finds of extinct fauna, the revolutionary insight into Stone Age cultures and species such as *Australopithecus* (*Zinjanthropus*) *boisei* and *Homo habilis* has turned out to provide accepted and important knowledge, the two truly impressive things about the career of Louis Leakey are how he succeeded in initiating projects and inspiring persons to do important scientific discoveries of their own, and last but not least: that he so effectively popularized his research. By his public writing, lectures, and broadcasts, Louis Leakey had single-handedly communicated the knowledge of evolution to millions of people, not as a dry scientific theory, but as a fact illustrated by his own magnificent discoveries of fauna, hominids, and geological and cultural sequences. Louis Leakey made prehistory come to life, and secured that both the public and science eagerly turned their eyes to the African continent for the answer to the riddle of man's evolution.

This chapter has pointed to the fact that science is made by humans, and that humans make mistakes, and Louis Leakey's science was no exception. Louis Leakey may not have found out the right answer to how humans had evolved himself but he—as the headstone says—pioneered projects and inspired minds and thus secured that the questions about the course of evolution continue to be explored in ever more sophisticated and interesting ways. A reviewer of Morell's book describes Louis Leakey quite correctly with the words: "That pioneer colonialist, archaeological empire-builder and charismatic philanderer who, through his lovers and offspring, succeeded in founding modern palaeoanthropology" (Knight, 1997, p. 388).

Among the minds he inspired were in fact many of those of his own family and other loved ones, and this legacy still lives on. Richard Leakey and his second wife Meave Leakey have since the 1970s been the driving forces behind several important hominid discoveries, and their daughter Louise continues this tradition. The outcome of their discoveries is several new species and confirmation of one of Louis Leakey's theories: that many hominid species were contemporary between 1 and 4 million years ago. The species *Homo habilis* has in every way been judged valid by other fossil discoveries, and though Louis Leakey's idea about *Homo habilis* as the direct ancestor to *sapiens* is still debated and many paleoanthropologists have pointed to the fact that two species might be represented in the *Homo habilis* fossil material, Louis Leakey's claims about its importance have been proved right. His view that no *Australopithecus* known at that time were ancestral to *Homo* is now shared by many scientists, since older and less specialized *Australopithecus* species have been excavated, among these *Australopithecus afarensis*. What was mockingly described as the Leakey Syndrome—the tendency to view your own fossils as the direct ancestor to *Homo*—has in the last decade infected other scientists as well. In addition, Louis Leakey's splitter tendency has now been adopted by many, and the number of controversies as well as the number and age of declared hominid species is soaring. This also points to the fact that pride and prejudice is a part of every science, but perhaps even more so in paleoanthropology, which strives to explain the origins of humanity. And no doubt a lot of what and where the science of paleoanthropology is today is due to the life and legacy of Louis Leakey.

FURTHER READING

Cole, Sonia. *Leakey's Luck: The Life of Louis Seymour Bazett Leakey 1903–1972* (London: Collins, 1975).

Leakey, L. S. B. *White African: An Early Autobiography* (Cambridge: Schenkman, 1966).

Leakey, L. S. B. *By the Evidence: Memoirs 1932–1951* (New York: Harcourt Brace Jovanovich, 1974).

Lewin, Roger. *Bones of Contention: Controversies in the Search for Human Origins* (Chicago: University of Chicago Press, 1997).

Morell, Virginia. *Ancestral Passions: The Leakey Family and the Quest for Humankind's Beginnings* (New York: Simon and Schuster, 1995).

LUCY

Holly Dunsworth

When their neighbors ask them what they do for a living, paleoanthropologists need only drop a single name: Lucy. Most people who have been exposed to mainstream news media since the mid 1970s are at least familiar with the internationally adored fossil human ancestor. Thus, no matter whether a paleoanthropologist has ever performed research on Lucy, her name can be used like a brand for the science of human origins and evolution.

Innumerable scientists, writers, and reporters claim that Lucy is the world's most famous fossil hominin. However, the petite skeleton is not even the most beautifully preserved or the most complete hominin individual on record. Since Lucy's discovery over thirty years ago, thousands of hominin fossils have been discovered, a handful of which rival and even surpass her scientific significance. Yet Lucy remains the most celebrated human ancestor and the benchmark by which all subsequent finds are measured.

Lucy's formula for stardom comprised several critical components. First, at the time she was found, she was the oldest decent specimen of a fossil hominin. Hominins (also referred to as "hominids") are humans and all of our extinct fossil relatives that lived since the human lineage split from that leading to chimpanzees. Lucy is dated to about 3.2 million years ago (Mya) and prior to the time of her discovery there were only scraps of very early hominin fossils that approached the estimated split (based on molecular clocks of genetic sequences) between the human and chimpanzee lineages around 6 Mya. These included a piece of arm bone from about 4 Mya, a fragment of a jaw from over 5 Mya, and a single molar tooth from 6 Mya, all from East Africa.

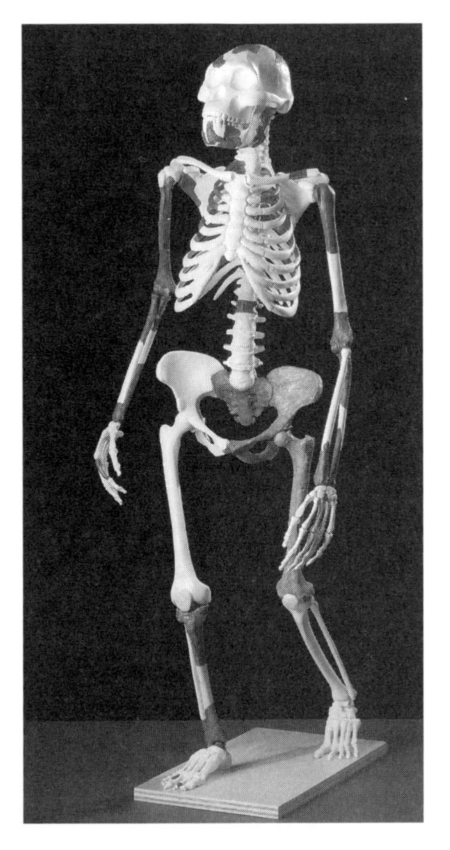

A complete reconstruction of the Lucy skeleton. The dark parts are the fossils. The lighter parts are new. (Photo care of the Cleveland Museum of Natural History.)

Second, with nearly 40 percent of her skeleton preserved, Lucy debuted as the most complete early hominin skeleton specimen on record. Using mirror imaging to fill in the missing bones on either side, over 70 percent of her skeleton could be reconstructed. There were specimens of Neandertals that were more complete but they were much younger, or much more recent in geologic terms.

Third, Lucy had a never-before-seen mosaic of ape- and human-like traits and she played a major role in the designation of an entirely new species that included the earliest hominins on record. This species, *Australopithecus afarensis*, eventually became widely considered, as it is today, the common ancestor to all later hominins, including humans. Therefore, Lucy was symbolic of the complete makeover of the human family tree. Her identity was so intertwined in that paradigm shift that scientists who experienced it can split their careers into two time periods: B.L. and A.L., or Before Lucy and After Lucy (Gibbons, 2006a). Tim White, a prominent paleoanthropologist at the University of California, Berkeley who was a co-leader in Lucy's analysis, summarized in retrospect that "Lucy was a turning point. Lucy had a fundamental role in changing the structure of paleoanthropology in east Africa" (Gibbons, 2006a, p. 1,740).

Fourth, Lucy's feminine nickname humanizes her ape-like qualities, and it also commands affection and adoration. In one fell swoop, the dubbing of a pile of cruddy ancient bones "Lucy" opened up the hearts of the world—even those who doubt humanity's evolutionary roots. The nickname created a simultaneously sweet and matronly persona that made human evolution more palatable to the public. There is no proof, but there is little doubt, that if we only knew Lucy by her museum catalog number A.L. 288-1 she might not be automatically included in volumes such as this one.

Finally, Lucy's discoverer, Donald Johanson, is a charming, appealing, and effective spokesperson for human evolution. The scientific importance of Johanson's 1974 find, combined with the publicity that he helped generate, almost instantaneously catapulted he and Lucy to international superstar status, far beyond paleontological standards of notoriety. His natural ability to promote his work and the science of paleoanthropology as a whole contributed largely to the adoption of Lucy as the mascot for the field of human origins and evolution.

LUCK BE A LUCY

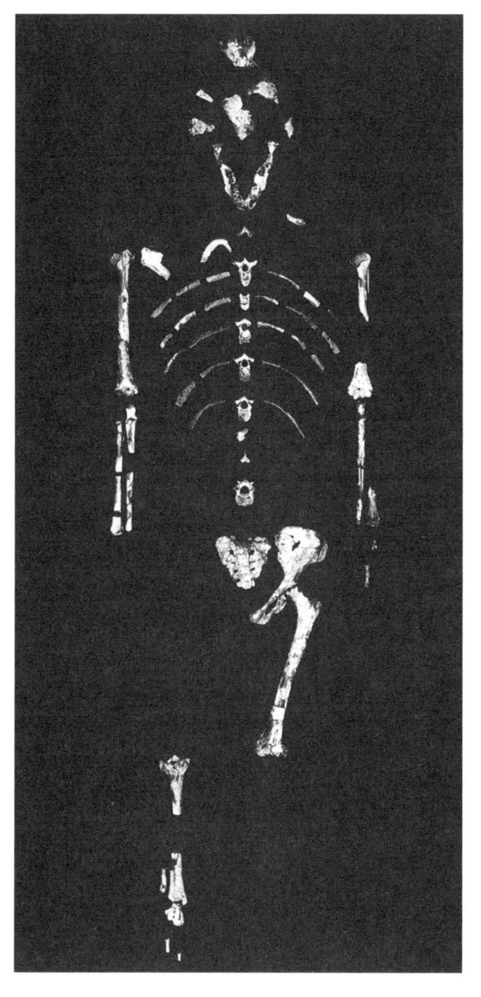

Here's Lucy! These are the actual fossils found by Tom Gray and Donald Johanson in 1974. (Photo care of the Cleveland Museum of Natural History.)

Lucy hails from the Afar region of Ethiopia which is the triangular depression located just west of the Horn of Africa at the northern terminus of the Great Rift Valley. It is a zone characterized by volcanic and tectonic activity where the foundation of East Africa is literally separating itself from the rest of the continent. The Afar is an inhospitable place to do paleontology because of the extreme heat, lurking carnivores, deadly malaria, and ethnic gunfights that have plagued the region throughout history.

French geologist Maurice Taieb did not see the Afar in those foreboding terms. Taieb had been exploring and mapping the Afar region and he found vertebrate fossils of species that were known to date to between 4 and 3 Mya at a site called Hadar. His excitement about the prospects of finding fossil hominins at Hadar led him to invite a young Don Johanson to join him for fieldwork. Johanson was technically still a graduate student at the University of Chicago, but he had already earned an appointment at the Cleveland Museum of Natural History. The two met while working on a joint French and American expedition at a site in southwestern Ethiopia along the Omo River. The age of the fossiliferous sediments at Hadar were particularly pertinent to paleoanthropology at that time since very little had been collected or securely dated prior to 2.5 Mya.

Hadar is about 100 miles northeast of Addis Ababa. Its endless expanse of badlands is difficult to access and is an unforgiving place to search for fossils. Remarkably, Hadar is the kind of place that paleontologists drool over because of the layer-cake geology that is full of volcanic ashes, which are used for absolute dating, and because of the fossils seeming to ooze out from the sediments. Johanson's first impression of Hadar was that of an intriguing dream-like place.

Taieb, Tom Gray (a graduate student in archaeology at the Case Western Reserve University in Cleveland), and a small team of colleagues (collectively called the International Afar Research Expedition [IARE]), embarked on the first short reconnaissance expedition to Hadar in 1972 and then returned in 1973 for a proper field season. Finding hominins in the first two forays is quite a feat and more than most people expect to accomplish at a cold site, but that is exactly what they did. Johanson felt his career was riding on finding a hominin at Hadar and by the end of the two expeditions the IARE had collected a bit of skull, three femur fragments, and a matching tibia fragment for one of the femurs that comprised a knee joint adapted for bipedalism. Those bits of fossils were promising enough to spark another season in 1974.

In his book *Lucy: The Beginnings of Humankind* (1981), Johanson recounts the events leading up to his introduction to Lucy in his third year at Hadar.

> I wasn't eager to go out with Gray that morning. I had a tremendous amount of work to catch up on . . . I *should* have stayed in camp that morning—but I didn't. I felt a strong subconscious urge to go with Tom, and I obeyed it. I wrote a note to myself in my daily diary: *Nov. 30, 1974. To Locality 162 with Gray in AM. Feel good.* (Johanson and Edey, 1981, p. 14)

Johanson describes the constant nagging dilemma of fieldwork. Fossils literally pile up with every day of collection and if they are not processed in step with collection (i.e., with detailed identification and labeling, and often sieving and screening of bags of sediment), then the guilt builds up along with them. The longer the lag between collection and processing, the greater the risk of introducing error. But it is difficult to resist the call from the paleontological playground every morning in favor of mundane lab work back at camp. Often paleontologists will wait for a rainy day to face the backlog of chores. The temptation to explore, survey, collect, and excavate every day is so great that sometimes the laundry makes the decision to stay back: Once paleontologists run out of it they can either get creative with their wardrobes or they can stay back to do the wash and tie up all the loose ends in the lab.

Because of a gut feeling, Johanson chose the field over the lab that extraordinary morning. Given that hominin fossils are some of the most difficult to find, he knew a good bit of luck would be needed to find anything at this site. Having already spent three years at Hadare, and having made a number of important discoveries he felt that his lucky streak would continue. On that fateful morning he had a feeling that listening to superstition would pay off. Some paleoanthropologists, even distinguished ones, spend their lives looking and never find a hominin fossil and Johanson was poised to avoid that curse.

Johanson describes how the morning he found Lucy seemed like any other morning, yet his lucky feeling had not produced a thing. He and Gray mapped some localities and surveyed some gullies where erosion would have exposed fossils only to find a few scraps of fossil horse, pig, antelope, and a part of a monkey jaw. That was the extent of their hard work on that lucky whim. Refusing to let go of the good vibes, Johanson explains that he decided to revisit a nearby gully that had been worked at least twice already by other people who had found nothing of interest.

The birth of an icon, as it turns out, was to be the result of one last detour, by one irrepressibly hopeful man, on the way back to camp after a rather dull morning. The gully that Johanson and Gray resurveyed had apparently no fossils in it. Disappointed, they turned to leave but just then something caught Johanson's eye. It was a bit of hominid arm. Gray looked at it and thought it was too small to be human and that it more likely belonged to a monkey. The two scientists debated for a moment, but their deadlock was broken when Johanson noticed a bit of skull lying next to the arm. Their interest suddenly exploded into excitement as they realized there were hominin fossils scattered all around their feet. They could not believe their good fortune. It was an unprecedented find. Despite the grueling 110-degree African heat they shouted for joy and danced around until they realized they might crush one of the fossils with their feet. Stepping back they had to resist the temptation to start scooping up the fossils. Carefully recording a fossil's location is almost as important as collecting the fossil itself. Just grabbing a fossil off the ground can damage it and not knowing where it was found makes it difficult to place into context. Their professionalism exerting itself, Johanson and Gray delicately removed a couple of pieces, carefully documented their position, marked the overall site, and then climbed back into their truck. As they approached camp Gray could no longer control his excitement and honked the truck's horn to get everyone's attention.

Donald Johanson, while certainly excited, was a bit more circumspect than Gray. He knew that just because their fossils looked like a hominin did not necessarily mean they had one. The bones would have to be studied in detail before any conclusion could be made. Their joy could quickly turn to disappointment if it turned out to be a monkey after all. Going public about it before they knew exactly what they had could also damage their reputations. Famous paleoanthropologists have been known to misidentify things like shards of fossil tortoise shells as hominin skull fragments. A major gaffe like that is no way to kick-start a fledgling career. It is safe to bet that the main source of Johanson's pragmatism was his inexperience. He had toured the collections of the United States, South Africa, and East Africa—both modern specimens and fossils. But he had not acquired a lifetime of knowledge like his senior colleagues. Without comparative material in the field, it was in his best interest to take small cautious steps before crying "hominin." But after the crew collected what turned out to be a partial skeleton, it was obvious that they had a creature that was human-like with some ape-like characteristics. They had a hominin.

The little skeleton was christened that first evening, just after the first bits of the skeleton were collected. The crew celebrated and gleefully anticipated what else they would find over the coming days of excavation. There was drinking, dancing, and singing. The Beatles' song "Lucy in the Sky with Diamonds" was playing over and over on the tape recorder at full volume as a form of celebration. Johanson would later recall that it was unclear just when and how it happened, but over the course of that night people on the team had begun referring to the fossil as "Lucy." It would later receive its official designation as A.L. 288-1 (Afar Locality #288-1).

In the midst of such a story, it is common for the players to imagine they are somehow blessed. Johanson describes in a later account in *Ancestors: In Search of Human Origins* (1994) how even some of the most scientifically minded people cannot help but feel like the stars have aligned perfectly or that the gods are smiling down on them when they discover exciting hominin fossils. Johanson never lets go of this lucky thread twenty years after he found Lucy. He claims that it was chance and good fortune which led him and Tom Gray to find Lucy.

The battered skeleton was an earth-shattering find and is still one of the very few skeletons of early hominins in existence. Before people intentionally buried the dead, scavengers and environmental processes usually scattered hominin skeletons making the discovery of complete ones a rarity. The chance that a bone survives carnivores and decomposition to become a fossil is very slim and then the chance that someone will actually find the fossil is even slimmer. Lucy's species was probably rare in the first place. Most human population sizes today are an extremely inappropriate model for humans in the past, let alone ape-like hominins. Only recently have humans become insect-like in their world colonization. Back in Lucy's day, population sizes were probably comparable to chimpanzees, maxing out at about 150 members per group. Lucy's kind would have covered very little of the earth. Perhaps Johanson was lucky after all.

I LEARN LUCY

Despite the starstruck reaction she received immediately upon her discovery, Lucy got quite a low-class treatment on her first trip across the Atlantic. Johanson wrapped her bones in toilet paper—a tactic used by paleoanthropologists to protect fossils during transport. He nestled the mummy into a carry-on suitcase and tucked her under his feet on the plane from Addis Ababa, Ethiopia, to Cleveland, Ohio.

Practically the moment the two stepped off the plane, Johanson and his colleagues—including locomotion and functional anatomy expert Owen Lovejoy of Case Western Reserve University—had Lucy in the lab at the Cleveland Museum of Natural History.

More than any hominin fossil before her, Lucy offered scientists great insights into human evolution. In fact, she offered so much that an entire volume of the American Journal of Physical Anthropology was devoted to her skeleton and to the rest of the Hadar fossils collected in the mid 1970s. Paleoanthropologists rarely have skeletons of individual hominins with associated bones. These specimens stand out as fossil decoder keys for the comparison, identification, and analysis of enigmatic, isolated, and fragmentary finds which are far more common.

Skulls and teeth are diagnostic for species, but isolated hominin limb bones are often left unidentified. However, with Lucy, many body parts are present and can be linked together to draw a more complete picture of the species. Associated body parts within a skeleton are also invaluable because of the general relationships that are known to exist between regions of the mammalian body. For instance, the relationship between brain size and body size tends to align with intelligence. Species that have larger brain-to-body-size ratios tend to have higher cognitive abilities. Lucy's estimated brain-to-body-size ratio is similar to that of a chimpanzee, which is very high compared to most mammals.

After Johanson spotted that first bit of right ulna (one of the long bones of the forearm), and then a bit of skull bone, some of a femur, some ribs, the pelvis, and the lower jaw, he and Gray had already collected an astonishing specimen. Then after two weeks of careful surface screening and excavation of the site, they had accumulated several hundred fragments of bone that represented Lucy's 40 percent complete skeleton. Johanson's estimate, however, neglects to include the missing tiny bones of the wrists, ankles, fingers, and toes which would reduce the level of completeness considerably. When those bones are taken into account, Lucy is more like 20 percent complete.

Regardless of the correct numerical percentage of preserved bones, Lucy was and still is a remarkably complete specimen. It was clear that she composed one single individual because not one of her several hundred bone fragments is duplicated. If just one of the bones or bone fragments had a twin, then Johanson and his colleagues would have had to assume there were at least two individuals represented in the fossil cache. Furthermore, the bones were all of the same developmental age and they were the same size. That is, the short length of the femur indicated that the body was small, which is consistent with the size of the teeth. Nothing about the sizes of the various bones contradicted those of other bones within the limits of a single individual.

It is difficult to determine or even to guess how Lucy died. There is no preserved trauma to her skeleton and there are myriad ways a hominin can die that leave no skeletal clues. The lack of postmortem carnivore tooth marks, made by either predators or scavengers, suggests she was not killed or eaten by such creatures. Unlike the chewed and crushed skeletons of many hominins and other fossilized animals, Lucy has only a single puncture wound on the top of her left pubic bone that occurred around the time of her death (as indicated by the style of fracture typical in fresh bone, and the lack of subsequent healing). If the wound to her pelvis occurred after her death it may or may not have anything to do with the cause of death.

Because Lucy's third molars, or "wisdom teeth," were erupted and even worn slightly, she can be designated a fully grown, but young, adult. Furthermore, all of her bones were fused at the growth plates, that is, all of the ends were fused onto the shafts, which is another sign of maturity. Lucy's cranial sutures were also closed indicating her skull was fully developed. Her geologic age is estimated to about 3.2 Mya, according to the volcanic ash in the sediments where she was found, which can be dated using the 40Ar/39Ar (Argon-Argon) dating technique.

Using the long bones and the size of the teeth for calculation (based on the known relationships in apes and in humans), Lucy is estimated to have stood about three and a half feet tall in life and weighed about sixty to sixty-five pounds.

A.L. Woman

Because of its diminutive size, Johanson's team assumed, right there in the field, that their fossil was female. The name Lucy went naturally with the conspicuously petite bones, but back in the laboratory A.L. 288-1 had to be sexed according to more rigorous scientific standards. The best skeletal indicators of sex are located in the skull and the pelvis. These are the regions forensic anthropologists use to determine whether an unidentified homicide victim is male or female. Unfortunately Lucy did not have much of her skull left, but she did have a nearly complete half of a pelvis which Owen Lovejoy mirror-imaged in order to reconstruct the entire thing. Unfortunately there was not a single pelvis on record at the time that was decidedly male or female with which to compare Lucy's. A separate team of researchers, outside of the Hadar group, argued that despite the small size, the shape of the pelvis made the birth canal too small to fit a newborn's head, so it must have been male. They offered a new nickname for A.L. 288-1: Lucifer. However, the motion to change Lucy's sex never earned support. Most of the other hominins collected from the Hadar sites are large and Lucy is actually one of the smallest, if not *the* smallest adult *A. afarensis* on record. Lucy's species is highly sexually dimorphic with males being nearly twice as large as females—a condition surpassing that of chimpanzees and approaching that of gorillas. Therefore, Lucy's size alone strongly supports her feminine status.

But does Lucy's sex really matter in the big picture? Beyond the pull of Lucy's scientific importance, Johanson thought her popularity had a lot to do with her femininity.

A piece of jaw or a fractured skullcap has no gender, and if we endow them with one, it will tend automatically to be male. In our society male is still the undifferentiated standard, the "generic brand." Lucy becomes all the more recognizable for being female. Whether or not one believes, as I do, that her species was a direct ancestor of humankind, her sex grants her the rudiments of a human identity. And I think it also taps a very deep well in our collective imaginations. In an elusive but powerful sense she represents the Mother, Gaea, Isis—or whatever history has called the fertility that lingers at the beginnings of our consciousness (Johanson and Shreeve, 1989, p. 30).

LUCY IN THE TREES WITH APES?

The mosaic of ape- and human-like attributes in Lucy's body provided more rock solid evidence against the notion that was once used to uphold the Piltdown hoax; the old idea that big brains evolved before bipedalism. Although Lucy's entire skull was not preserved, enough remained to show that her brain size was comparable to a chimpanzee's yet her body already had specializations for upright locomotion. Supporting previous hominin fossil discoveries, Lucy also confirmed Darwin's (1871) hypothesis that the earliest ape-like human ancestors would be found in Africa.

Despite bolstering two prevailing views of human evolution, Lucy opened up a whole new line of questions: What was the transition from quadruped to biped like? How arboreal was Lucy even though she could clearly walk upright? The argument that she started has yet to be resolved.

There are features that hominins share with one another that differ from chimpanzees and humans and also distinguish hominins from ancestors of chimpanzees. Human-like traits appear at different times and change at different rates through the course of hominin evolution. For example, brain size increases through time. Canine size decreases through time. Sexual dimorphism in body size decreases but overall body size increases. Aside from the changes in the skull associated with dramatic brain size increase, the most conspicuous traits that evolve in hominins are involved in the transition from quadrupedal walking and climbing in the trees to bipedal walking and running.

A common method for determining bipedality, particularly when there is no postcranial material (i.e., bones of the skeleton excluding the skull and teeth), is to look at how the vertebrae attach to the base of the skull. The foramen magnum (literally "big hole") where the spinal cord exits the skull is tucked underneath the head of a biped because the skull rests atop the body. On the contrary in quadrupeds, like chimpanzees, the foramen magnum is located further back on the skull because the head sits out in front of the body. Unfortunately for Lucy, with her fragmented and poorly preserved bits of skull, this trait could not be determined. Luckily other reliable indicators of bipedalism in the postcranial skeleton were preserved and could be analyzed.

The evolution of bipedalism required muscles and bones to adapt to the redistribution of forces from walking on all fours to balancing over only two legs. Lucy's pelvis is more like a human's than a chimpanzee's and has several adaptations for bipedality. Human pelves are basin-shaped. Our hips, or ilia, are on the sides of our bodies. Chimpanzee ilia are long and flat and located on their lower backs. Because our ilia are short, round, curved, and rotated around to the sides they are able to anchor muscles in positions that are crucial for bipedal walking. Once the center of gravity was shifted from in front of the pelvis to above it, the muscles of the hips and legs had to change to accommodate the shift and to aid in the change in locomotion. Furthermore, during bipedal walking, most of the striding gait takes place on only one leg at a time. Because of the need to balance (and to avoid falling over to the side that is unsupported) many of the muscles of the hips (e.g., the gluteal muscles) needed to move to the sides and also to enlarge in order to hold up the trunk by contracting during walking. When they attempt to walk bipedally, chimpanzees cannot stride effortlessly like humans; instead they waddle and struggle to stay balanced.

Changes in the hindlimb must take place when a quadrupedal creature becomes bipedal and Lucy has some of these adaptations in her legs. Lucy's distal femur (the end of the thighbone at the knee) shows that her femur was angled toward the body's imaginary midline running down between the legs. This adaptation allows for better balance while walking since much of the stride takes place over only one foot. By angling the femur inward over the tibia (the shinbone) the foot is placed directly under the center of gravity, requiring very little work to keep balance during gait. Chimpanzee legs, in contrast, articulate nearly completely straight up and down at the knee.

There is a danger to angling the femur down and inward, the forces can be so great that the patella (kneecap) may slide toward the side of the knee that is close to the midline of the body. To prevent this potentially debilitating injury from occurring, there is a prominent ridge or lip on Lucy's femur at the knee that blocks the patella from dislocating medially. Furthermore, Lucy's femoral condyles—which are the two bulbous areas of the femur that comprise its part of the knee joint—are large, which is an adaptation for experiencing larger forces as all the body weight is shifted to two legs instead of spread out over four like in our quadrupedal ancestors and cousins.

The feet also change to bear the added weight and to adapt to the mechanics of walking and running. Lucy's talus, the ankle bone that shares a joint with the tibia and also with the bones of the big toe, has anatomy that indicates that the big toe was in line with the rest of the toes. Chimpanzees have a divergent, thumb-like big toe, and Lucy had already compromised grasping abilities of the foot for efficient bipedal locomotion. There is also evidence in Lucy's trunk that she was bipedal. Her vertebrae fit together to form an S-curved backbone, which is the human strategy for holding the trunk over the hips and legs effectively during upright locomotion. Chimpanzees have a C-curved spine when they stand erect.

However, other anatomy of Lucy's trunk is ape-like. Her rib cage is shaped like an upside-down funnel as opposed to more of a barrel-shape like humans. Other ape-like anatomy includes her long curved fingers and toes and her long arms compared to her body and to the length of her legs. Her upper arm morphology has also been used to argue arboreal abilities.

The transitional anatomy and the abundance of preserved material over which to argue make *Australopithecus afarensis* prime for a heated debate over its locomotion. Both sides agree that Lucy's anatomy, and that of the other specimens of her species, contains strong evidence for bipedality. But the crux of the argument lies in the degree of bipedality. One side, led by Owen Lovejoy, argues that Lucy and her kind walked habitually on two legs just like we do today and that the arboreal anatomy evident in her skeleton is just evolutionary baggage. The other side, which includes Jack Stern and Randall Susman of SUNY-Stony Brook, attests that Lucy was still highly adapted for climbing and swinging about the trees.

ICONOFOSSIL

Supporting Lucy's iconic status is her iconoclastic nature. Of course, by definition she is "clastic" since her bones turned into rocks over millions of years of geologic and geochemical processes. But she is also an iconoclast because of her pivotal role in breaking long-held assumptions about human evolution.

Lucy was incredibly pertinent at the time of her discovery because she provided information about a time in human evolution that had long been the subject of speculation and hypotheses. In the early 1970s the hominin fossil record of the Plio-Pleistocene age was a jumble with little to pull the bits and pieces together into a recognizable pattern. The discovery of Lucy at that time became a catalyst for developing a more focused picture of the field.

When Lucy was discovered in 1974, there were very few fossils older than 2.5 million years old. After 2.5 Mya, the human genus *Homo* appears in the fossil record. Fossils belonging to *Homo* are more human- than ape-like. The genus that predates *Homo* is *Australopithecus* and it spans from about 4.2 to 1.0 Mya. (Here, *Australopithecus* includes the robust group that is sometimes put into a separate genus called *Paranthropus*.)

In the mid 1970s there were few *Australopithecus* fossils from East Africa aside from the famous robust *Australopithecus* skull found by Mary Leakey at Olduvai Gorge, OH 5, known as "Zinj" (short for the original genus it was given, *Zinjanthropus*) in 1959. It was obvious that OH 5, with its enormous molars, five times the size of modern human molars, and with its specialized features of the skull for large chewing muscles, most likely for chewing tough fibrous fruits and vegetation, was not an ancestor of humans, but rather an evolutionary side branch on the hominin tree. OH 5 was merely our vegetarian cousin.

Down in South Africa, *Australopithecus* had been present since 1925 when Raymond Dart created the genus for the Taung Child—the beautifully preserved partial cranium with face and teeth and a natural brain endocast of an ape-like hominin infant. The fossil baffled scientists for many years because it had a small, ape-like brain size but had human-like features in the teeth. Dart considered the Taung Child and its species *Australopithecus africanus*, to be the direct ancestor to humans. Soon thereafter, more *Australopithecus* fossils from the *A. africanus* species and also a more robust species *A. robustus* were discovered.

In 1936, a cranium nicknamed "Mrs. Ples"—for its original genus *Plesianthropus* (which means "almost human")—came out of Sterkfontein Cave and was the first adult *Australopithecus* to go in the record. These South African caves are not shelters that hominins occupied, but rather these are ancient underground limestone caves where bones accumulated probably from carnivores, like big cats, and other creatures, like birds of prey. By the mid 1950s there was enough material in South Africa for John Robinson to differentiate between two types of hominin lineages living in the African Plio-Pleistocene: gracile (*Australopithecus africanus*) and robust (*Australopithecus robustus*) (Broom and Robinson, 1952). It was hypothesized that the former was ancestral to *Homo* and the latter was an evolutionary side branch that went extinct. The species containing OH 5, *A. boisei*, would fall under the latter category.

When there are few fossils for comparison and when fossils at the root of an evolutionary lineage are discovered, it is difficult to assign them to a species. By the mid-1960s, when the Leakey expeditions had collected enough evidence from Olduvai Gorge, Louis Leakey, Philip Tobias, and John Napier created and described the species *Homo habilis* (literally "handy man") which included several specimens, a few of which were crania, that many people considered to be *Australopithecus* at the time.

The genus seemed to begin around 2 Mya at the time and today it is pushed back to about 2.33 Mya with an upper jaw from Hadar. Leakey and colleagues defined the genus *Homo* by the use of tools (which in archaeological terms means the bones are associated with tools) and by a minimum brain size requirement of 600 cubic centimeters. Of course, the earlier in a lineage one looks, the more primitive a hominin will be, so the earliest *Homo* are difficult to distinguish from *Australopithecus*, even today with many more specimens on record.

It was in this theoretical climate, with the earliest *Homo* finally being defined in East Africa and with a variety of *Australopithecus* in South Africa but only the robust form in East Africa, that Johanson and his colleagues began collecting hominins at Hadar. Even though they were collecting hominins older than 3 million years, they first considered them to belong to *Homo*. Their fossils did not resemble the large toothed *A. boisei* (OH 5's species), so the next reasonable expectation was *Homo*. When they started to find large and small hominins, they thought that the larger ones were *Homo* and the smaller ones were *Australopithecus*. The situation got even messier in 1975 when they collected what would become known as the "First Family": a collection of 216 fossils from a minimum of 13 individuals from the site A.L. 333/333w. The group—comprising males, females, and juveniles—probably died together in a catastrophic event. Researchers have since returned to the site and found even more remains. Now the estimate for minimum number of individuals is up to seventeen (nine adults, three adolescents, and five juveniles.

It was during a visit to the National Museums of Kenya in Nairobi that Johanson met his closest collaborator in the Hadar fossil analysis, Tim White, who was just finishing his doctorate at the University of Michigan. They began working in Cleveland in 1977 and struck up such a productive and stimulating partnership that Johanson even compared them to Watson and Crick, the famous pair of scientists that won the Nobel Prize for discovering the structure of the DNA molecule. The two worked diligently on making sense of all the puzzling Hadar fossils and putting them into context with the new *Australopithecus* and early *Homo* fossils coming out of Kenya and Tanzania at the time.

Resolution within *Australopithecus* came in 1978 when Johanson, Tim White, and Yves Coppens (a French paleoanthropologist on the Hadar expeditions) announced the new species *Australopithecus afarensis* which means "southern ape-man from the Afar." They lumped all of the Hadar and the fossils of an even older age that Mary Leakey had collected from Laetoli, Tanzania, into the species. Including the Tanzanian fossils in with Lucy meant that a hominin like Lucy left the trail of footprints preserved at Laetoli.

Johanson, White, and Coppens hypothesized that the large ones were the males and the small ones were the females of the species. It was therefore a sexually dimorphic species approaching gorillas in magnitude with the largest males being nearly twice as large as the females. Of course, as with any new species birth, *A. afarensis* caused a stir. Was there enough similarity between the Laetoli fossils at 3.6 Mya and the Hadar ones to lump them together? They are separated by a thousand miles, after all. Was there convincing evidence to distinguish them from *A. africanus* in South Africa? Was their inclusion of the Laetoli material necessary or was it a ploy to associate Hadar with the early dates at Mary Leakey's site? Why choose the type specimen from Tanzania but name the species after the Ethiopian region?

Johanson and colleagues were even so bold as to place *A. afarensis* at the root of the hominin family tree. Any time a paleoanthropologist claims that their fossils are ancestral to all others, they are met with opposition. White does not mince words about arguments that continue to linger over *A. afarensis*. He writes that attempts to separate the species and link only certain specimens or subgroups to later hominin taxa and to push the rest onto a side branch (or branches), "have proven to be contradictory to themselves. Most often they have been found to be based on fallacies in statistical manipulation, or errors of basic observation or logic" (White, 2002, p. 414).

The creation of *Australopithecus afarensis* and the assertion that it was an ancestral species to *Homo* was one of the most controversial hypotheses of human evolution in the last fifty years but it actually held. Even after political strife that prevented the American and French teams from working in Ethiopia for nearly a decade, *A. afarensis* remained solid, sitting front and center on the hominin family tree.

In the early 1990s, crews were able to resume work at sites in the Afar region and the work there is continuing to greatly augment the *A. afarensis* fossil record. Almost immediately after Ethiopia opened back up, a team comprising Bill Kimbel (currently a professor at Arizona State University with Johanson), Johanson, and Yoel Rak (an expert on the *Australopithecus* face) pieced together the first skull of *A. afarensis,* from Hadar (A.L. 444-2).

Today *A. afarensis* is a very well-known hominin species that is found at sites in Chad, Ethiopia, Kenya, and Tanzania. The fossil record for the taxon spans between 4 and 3 million years ago. With over 362 specimens from Hadar alone, much is known about the transitional ape- and human-like morphology of the species. Its skull, jaws, and long bones of the limbs are too ape-like to include it in our own genus *Homo*, but it does have primitive features shared by early *Homo* and even humans that indicate it was most likely ancestral to many if not all of the hominin species to subsequently evolve, at least in East Africa. The cheekteeth or chewing teeth (the molars and premolars) are large and very primitive compared to *Homo* and modern humans. However, the canines are smaller and much more human-like than those of apes.

BEYOND LUCY

Plenty of fossil specimens from other hominin species have been touted as the next Lucy or have threatened to replace Lucy as the world's most famous hominin. All of these are exciting in their own ways but none of them have usurped the title of most celebrated or most famous hominin.

The most complete skeleton on record of a *Homo erectus* is the "Nariokotome Boy" better known in many circles as the "Turkana Boy" (museum catalog number: KMN-WT 15000). Both nicknames refer to the geographic region where Kamoya Kimeu—a highly successful member of the "Hominid Gang" working out of the National Museums of Kenya—first discovered a bit of the boy's forehead on the surface at the site of Nariokotome on the west side of Lake Turkana, Kenya. The ensuing excavation that began in 1985 led to the recovery of a nearly complete skeleton. The incredible level of preservation enabled a large group of scientists, led by Alan Walker and Richard Leakey (who had also led the excavation), to analyze various aspects of the boy's paleobiology. He already stood 5'3" tall at an estimated age of only eight years and he would probably have reached well over six feet if he had lived into adulthood. That is a considerable shift toward the human body size from the prior small-bodied hominins of the Pliocene, like Lucy. The Nariokotome Boy also lacked any lingering arboreal adaptations and possessed all the adaptations for fully committed and efficient bipedalism of a modern human.

About a decade after the Nariokotome Boy was discovered, remains of some of the earliest known hominins were reported out of Aramis in the Middle Awash region of Ethiopia. Tim White, Gen Suwa, Berhane Asfaw, Johannes Haile-Selassie, and their colleagues placed the fossils into *Ardipithecus ramidus* which means "root ape" in both Latin and in the native language of the region. They date to about 4.4 Mya and the species now includes other specimens as old as 5.7 Mya. There is a fragmentary skeleton waiting to be fully published, but for now there are mostly bits of skull, teeth, jaws, and a few postcranial remains on record. So, although *Ardipithecus* is older than *A. afarensis*, there are no specimens that rival Lucy's completeness. Currently there is enough of a fossil record for *Ardipithecus ramidus*, *Australopithecus anamensis* (the most primitive *Australopithecus* species known from the southern end of Lake Turkana and from Ethiopia, dated to about 4.2 Mya), and *A. afarensis* to link them anatomically into an evolutionary lineage.

In the same decade that White's team first found *Ardipithecus* fossils in Ethiopia, Ron Clarke of the University of the Witwatersrand in Johannesburg, discovered an astonishingly complete skeleton in Sterkfontein cave in South Africa that tentatively dates to around 3 Mya, which is *A. africanus* times. The skeleton is known as "Littlefoot" (STW 573) because its beautifully preserved foot was discovered early, before the rest of the skeleton was dug out of the rock to reveal not only complete long bones but an incredibly preserved skull. It is perhaps the best-preserved *Australopithecus* skull on record. Clarke is still patiently removing the rock-hard matrix from the specimen with airscribes and chemicals. If he moves slowly enough he diminishes the risk of destroying fragile parts of the bones. It is clear based on what Clarke has thus far revealed that STW 573 will eventually become the most complete specimen of an adult *Australopithecus* on record. Once it is completely cleaned it will also be easier to compare it against *A. africanus* and *A. afarensis* fossils, since some scientists have already suggested, based on the skull's affinities to those from Hadar, that it could be the first evidence for *A. afarensis* to have lived all the way down in South Africa.

Brigitte Senut, Martin Pickford, and their team from the Community Museums of Kenya (a completely separate entity from the National Museums of Kenya) announced the so-called Millennium Man from 6 Mya in 2001. These have the earliest postcranial fossils of hominins and they include a couple fragments of leg bones and a few teeth of the species *Orrorin tugenensis* from the Tugen Hills of the Lake Baringo region of central Kenya. The bipedal status of *Orrorin* is currently under intense debate because the scrappy fossils do not offer clear-cut, convincing evidence of upright locomotion. The other controversy has to do with the new evolutionary picture Senut and Pickford paint based on their new species. They placed *Orrorin* as the ancestor to *Homo* and relegated fossils like Lucy to a dead-end side branch removing them from the lineage leading to humans. They based their new view of human evolution on their claimed advanced state of bipedalism in *Orrorin* that they argue was even more adapted to upright walking than Lucy was, nearly 3 million years later in time.

Perhaps even earlier in the hominin lineage than *Orrorin* is the nearly complete fossil cranium from Chad. This sole representative of *Sahelanthropus tchadensis* was discovered by French paleoanthropologist Michel Brunet and announced just after *Orrorin* was (Brunet et al., 2002). The *Sahelanthropus* cranium came from the dry Lake Chad Basin near Toros-Menalla. The species name means "Sahel-ape from Chad" but the fossil was nicknamed "Toumai" meaning "hope of life" in the local language. Although *Sahelanthropus* may be of equivalent age to that of *Orrorin* at about 6 Mya, it is also possible that the Chad hominin is as old as 7 Mya. There are no volcanic rocks for dating the site with absolute methods, so relative methods must be used but they have not so far been able to pin down anything more refined than between 7 and 6 Mya. Another point of contention is centered on the hominin status of the skull. Because there is no postcranial skeleton, the anatomy of *Sahelanthropus* does not show direct evidence of bipedalism. However, its discoverers emphasize its shared dental and cranial characteristics with later hominins. But because there is no skeleton to go with the skull, it is not yet known if this species was bipedal.

Although it is still unclear as to which of the three taxa, *Ardipithecus*, *Orrorin*, and *Sahelanthropus*, are hominins and which are not (and in that case, whether or not they are chimpanzee ancestors), there is one thing that all of their discoverers agree upon: The sites are reconstructed to have been woodlands and forests, not savannahs. This means that the early hypothesis that the first human ancestors emerged when they moved out of the trees and onto the drying East African savannah is unsupported. The earliest hominins were still, in fact, living among the trees, presumably like chimpanzee ancestors.

On the other end of the hominin spectrum, the recent end, there is a healthy contender for Lucy's limelight: *Homo floresiensis* or the so-called hobbits. Judging by the overwhelming attention *Homo floresiensis* has garnered at professional meetings and by the flurry of publications already stemming from the discovery, *Homo floresiensis* has been the greatest paleoanthropological find of the twenty-first century thus far.

In 2003, a team of Australian and Indonesian researchers led by Peter Brown of University of New England in New South Wales, Australia, dug down into the floor of a cave called Liang Bua on the island of Flores. They uncovered a skull with an associated skeleton (LB 1) and also the skeletal remains of at least eight other small individuals. These diminutive hominins stood about three feet tall and had very small ape-sized brains so they were given the new species name, *Homo floresiensis*. Their physical description may sound initially like Lucy's, but the teeth and skull resemble early *Homo* and complex stone tools were associated with them in the cave along with evidence that they hunted pygmy elephants. Also, the site dates to between 18,000 and 13,000 years ago, which is far later than any *Australopithecus,* the last of which died out by 1 Mya. Intriguingly, the morphology of the postcranial skeleton is just now being analyzed and parts like the wrist have affinities to *Australopithecus afarensis*.

The hobbits are currently the number one fascination in paleoanthropology and, of course, there is a battle over how to interpret them. One side argues that the small skull is pathological, not indicative of the evolution of a new species. After all, there are no fossil hominins with brain sizes that small past about 1 Mya. The proponents of pathology posit that LB 1 had a condition like microcephaly where the brain does not develop properly and remains dangerously small into adulthood.

Their opposition points out that microcephalics often have a misshapen cranium and face and that LB 1 does not. At present, *Homo floresiensis* is hypothesized by those against the microcephaly diagnosis (which is the current majority of the field) to have been a descendent of Indonesian *Homo erectus* that got isolated on Flores. Then after years of island living, the hominins succumbed to the ecological pressures by dwarfing in body size like other large mammals on islands have been documented to do (e.g., elephants and hippos). These incredible finds since Lucy's discovery are just a few examples of those that have threatened to steal her glory but have not come close to achieving her iconic status.

Getting a Handle on Lucy

Is Lucy's fame due in part to her nickname? Compare and contrast calling Johanson's fossil *Lucy* as opposed to A.L. 288-1 or even the species name "*Australopithecus afarensis.*" The Latin taxon is difficult to pronounce and the specimen number sounds more like a robot than a humanoid. Even less catchy is Lucy's nickname in the local Amharic language, which is "Denkenesh" (Johanson, 2004). Perhaps Lucy's fame was aided by her name's familiarity. In the minds of many Americans, Lucy is a name for plucky, headstrong, and lovable characters like Lucille Ball in the 1950s television program *I Love Lucy* and Lucy van Pelt in Charles Schultz's cartoon *Peanuts.* Assigning nicknames to fossils has long been a habit in paleoanthropology which is probably due to both the human tendency to anthropomorphize beloved things like pets, boats, and cars, and also to the strong duty we feel to identify the dead.

Robert Broom and John Robinson gave us "Mrs. Ples" in 1947. The fossilized but fully modern human skeleton from La Chapelle-aux-Saints, France, is endearingly called the "Old Man" since there are few specimens of prehistoric hominins that lived as long as he did. One beautiful *Homo erectus* cranium (D 2700) from the site of Dmanisi in the Republic of Georgia was dubbed "Poor Marc" in droll honor of paleoanthropologist Marc Meyer. During the entire summer field season of 2002, he unwittingly dug down toward the skull. Then, the very next day after he returned to graduate school at the University of Pennsylvania, the excavator who took over for him removed a small layer of dirt to find the prize that would eventually grace the cover of *National Geographic.*

Louis and Mary Leakey were fond of nicknaming their fossils with monikers like "Nutcracker Man," "Dear Boy," and "Zinj" for OH 5, and then "George" (OH 16), "Cinderella" or "Cindy" (OH 13), and "Twiggy" (OH 24) for the *Homo habilis* specimens they also collected at Olduvai Gorge. Richard Leakey and Alan Walker chose not to continue this tradition and to stick to museum numbers as their teams discovered fossils from Koobi Fora and elsewhere in Kenya in the 1970s and 1980s. However, nicknames still found their way onto some of the fossils they discovered. For instance, the skull numbered KNM-WT 17000 is much better known as the "Black Skull" for its dark color.

LUCY'S BABY

Lucy's notoriety is due to many things and the rarity of nearly complete skeletons plays a major role. But even rarer than adult hominin skeletons are those of infants. Their small bones are fragile, which means that if by chance they make it to the fossilization stage they are hardly ever found. Special preservation circumstances in the sandstone at a site called Dikika in Ethiopia afforded a joyous exception: a nearly complete skeleton of an infant *A. afarensis*.

Dikika is just across the Awash River from Hadar. There, just ten kilometers from where Lucy was found over thirty years ago, native Ethiopian Zeresenay "Zeray" Alemseged of the Max Planck Institute for Evolutionary Anthropology in Leipzig, Germany, and his team (The Dikika Research Project) found the spectacular fossil in late 2000. Upon its announcement, the press touted the newborn specimen as "Lucy's Child" and "Lucy's Baby" (DIK–1–1). It is the most complete specimen of a very young hominin ever discovered, and the youngest to boot. Parents always want better for their children and that is precisely what Lucy got. Don Johanson said that, "If Lucy was the greatest fossil discovery of the twentieth century, then this baby is the greatest find of the twenty-first thus far" (Wong, 2006).

Geologists have reconstructed the ancient burial conditions and determined that the baby's corpse was covered quickly with sediment from low-energy (i.e., not turbulent) flooding conditions in a small river channel near where a river emptied into a lake, probably a delta environment frequently visited by elephants, rhinoceroses, and hippopotamuses. Argon-argon dating of the volcanic ash contained in the fossil-bearing sediments puts the baby at 3.3 Mya which is within the known time span of *A. afarensis*.

Certainly the Taung Child reported first by Raymond Dart in 1925 came before the Dikika baby with its beautifully preserved facial skeleton and even a natural cast of the brain. But, the Dikika baby offers much more to paleoanthropology than Taung could. After years of patient preparation using dental picks, Alemseged extracted much more than he imagined to be encased in that sandstone rock. Only the skull and a few of the bones were clearly present upon discovery, but once the rock was removed, much more of the infant was revealed.

The braincase of the Dikika baby was preserved intact so the cranial capacity can be calculated at about 330 cubic centimeters which is roughly equivalent in size to a three-year-old chimpanzee's. Remarkably, even the fragile bones of the face are preserved, and it is the face that links this baby to adult *A. afarensis* with its morphology that is both ape- and human-like. The postcranial skeleton shows off the scapulas (shoulder blades), clavicles (collar bones), ribs, vertebrae, humerus (upper arm), fingers, patellae (knee caps), femurs (thigh bones), tibias (shin bones) from both legs, and an almost-complete foot. The hyoid, the small horseshoe-shaped bone in the neck, was even preserved. In total over 50 percent of the skeleton is present. Following the trend in *A. afarensis* and especially visible in Lucy, the lower half of the body is human-like, but the upper parts are ape-like. The well-preserved upper body should shed bright light on the debate about how arboreal *A. afarensis* was.

It is clear from computed tomography or CT ("cat") scans of the face that the Dikika baby is at a similar stage in tooth development and eruption to that of a three-year-old chimpanzee. It is highly unlikely that *A. afarensis* grew at exactly the same rate as chimpanzees, but for now that age will remain the best estimate until further growth studies can be done. Those studies are probably underway already, thanks to this beautiful toothy specimen. Regardless of the pace at which the baby grew, it died with the teeth of an infant that is still of nursing age, which is ironic considering Dikika, which was named for a curiously shaped hill, means "nipple" in the local Afar language.

EVEN BETTER THAN A REPLICA

All rock stars and international celebrities embark on world tours. Lucy is set to take on her second trip "across the pond" in 2007. Along with about 200 other Ethiopian cultural artifacts such as prehistoric stone tools, Lucy's first stop on her six-year tour of America will be in September 2007 at the Houston Museum of Natural Science in Texas, with other appearances thus far scheduled in Washington, New York, Denver, and Chicago. Several other U.S. cities are said to still be in negotiations with Ethiopian officials and have not yet been named.

Lucy visited the United States once before. Just after her discovery in 1975, Don Johanson took Lucy to the Cleveland Museum of Natural History for five years and returned her to Ethiopia in 1980. Since then, Lucy has rested in a vault, specially created for her, in the Paleoanthropology Laboratories of the National Museum of Ethiopia in Addis Ababa. Many museums, universities, and other such institutions have high-quality replicas or casts of Lucy's bones made from molds taken of the original fossils. These casts can be used for scientific research, study, and teaching and for museum exhibits since they are nearly identical in size and shape to the specimen and are often artistically rendered to mimic every bit of color detail on the original fossils.

Instead of throngs of screaming fans greeting Lucy upon her arrival, there are and will continue to be the public outcries of scientists protesting her trip away from her Ethiopian home. Lucy is now at the center of what will certainly continue to be a long, drawn-out fight over the transport and display of original hominin fossils outside of their secure vaults in their native countries.

Taking fossils home for study and scrutiny was standard procedure until relatively recently. Johanson was one of the last to utilize the privilege. In 1998 three dozen scientists from twenty-three countries signed an agreement, in affiliation with UNESCO (United Nations Educational, Scientific, and Cultural Organization), which stipulated that fossils would remain in their country of origin and would not be transported from their home without overwhelming scientific reasons. It was expected that foreign scholars would travel from all over the world to work on the specimens in their native countries and by doing so they would generate local interest and encourage and facilitate homegrown science.

But Ethiopian officials who negotiated the million-dollar deals with the American Institutions insist that Ethiopian museums will benefit from the profits generated from visitors to Lucy's exhibit. Since King Tut put a bright and lucrative spotlight on Egypt, Ethiopia is hoping Lucy will do the same. It is a bold tactic used by a country that wishes to be known in the minds and hearts of the world as the cradle of humankind, above Kenya and above South Africa. Lucy already brought Ethiopia to the forefront of the hominin evolutionary map in the eyes of the professionals. There is an enormous fossil record for human evolution and the evolution of countless other organisms in Ethiopia thanks to the herds of paleontologists who have flocked there and who have come from there. Ethiopians are proud of their prehistoric heritage, have named cafés after Lucy and have featured her on postage stamps. Lucy's tour provides a chance for the rest of the world to take notice of their pride and joy.

The top two natural history museums in the United States—which also have highly-respected research groups dedicated to the study of human origins and evolution—are choosing not to participate. The National Museum of Natural History at the Smithsonian in Washington, D.C., led by director of the Smithsonian's Human Origins Program Rick Potts, is standing out as one of institutions that are boycotting the Lucy exhibit. Likewise, the American Museum of Natural History in New York is refusing to show Lucy's real bones.

The scientists that oppose Lucy's tour advocate that hominin fossils should not leave their vaults unless there is compelling scientific reason to do so. High-quality casts are sufficient for most types of study. But scientists like John Kappelman at the University of Texas, Austin, are planning to scan Lucy with high-resolution computed tomography (microCT or "cat" scanning) which is a technology that is not yet available in Ethiopia. Kappelman also emphasizes Lucy's wow-factor in an America where evolution is becoming a dirty word. "Millions of Americans will look at this original material. That could change the way many people think about human evolution" (Gibbons, 2006c, p. 575). While Lucy is busy wowing the public, however, she will be unavailable for scientific research for most of the time between now and 2013 when she is scheduled to fly home. Lucy's American home-away-from-home will not even welcome her back. Head of the Cleveland Museum of Natural History, Bruce Latimer, who did pioneering work on the foot bones of Lucy, said that, "There is only one Lucy. If something should happen to her, she's irreplaceable" (Gibbons, 2006c, p. 574). So far there have been no reported reactions from Don Johanson.

LUCY'S BETTER HALF ON HER BEHALF

Based mainly on small brain size and lack of distinguishing bony links to speech and language, the paleoanthropological community assumes that Lucy could not speak and did not have language like we do. So even if she had not been dead for 3.2 million years, she would have needed someone to speak for her upon her discovery and, like any international superstar, she had a superstar promoter. Lucy's bones spoke volumes on their own, but the public relations bonanza spurred by Don Johanson helped her achieve an astonishing level of notoriety.

Johanson is a storyteller. Detailed accounts of the day-to-day dialogue and adventure for many of his fieldtrips make their way into his books. He is one of the few paleoanthropologists working today that succeeds in both the scientific and popular realms. Just after finding Lucy and writing a book about her, Johanson opened the Institute of Human Origins (IHO), in 1981. First based at the University of California, Berkeley, then moved to Arizona State University, the IHO is a think tank for all things of human origin and evolution. As the Virginia M. Ullman Chair in Human Origins, Johanson is both a full professor in the School of Human Evolution and Social Change at Arizona State University in Tempe, Arizona, and also still the director of the IHO. Early on in his career the increasingly burdensome demands of fund raising, lecturing, hiring staff, and administrating encroached on the time he could spend advancing the very science that he was publicizing.

The popularization of science, like Johanson is very good at doing, is often frowned upon by dedicated scientists even though he did a great service to anthropology by relating human origins and evolution to the masses. The fame and celebrity status he achieved may be unfairly seen as his motivation. Perhaps he simply realized that he served well as a coach or commentator rather than as a starting pitcher. In his writings, Johanson acknowledges the critics and admits to struggling with the different sides of his career.

> But the celebrity Lucy brought me had a bitter side as well. Along the way, I would also be called a prima donna, a slick operator, a publicity hound. I lost friends, including some of the closest colleagues in the field, whose interpretations of humanity's origins were thrown into serious doubt by Lucy and her Hadar companions. (Johanson and Shreeve, 1989, p. 23)

It seems, however, that Johanson found a comfortable niche and thrived in the public eye and the face of human evolutionary science. After finding a partial skeleton of a missing link few graduate students would have the wherewithal to write a press release right there in the field like Johanson did. Other details, like his interaction with an airport customs official, reveal Johanson's natural inner publicist. He describes his stop-over in Paris on the way home to Cleveland from Addis with Lucy in his carry-on suitcase. A customs official wanted to know what he was carrying in his bags. Johanson told the man that they were fossils from Ethiopia. To his great surprise the official dropped his suspicions and smilingly waved him on. The official already knew about Lucy from local newspaper accounts.

The average paleontologist probably would have felt disarmed to learn the world was aware of their fossils, and even knew them by name, before they were even analyzed in any great detail. The incident with the Ethiopian customs official was the beginning of Johanson's realization that even to the general public Lucy was more than just another fossil. Not only was Lucy now the most famous hominid fossil since the discovery of the Neanderthal Man, but Johanson himself was now famous. He had been launched from being an obscure junior paleoanthropologist to being a superstar. His rise to fame would be quick indeed.

Hand in hand with Lucy, Johanson's fame rose to iconic status. He and members of the Leakey family are perhaps the only paleoanthropologists that laypeople can readily identify by name or face. Yet, there are numerous paleoanthropologists that have been part of expeditions that have discovered more hominin fossils, that have been involved in just as tumultuous controversies, and that have made more expansive and/or significant contributions to the science of human evolution than Johanson.

Johanson's efforts to familiarize the public with human evolution have earned him more notoriety than most. He has co-authored six books—one of which, the best-selling *Lucy: The Beginnings of Humankind* that he co-authored with Maitland Edey, won the 1981 American Book Award in Science. Johanson writes about fossils like they are made of gold or like they are precious jewels in a pirate's treasure chest, describing dusty, old bones of dead ape-people with adjectives like "glittering" and "fabulous." His sparkly writing style translates well onto the television screen where he hosted and narrated a three-part PBS/NOVA series entitled *In Search of Human Origins*. He has remained on top of the latest technology as well. His current, state-of-the-art Web site is the premier Web site in the world on human evolution: www.becominghuman.org. Clearly, without Johanson's hard work and efforts, let alone his so-called "lucky" discovery in the first place, Lucy would not be an icon.

On the Origin of Controversies

Paleoanthropologists are generally well-behaved hominins just like everyone else. Yet the occasional public feuds—the accusations of metaphorical backstabbing and literal site and fossil stealing—earn the big headlines. It is common for the people involved in such rumpus to claim that the field as a whole is highly contentious or that it is fraught with infighting. Usually such a stereotype is blamed, in part, on the personality types that are drawn to paleoanthropology where people are passionate, ambitious, sharp, independent, and adventurous. These are usually the types of characteristics that lead to success out in the field when searching for fossils, but they also foster rivalry between scientists. Competition is also inherent to the basic logistics of fieldwork. Governments normally only dole out one research permit per region of land and funding agencies can only support a small number of research proposals.

The bitter feuds that the media loves to track have a tendency to mislead the public's perception of paleoanthropology. Stock claims that the latest fossil finds are "highly controversial" indicate that the fossils themselves are under some sort of fraud investigation, when paleoanthropologists can usually agree that fantastic front page fossils are indeed genuine (Piltdown excluded). The true controversies are over whether or not the fossils are humans, or direct human ancestors, or the earliest hominins, or whatever role the discoverers claim they played on the evolutionary stage. The controversies are also over whether or not other scientists are given the opportunity to study the fossils. Fairly objective versions of the famous feuds in paleoanthropology, including those fueled by events around Lucy's discovery, are recounted in *Missing Links* by John Reader (1981), *Bones of Contention* by Roger Lewin (1997), and *The First Human* by Ann Gibbons (2006b).

It is not clear which side to believe in any given contest, nor is it wise to pass judgments as an outsider on such incidents. However, it is worth noting that today many of the people originally involved in field or laboratory research on the Hadar fossils, including Don Johanson, do not (and will not under most circumstances) speak to one another. In fact, geologist and paleontologist Jon Kalb (now of the University of Texas, Austin) who was drummed out of fieldwork in Ethiopia wrote a memoir,

Adventures in the Bone Trade (2001), describing how he believes that because he was accused of being in the CIA, he was unable to obtain NSF funding for Ethiopian fieldwork he began in the 1970s.

A DAY IN THE LIFE OF LUCY

The day that Johanson discovered Lucy was an extraordinary day but the day Lucy died, or any day in her life for that matter, was probably rather ordinary. Modeled after chimpanzees, a typical day in Lucy's life would have involved caring for her offspring and feeding. Tool use has been documented in a variety of mammals and birds but the modification of natural materials for use as tools is unique to primates. Like modern chimpanzees it is possible that Lucy stripped and shaped tree branches and twigs for use as termite fishing wands or even to make spears for jabbing at and eventually eating small animals. Lucy would have taught these skills to her offspring. She also could have used the hammer-and-anvil technique to crack open foods like hard nuts. Chimpanzees in the Taï Forest in the Ivory Coast have been observed to practice nut-cracking behavior and the process leaves behind telltale evidence on the rocks. Presently, archaeologists are honing their ability to identify evidence for nut-cracking behavior in ancient chimpanzee populations and it may soon be possible to detect nut-cracking at Plio-Pleistocene hominin sites under the right preservation conditions.

Although they probably used stone tools, Lucy's kind left no traces of stone tool manufacture, or the modification of rocks by chipping and flaking them into specific types of implements. Regular occurrence of hominin stone tool production and use does not appear until about 2.5 Mya with later *Australopithecus* or early *Homo*.

Without associated evidence of a stone tool culture and butchered and bashed animal bones, it is assumed that meat and bone marrow were not crucial components of the *A. afarensis* diet. But they are not ruled out completely. Male chimpanzees are known to obtain meat without the use of stone tools and most documented cases of hunting involve killing colobus monkeys. Lucy ate a varied diet comprised of a combination of termites, fruits, nuts, grasses, leaves, and underground "storage organs" like roots, corms, and tubers.

As a young adult female Lucy probably had at least one baby which she probably left behind when she died. She would have spent most days in a small unit alone with her offspring, both male and female. She may not have seen a single adult male all day. If she did come across a male, he would most likely be in a group with other males that her male offspring would probably join one day.

A. afarensis infants were more of a burden than infant chimpanzees because they could not grasp onto their mothers with their bipedally adapted feet. Lucy would probably have had to carry her babies as she foraged since we assume that *A. afarensis* was not sophisticated enough to construct baby slings. However, *A. afarensis* did retain long arms which may have helped the babies hold on better to their mothers, like chimpanzee infants do with theirs. If Lucy was carrying her babies as she foraged, her food-gathering ability would have been compromised. Thus, greater assistance from a reliable male (i.e., the father or a male who may think he is the father) would benefit both her and her offspring. With the ever-increasing fossil record for *A. afarensis* and a greater understanding of its behavior and life history, Lucy will continue to be in the spotlight because she has the potential to represent a revolution in the evolution of human motherhood.

Despite all of the Earth Mother connotations surrounding Lucy, she should not be considered a woman. She was a small-bodied, small-brained bipedal ape. She may or may not have been covered in hair, depending upon when the thermoregulatory adaptation of body hair reduction (linked to sweating) evolved. There is no evidence to suggest that she could speak like us, so we assume that she probably vocalized like chimpanzees. Lucy could walk like us, but she was probably much more acrobatic in the trees than us. It then follows that she would have nested overnight in the trees as well. If we could meet Lucy today, we would probably consider her a delightfully gifted ape.

To outsiders it must seem like paleoanthropology stopped the day Lucy appeared, but it has thrived with every passing day at an increasingly exciting pace. The hominin fossil record will continue to expand in the coming years. Once found, the elusive remains of early chimpanzee ancestors will revolutionize our currently nebulous understanding of how the human and chimpanzee lineages diverged. But despite all of the important fossils discovered since 1974 and despite what we anticipate will be found in the future, Lucy will always be there. She will always and forevermore be the ultimate ancestor to all icons of human evolution to come.

FURTHER READING

Gibbons, Ann. *The First Humans* (New York: Doubleday, 2006). This is an entertaining and insightful collection of tales about the claims for the discovery of the earliest human ancestor over the last fifteen years. Gibbons is a premier science writer and the reporter on paleoanthropology for *Science* magazine.

Hartwig, Walter C. (ed.). *The Primate Fossil Record* (Cambridge: Cambridge University Press, 2002). Tim White's chapter, "Earliest Hominids" (pp. 407–417) includes a brief history of discovery and debate of *Australopithecus* and *Ardipithecus* with discussions of even earlier hominins just beginning to be announced and analyzed at the time. This is a technical book compared to the popular works listed above and below, but it is a good entry point for anyone wishing to get into the nitty-gritty of the primate fossil record.

Johanson, Donald, and Edey, Maitland. *Lucy: The Beginnings of Humankind* (New York: Simon & Schuster, 1981). Described on the cover as account of "how our oldest human ancestor was discovered—and who she was," this book is a firsthand account of Lucy's discovery in 1974 and the initial analyses. Although Lucy and her kind (*Australopithecus afarensis*) are no longer the earliest hominins in the fossil record, the history of paleoanthropology and the story of how Lucy was discovered, disputed, and analyzed still resonates today.

Lewin, Roger. *Bones of Contention, 2nd Edition* (Chicago: University of Chicago Press, 1997). The major controversies in paleoanthropology, including those of historical significance and those that persist today, are explained in this highly entertaining read.

Sloan, Christopher P. "Meet the Dikika Baby," *National Geographic* (November 2006): 148–159. The Dikika baby made the cover of this issue and the adorable fleshed-out reconstruction is a must-see.

"Out of Africa"

Christopher Stringer

In *The Descent of Man*, first published in 1871, Charles Darwin suggested that Africa was the most likely evolutionary homeland for humans because it was the continent where our closest relatives, the African apes, could be found today. However, it was to be another fifty years before the fossil evidence that was ultimately to prove him right began to be discovered. Before then Europe, with the Neandertals, "Heidelberg Man," and the spurious "Piltdown Man," and Asia, with "Java Man," were the foci of scientific attention concerning human ancestry. But the 1921 discovery of the Broken Hill cranium in Northern Rhodesia (present-day Zambia) and the 1924 discovery of the Taung skull (from South Africa) started the process of giving Africa its rightful paramount importance in the story of human evolution. By the 1970s a succession of fossils had established that Africa was not only the place of origin for the human line (i.e., the continent in which the last common ancestor of humans and chimpanzees lived) but was probably also where the genus *Homo* had originated. But where did our species *Homo sapiens* (modern humans) originate? This was still unclear in the 1970s and remained so until quite recently.

THE EVOLUTION OF *HOMO SAPIENS*

Discussing the origin of our species involves an understanding of the evolution of the special characters which living humans share, e.g., a more lightly built skeleton compared to our more ancient relatives, a higher and rounder braincase, smaller brow ridges, and a prominent chin. But we also need to examine the evolution of the characters that

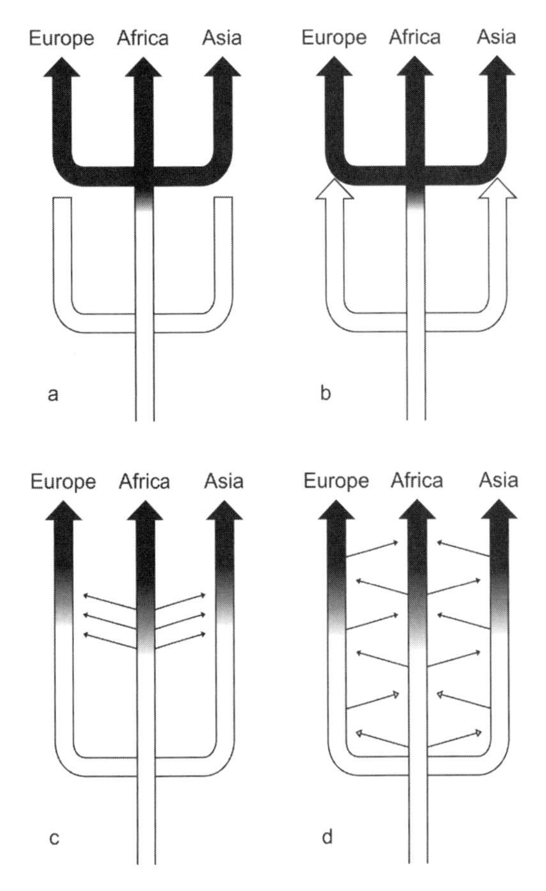

Four models for the evolution of Homo sapiens (from Stringer 2002): (a) Recent African Origin (Out of Africa); (b) (African) Hybridization and Replacement Model (see, e.g., Bräuer 1992); (c) Assimilation Model (see, e.g., Smith 1992); (d) Multiregional Evolution (see, e.g., Wolpoff & Caspari 1997).

distinguish different geographic populations today, the regional or "racial" characteristics, such as the more projecting nose of many Europeans, or the flatter face of most Orientals. In the last twenty years there have been two diametrically opposed views on how our species *Homo sapiens* evolved from its assumed early Pleistocene ancestor, *Homo erectus*, with many intermediate views between these extremes. The two extreme views differ quite radically over where and when we developed our special "modern" features, and when we began to evolve our regional differences.

MULTIREGIONAL EVOLUTION

Supporters of one extreme view, the Multiregional Model, say that *Homo erectus* gave rise to *Homo sapiens* across its whole range which, about 1 million years ago, included Africa, China, Indonesia, and, perhaps, Europe. According to this view, when *Homo erectus* dispersed around the Old World over a million years ago, it gradually began to develop both the modern features, and the differences that lie at the root of modern regional ("racial") variation. Particular features in a given region developed early on, and persist in the local descendant populations of today. For example, 500,000-year-old Chinese *Homo erectus* specimens had the same flat faces, with prominent cheekbones, as modern Oriental populations. Indonesian *Homo erectus* of 700,000 years ago had robustly built cheekbones and faces that jutted out from the braincase, characteristics found in modern native Australians. In Europe, another line of evolution gave rise to the Neandertals, who were the ancestors of modern Europeans. Features of continuity in this European lineage include prominent noses and midfaces. Recent supporters of multiregional evolution emphasize the importance of gene flow (interbreeding) between the regional lines, which prevented them from diverging and speciating, and allowed new traits to spread from one population to another across the inhabited world. In fact, they regard the continuity in time and space between the various forms of *Homo erectus* and their regional descendants to be so complete that they should all be regarded as representing only one species, *Homo sapiens*. Thus under the Multiregional Model, there is no "place of origin" of modern humans, since the species evolved from a network of ancestors across the inhabited world.

THE OUT OF AFRICA MODEL

The opposing view is that *Homo sapiens* had a restricted origin in time and space, with recent proponents focusing on Africa as the most important region. Some of them argue that the later stages of human evolution, like the earlier ones, were characterized by evolutionary splits, and the coexistence of separate species. They often recognize an intermediate species between *Homo erectus* and *Homo sapiens*, called Homo *heidelbergensis*. On this view, by about 600,000 years ago some *erectus* populations in Africa and Eurasia had changed sufficiently in skull form to be recognized as a new species, *Homo heidelbergensis*, named after a 500,000-year-old jawbone found at Mauer, near Heidelberg, Germany. Members of this species had a less projecting face, more prominent nose, and a more expanded braincase than *erectus* fossils. *Homo heidelbergensis* is known from Africa, Europe, and possibly Asia between about 600,000 and 300,000 years ago. From the "Out of Africa" viewpoint, from about 500,000 years ago, *heidelbergensis* began to give rise to two descendent species, *Homo sapiens* and *Homo neanderthalensis*, the former evolving in Africa, and the latter in western Eurasia. About 100,000 years ago, the African stock of early modern humans spread from the continent into adjoining regions and eventually reached Australia, Europe, and the Americas (probably by 50,000, 40,000, and 15,000 years ago respectively). Regional (racial) variation only developed during and after the dispersal, so that there was no continuity of regional features between *Homo erectus* and present inhabitants in the same regions. Like the Multiregional Model, this view accepts that *Homo erectus* evolved into new forms of human in inhabited regions outside Africa, but argues that the non-African lines became extinct without evolving into modern humans. Some, such as the Neandertals, were apparently replaced by the spread of modern humans into their regions. For this reason the Out of Africa Model is sometimes also known as the Replacement Model.

INTERPRETING THE FOSSIL DATA

As discussed above, the two extreme models of modern human origins provide quite different expectations from analysis of the fossil data. The Multiregional Model posits evolutionary continuity across the inhabited world, such that each inhabited area would show a relatively complete and potentially gradual sequence of anatomical changes from non-modern to modern humans. There would also be continuity of regional traits—"racial" features—that would evolve *alongside* the modern ones over a long period of time. In contrast, the "Out of Africa" model would be supported from the fossil record if there was only *one* place—Africa—which showed a complete evolutionary sequence from non-modern to modern humans. Outside of Africa, there would be no continuity, either in the transition to *Homo sapiens*, or the presence of regional/"racial" characteristics. These latter traits would only have evolved after the spread of modern humans began about 100,000 years ago, and especially in the last 50,000 years, as many populations reached their ultimate homelands in places such as Australia, Europe, and the Americas, while others remained behind in Africa.

Of course there are various theoretical models lying between these two extremes which allow *Homo sapiens* to originate in one region, such as Africa, and then spread out while interbreeding to a greater or lesser extent with existing populations elsewhere. For Bräuer (1992) there was probably only a brief phase of such intermixture following an African origin and dispersal, while for Smith (1992), the process was more gradual and longer-term, a view known as the Assimilation Model. Given the paucity of the fossil record from many parts of the world, we should not expect complete answers to some of these complex questions. Nevertheless, the last fifteen years has seen significant advances in our knowledge, in terms of new fossil or archaeological discoveries, novel methods of study, and progress in our ability to date important evidence accurately.

Evidence now suggests that only Africa has credible transitional fossils between pre-modern and modern humans, and it is here that *Homo heidelbergensis* was apparently transformed into *Homo sapiens*. Fossil remains of modern type have been discovered in African sites such as Border Cave and Klasies Caves (South Africa), Guomde (Kenya), and Omo-Kibish and Herto (Ethiopia). The southern African remains are between 70,000 and 120,000 years old, while the Guomde skull and femur and the Omo and Herto fossils are probably all more than 150,000 years old. Although they can be classified as fundamentally modern in anatomy, these people still retained some primitive features and were not necessarily fully modern in behavior. The transition from *Homo heidelbergensis* to these early *Homo sapiens* may be represented by fossils of even greater antiquity, such as the Florisbad skull from South Africa (about 260,000 years old), and less well-dated material from sites like Ngaloba (Tanzania), Eliye Springs (Kenya), and Djebel Irhoud (Morocco).

How the evolution of modern people proceeded in Africa, and why the transition happened at all, is still uncertain. There is some evidence of an increasing sophistication in African Middle Stone Age industries, which could have promoted evolutionary change, including the beginning of symbolism and art. Alternatively, geographical isolation following severe climatic changes may have been responsible. As discussed below, genetic evidence indicates that all living people are closely related and share a common ancestor who lived in Africa during the last 250,000 years, but it is not yet clear where the ancestral population or populations lived. While the fossil record points to East Africa as the most probable location, this is also the region with the best fossil record for this period—while artifacts show that humans were also living in central and western Africa at this time, not a single fossil is known to show what these mysterious populations looked like, nor what role they may have played in our evolution.

From about 125,000 years ago, and possibly earlier, *Homo sapiens* began to emerge from Africa, first entering the adjoining regions of western Asia. A number of modern human skeletons, apparently deliberately buried, have been excavated in the Israeli sites of Qafzeh and Skhul, dating between about 100,000 and 120,000 years. While their stone tools suggest they had a way of life superficially little different from that of their Eurasian neighbors, the Neandertals, there is some evidence from the structure of their skeletons that they exploited their environment more efficiently, and the sites contain evidence of symbolism in the form of grave goods, red ochre pigments, and seashell beads.

The dispersal of early modern people (the Cro-Magnons) into Europe probably occurred about 40,000 years ago, and there is more evidence of behavioral differences between Neandertals and early modern people by this time. Although most Neandertals are associated with Middle Paleolithic (Middle Old Stone Age) or Mousterian tool industries, the Cro-Magnons are associated with Upper Paleolithic industries. These contain narrow blade tools of stone, which could be used to work bone, antler, and ivory, and even to produce engravings and sculptures. However, some of the last Neandertals in Europe also briefly developed tool industries with Upper Paleolithic characteristics, suggesting possible contact and even intermixture with the contemporaneous Cro-Magnons. The disappearance of the Neandertals may have been related to a combination of factors—the rapidity of climatic oscillations at this time and consequent unstable environments, coupled with the presence of newcomers who were perhaps more flexible in adapting to these rapidly changing environments.

While populations who remained in Africa continued to evolve and differentiate, the ancestors of present-day Europeans, Asians, and the native populations of the American and Australian continents probably shared common ancestors inside or outside Africa within the past 60,000 years. The early modern people who reached Australia and New Guinea by at least 50,000 years ago must have needed boats or rafts, since even when sea level was at its lowest because of water locked up in the expanded ice caps of the last glaciation, there were never land bridges leading there, only chains of islands. On their way to Australia they perhaps encountered surviving archaic humans such as the controversial dwarfed humanoid form known as *Homo floresiensis* (nicknamed "the Hobbit"), from the island of Flores, which may have survived there until as recently as 12,000 years ago. For people to reach the Americas, it would then have been possible to walk from Asia across the land bridge between Siberia and Alaska, or to travel along a more southerly coastal route by island hopping. However, it is still uncertain whether this had happened before or after 15,000 years ago, and there is evidence that distinct populations may have been involved. It was apparently only during these final stages that modern regional differences evolved through a combination of continuing environmental adaptation, human selection through cultural preferences, and the growing degree of isolation as small numbers of pioneer populations spread out over ever-increasing distances.

Thus the majority of the data seems to support the "Out of Africa" model, but some gene flow resulting from sporadic hybridization between peoples such as the Cro-Magnons and Neandertals in Europe, or early modern people and descendants of *erectus* in the Far East, could also have occurred. To investigate this possibility, a better fossil record from the time of the dispersal phase of *Homo sapiens* will be needed for each area of the world, together with improved dating techniques and more detailed methods of study to extract the maximum information from the evidence. The further development of genetic analyses applied to living humans and, where possible, to the fossils themselves will also greatly help our ability to reconstruct this fascinating stage in the evolution of our species.

GENETIC DATA ON MODERN HUMAN ORIGINS

Genetic data have assumed increasing importance in reconstructions of recent human evolution over the last twenty years. A great deal of inferential information can be reconstructed from the DNA of living humans, and in the last ten years DNA from Neandertal fossils has also become available. Because DNA is repeatedly copied, most pertinently when it is passed on from parents to their children, copying mistakes are made, and if the changes are not lethal, these mutations are then also copied on. They thus accumulate through time and allow us to follow particular lines of genetic evolution, and also to estimate the time involved in their accumulation. For our purposes, there are three kinds of DNA that can be studied. The first is the DNA which makes up the paired chromosomes contained within the nucleus of our body cells—called nuclear or *autosomal* DNA. This DNA contains the blueprints for most of our body structure, and we inherit a combination of it from both of our parents. Autosomal DNA also contains many segments of so-called junk DNA, which do not code for features such as eye color, blood group type, or proteins. They nevertheless get copied along with the coding DNA and mutate through time, and can therefore also give us information on evolutionary relationships. The second type is *Y-chromosome* DNA, which lies on the chromosome which determines the male sex in humans. The DNA on this chromosome can be used to study evolutionary lines in males only, without the complication of inheritance from two parents which comes with the study of normal autosomal DNA. The third type is *mitochondrial* DNA (mtDNA), which is found outside the nucleus of cells, in the mitochondria. These are little bodies which provide the energy for each cell. Their DNA is passed on in the egg of the mother when it becomes the first cell of her child, and little or no DNA from the father's sperm seems to be incorporated at fertilization. This means that mtDNA essentially tracks evolution through females only (mothers to daughters) since a son's mtDNA will not be passed on to his children. This type of DNA seems to mutate at a much faster rate than autosomal DNA, allowing the study of short-term evolution.

Perhaps the biggest single impact of genetic data on research on modern human evolution came in 1987, with the publication of a study of mtDNA variation in modern humans. About 150 types of mtDNA from around the world were investigated, and their variation determined. Then a computer program was used to connect all the present-day types in an evolutionary tree, reconstructing hypothetical ancestors. In turn, the program connected these ancestors to each other, until a single hypothetical ancestor for all the modern types was created. The distribution of the ancestors suggested that the single common ancestor must have lived in Africa, and the number of mutations which had accumulated from the time of the common ancestor suggested that this evolutionary process had taken about 200,000 years. This, then, was the birth of the famous "mitochondrial Eve" or "lucky mother," since the common mitochondrial ancestor must have been a female. These results seemed to provide strong support for the "Out of Africa" model of modern human origins since the research suggested that a relatively recent expansion from Africa had occurred, replacing any ancient populations living elsewhere, and their mtDNA lineages. However, the work was soon heavily criticized. It was shown that the kind of computer program used could actually produce many thousands of trees which were all more or less as economical as the published one, and not all of these alternative trees were rooted in Africa. Moreover, other workers criticized the calibration of the time when "Eve" lived, while yet others questioned the constitution of the modern samples analyzed (for example, many of the "African" samples were actually from African Americans). The team involved in the original work admitted that there were deficiencies in their analyses, but they and many other workers have continued to use mtDNA to reconstruct recent human evolution. The more detailed results obtained since suggest that even if the 1987 conclusions were premature, they were essentially correct, and that a recent African origin for our mtDNA is highly probable (Ingman et al., 2000). Nevertheless, most workers agree that mtDNA, while very useful, is only one small part of the genetic evidence needed to reconstruct our evolutionary origins.

In the case of Y-chromosome DNA, instructive results have also been obtained. Recent research suggests that variation in Y-chromosome DNA is very low, and the hypothetical "Adam" for present day males also lived in Africa (Underhill et al., 2001). Y-chromosome "dates" for the most recent human common ancestor and for African dispersal events are generally younger than those obtained from mtDNA, and it is still unclear to what extent this is a question of real events, differing female and male demographic histories, or methodological factors.

The analysis of autosomal DNA and its products (most of our body proteins, enzymes, antigens, etc.) has a much longer history in evolutionary studies than that of mtDNA. For example, it was a study of ape and human albumins which led, thirty years ago, to the first suggestion of a late divergence between humans and African apes. Now, studies are able to use combinations of data from many different gene systems, or they can look at variation in a particular segment of DNA. These data have been used to compare the DNA of human populations in ever greater detail, to estimate coalescent (last common ancestral) dates for various gene systems, to reconstruct ancient demographic patterns, and to develop phylogeographic studies to map ancient dispersal events. While most of these data support a recent African origin for recent humans and their genetic diversity, they do not all. While the data are growing in power and resolution, they still cannot yet resolve the precise time and place of our origins, nor establish whether there was only one or perhaps several significant dispersals of *H. sapiens* from Africa during the later Pleistocene. Many autosomal DNA studies have given strong support to the Out of Africa Model, but there are exceptions, leading to the suggestion that some gene systems (perhaps as much as 10 percent) in *Homo sapiens* have deeper roots outside Africa than within. This has led Templeton (2002) to propose a model of multiple dispersals from Africa and continuing gene flow, following the expectations of the Assimilation Model. This would seem to falsify a pure Out of Africa Replacement Model, though some workers have cautioned that more extensive sampling within Africa may yet reveal the real roots of these apparent genetic anomalies.

Genetic data have even become available from Neandertal fossils. The recovery of the first mtDNA from the Neander Valley skeleton (the type fossil of *Homo neanderthalensis*) in 1997 caused a sensation, and about a dozen Neandertal fossils have now yielded up this genetic material, analyses of which suggest that we (*Homo sapiens*) and the Neandertals began to take our separate evolutionary paths about 500,000 years ago (e.g., see Krings et al., 2000). In 2006, using a particularly well-preserved Neandertal fossil and massive improvements in analytical techniques and computing power, two international teams of scientists reconstructed genetic maps of the Neandertal autosomal genome, holding the promise of the reconstruction of a complete Neandertal genome in the next few years. The results confirmed the distinctiveness of the Neandertals, and supported previous estimates of the divergence time. But this research also promises breakthroughs in our understanding of their whole biology. Portions of DNA have already been recognized that originated from the Y-chromosome of the Croatian Neandertal sampled, indicating it was a male individual, and this will provide valuable information for comparison with the extensive data from modern human Y-chromosomes that indicate a recent African origin for our species. Research will now extend to complete the whole genome of a Neandertal and to examine Neandertal variation through time and space to compare with ours. Having such rich data holds the promise of looking for the equivalent genes in Neandertals that code for specific features in modern humans, e.g., eye color, skin and hair type, cognitive and language skills, etc. It will be especially interesting to look for the microcephalin gene (a gene that contributes to the form of the brain) in Neandertals, as it has recently been suggested that a variant of this gene was introduced into *Homo sapiens* quite recently from a distinct source, perhaps through interbreeding with Neandertals. Having a Neandertal genome will also throw light on our own evolution, by allowing a three-way comparison of the genetic blueprints that produced Neandertals, and that today produce us and our closest living relatives, the chimpanzees. We should then be able to pin down unique changes in each genome to show how we came to be different from each other.

These breakthroughs come at a time when renewed claims are being made that Neandertals and early modern humans (Cro-Magnons) interbred in Europe about 35,000 years ago, and the new data will provide many more clues about whether this happened, and if so, on what scale. Both fossil and DNA data indicate that the Neandertals were a distinct lineage from modern humans, but a closely related one, and the level of morphological difference in the skeleton is comparable to those in recent primates and fossil mammals that demarcate distinct species. However, closely related mammal species may still be able to hybridize, so this was presumably possible between Neandertals and Cro-Magnons. How often it happened, and its importance to the bigger picture of modern human origins, are still unclear, with scientists taking quite different views. The essential question is whether when the populations met, they regarded each other as simply people, enemies, alien, or prey even. We simply don't know the answer, and the answer may have varied from one time and place to another, especially given the vagaries of human behavior. Regardless of the outcome of such fascinating questions, there is little doubt that Out of Africa is now the dominant paradigm in studies of the evolutionary origins of our species.

FURTHER READING

Jones, Steve, Martin, Robert, Pilbeam, David, and Bunney, Sarah (eds.). *The Cambridge Encyclopedia of Human Evolution* (New York: Cambridge University Press, 1992).

Mellars, Paul, and Stringer, Chris (eds.). *The Human Revolution: Behavioral and Biological Perspectives on the Origins of Modern Humans* (Princeton: Princeton University Press, 1989).

Stringer, Christopher, and McKie, Robin. *African Exodus: The Origins of Modern Humanity* (New York: Henry Holt, 1997).

Tattersall, Ian. *Becoming Human: Evolution and Human Uniqueness* (New York: Harcourt Brace, 1998).

Wolpoff, Milford, and Caspari, Rachel. *Race and Human Evolution* (New York: Simon & Schuster, 1997).

Intelligent Design

Brian Regal

By the mid-twentieth century anti-evolutionists had reached a dead end in their attempt to thwart the teaching of and belief in evolution. The evidence supporting evolution, much of it explained in this book, just kept piling up. A new approach was needed: an approach more comfortable with the scientific, genetics, computer modeling–based late-twentieth century than the scriptural and fossil-based nineteenth century. EnterIntelligent Design. Eschewing biblical references or any specific religious connotations—at least in public—Intelligent Design theorists attempted to develop a cosmology that was both physical and metaphysical through the artifice of biochemistry, higher mathematics, statistics, and genetics. They worked hard at sounding scientific but the mainstream scientific world saw this mostly as the cloaking of other intentions. The problem was that under all the DNA, genetic sequencing, complexity theory, lawyerly rhetoric, and other scientific-sounding jargon was the basic fact that Intelligent Design was the search for God (a situation exacerbated by Intelligent Design theorists publicly insisting that it was not). As such Intelligent Design theory steps outside the bounds of genuine scientific inquiry.

This chapter may be problematic for a book about the icons of evolutionary thought. This is because Intelligent Design (ID) is an icon of anti-evolutionary thinking. In the minds of many non-scientists ID is linked to evolution in some way as part of an ill-defined "controversy" or "debate" concerning its being taught in the American public school system. Clearing up this confusion is the reason this chapter is included here. The reality is that there is no controversy or debate in the world of evolutionary biology over ID. This is because evolutionary biologists do not consider ID a science, but a religious belief better discussed as metaphysics than biology.

This chapter sets out to explain what Intelligent Design is, what its operating principles are, who its promoters are, and why evolutionary biologists dismiss it as a form of religious creationism. What will also be shown is that this is a more subtle, nuanced, and complex issue than most evolutionists and even creationists realize. While it certainly is a form of creationism, it varies in profound ways from traditional creationism just as much as it differs from evolution, but for different reasons.

ANTECEDENTS

In its simplest form, Intelligent Design is the belief that the living world, and indeed the entire universe, is such a vastly complex and interwoven continuum that it could only have been designed by some higher metaphysical power and that this condition can be proven by an application of modern scientific techniques. Proponents argue that the level of complexity and design seen in the natural world is by itself proof that it could not have come into being by natural selection and that the only possible way to explain the universe is that an intelligent designer purposefully engineered and assembled it. ID believers also argue that their position does not use biblical references but is proven by science.

While Intelligent Design has become a much-discussed topic in the early twenty-first century—due mostly to its inclusion in the various latter-day monkey trials—it is not a new idea. The belief that God, or the gods, created the universe goes back to the oldest human civilizations and probably back further than that. All cultures around the world regardless of ethnicity, time period, or religious persuasion had and still have some type of creation story. Intelligent design is only the most recent variation in a long line.

One thing all those creation stories have in common is that they argue life has a reason and purpose. The formulized notion of teleology goes back to Greek philosophers like Plato and Aristotle. They argued that if all life had purpose it must have been designed to fulfill that purpose. If all living things have purpose then the entire universe must as well, therefore the entire universe was designed with a purpose. That means there must be a designer: in the Greek, Theos. The study of God is theology. (By definition then, if Intelligent Design is the search for the grand designer, it is theology not science.) In the Middle Ages the Christian philosopher Thomas Aquinas—a big fan of Aristotle—adapted teleology to conform to the Christian notion of God: a God who was separate from and superior to the Greek gods and any other gods for that matter.

An attempt to explain, not so much the origins of all life in general, but humans in particular was Matthew Hale's *Primitive Origination of Man* (1677). Hale (1609–1676) was an influential British lawyer who instead of scripture, or even nature, employed the logic and rhetoric of the law to work out the proof of creation (an approach which would echo in the late twentieth century). He wrote his book over a period of years as a kind of intellectual hobby. Hale began his argument by positing that the earth could not be infinite and therefore had to have had a beginning. If that were the case then all life on earth, including humans, had to have been created as well. Using what he considered the structure of a reasoned legal argument, rather than simple religious belief based upon scripture alone, he argued that the Christian view of creation was the only one that made logical sense. Examination of the natural world led one to believe that the laws that seemed to govern the operation of nature were divine in origin (an idea he shared with Isaac Newton). He accepted that some creatures appeared through spontaneous generation, but certainly not man. He also reasoned that angels were not powerful enough to have created humans. Hale did not use the example of microscopic animals for his argument—despite the fact that microscopes were readily available at the time—but turned instead to the smallest creatures he could examine without a microscope: insects. They were far too complex to have appeared by spontaneous generation. Only an "Intelligent Agent" could have designed and created them. (Hale may have been the first to use this expression.) He also did not hesitate to say who he thought that intelligent agent was. "I shall in this place," he said, "inquire what kind of Intelligent Efficient this was, for among Intelligent Beings there is one *Primum*, the Glorious God" (Hale, p. 331).

Natural theology—proof of God's creation from evidence in the natural world—took form in England in the late 1600s. It came out of an increased interest in studying nature for itself and dovetailed into the attempt to use knowledge of the structure and order of the natural world to prove God designed it for a purpose. A number of philosophers and theologians tackled the problem of natural proofs of creation. The mechanists, like Robert Boyle and his *Origin of Forms and Qualities* (1666), argued that the universe was a vastly complex machine with a first but not a final cause. As such the only way to account for this device was by an intelligent designer who put the whole thing together and started it working. The most influential was John Ray (1628–1705) the Anglican priest-turned-botanist. Ray did pioneering work in the cataloging and describing of plant life in a way similar to the Swiss systematist Carolus Linneaus. Ray worked at devising a scientific arrangement for living things that would not only bring order to the human understanding of nature but also show how that order was proof God had designed it.

His *Wisdom of God Manifest in the Works of the Creation* (1691) was widely popular and helped influence many later writers who wanted to use the apparent design of nature to do the same thing. The best-known theologian to tackle the problem of intelligent design of the late eighteenth century was William Paley and his book *Natural Theology* (1802).

Paley (1743–1805) argued, as Ray had, that the complexity and interwoven nature of living things was evidence of it having been purposefully designed by God. Paley's novel approach was to use the simple analogy of the pocket watch—an analogy which became an icon all its own. He set up a thought experiment asking his readers to imagine strolling down a country lane and coming across a pocket watch. The stroller would instinctively know the watch was a complex object with many interlocking parts which could not possibly have come together on its own. The complexity of the watch would make it obvious that it had been designed and constructed for a purpose by a watchmaker. Paley then said that nature was like the pocket watch, a complex machine constructed by a designer for a purpose. Therefore by looking at nature it was obvious that God had designed it purposefully. Paley's book made natural theology more widely known than ever and came to be called the argument from design. At the famous Oxford debate over evolution between the Reverend Samuel Wilberforce and "Darwin's Bulldog," T.H. Huxley, in 1860, the Reverend—coached by the powerful anti-evolutionist Professor Richard Owen—called upon the argument from design to discredit Darwin's *Origin of Species* with comical results. The same basic argument from design would be used throughout the twentieth and twenty-first centuries.

Not all religious thinkers discounted evolution. An example of an influential theologian who supported evolution is James McCosh (1811–1894), the reform-minded president of Princeton College. Born in Scotland, McCosh studied theology at Glasgow and Edinburgh in the 1820s. He began his ministerial duties in the midst of a split in the Scottish Presbyterian Church between moderates (those following the philosophy of the eighteenth-century Scottish Enlightenment which was pushing the Church away from its strict Calvinist origins) and the evangelicals who opposed them. McCosh came down on the side of the evangelical camp, and supported that group's attempt to limit the meddling of the state and wealthy elites in church affairs, which he feared might have short-term benefits but long term would be detrimental to religious freedom. Unable to reconcile their differences, the evangelicals split from the Church to form the Free Church of Scotland. McCosh was not however, a dogmatic evangelical. He wanted to reconcile the cold intellectualism of the moderates and the emotionalism of the evangelicals. In the style of William Paley he argued that the complexity of nature was a sign of its divine inspiration—though with an emphasis upon the human mind, its emotions, and will as evidence. For the Scotsman the workings of thc mind showed that God ran the world in an orderly fashion according to His laws.

In his teachings at Princeton McCosh focused on the human mind and its workings. He argued that there were absolutes of truth and meaning because humans could conceive of such things intuitively. He believed the human mind had certain fundamental ideas and principles built into it. This "primitive cognition" was the foundation upon which all common truths rest and was the basis for all religion. He felt intuition plus experience were how people learned about nature and thus were the divine laws by which nature operates. He insisted these beliefs be based upon factual evidence not speculation.

In addition to his views on intuition, James McCosh was also a pioneer in the teaching of evolution in America. Despite opposition from Princeton Theological Seminary head Reverend Charles Hodge, McCosh persisted. Hodge argued that Darwinism was a rejection of a God-created universe. He equated denial of design with a denial of God. Just because McCosh accepted evolution did not mean he was a pure Darwinian. As far back as 1856 he elaborated his ideas about the design of life borrowing from British naturalist Richard Owen's transcendental morphology and archetypes to argue God's handiwork, claiming there was unity and order in the universe. When *Origin of Species* appeared, however, McCosh saw the importance of Darwin's approach and adopted a teleological form of transmutation. He accepted the fact of evolutionary change, but argued it was a Divine Law. He also argued that students were in no danger from exploring Darwin's work. If evolution were properly understood, the Scotsman said, it would be seen as a mechanism of God and not antagonistic to revealed religion. McCosh was an evolutionary theist who was comfortable with the notion of an evolved body but argued man's mind was divinely inspired. Morality, instinct, and intelligence could only be accounted for by divine processes. Accepting what seemed an inevitable outcome, he held that man may or may not have evolved from brutes, but his soul was of higher origin.

McCosh had no time for unsubstantiated notions whether scientific or religious. He opposed atheistic evolutionary systems that attempted to take God out of the picture, but was equally opposed to those who leveled blanket condemnation at science and evolution. He warned against the "undiscriminating denunciation of evolution from so many pulpits" (McCosh, p. ix). He did not believe evolution led to infidelity or a breakdown in the social order. He took evolutionists who thought that mind was a result of material process to task when he told his students that it was absurd to think there was no such thing as the mind. Giving the devil his due, however, he warned that "the philosopher who believes that everything is mental and nothing material, takes good care in going downstairs not to slip and hit his head against a hard idea" (McCosh,

1870). As he did with Presbyterian schisms, McCosh tried to walk a middle way between science and religion to bring the two together, not push them apart.

It was not just biologists and theologians who saw the impact of evolution on domains outside of science. Some American intellectuals—like Charles Sanders Peirce, William James, and members of the Massachusetts-based "Metaphysical Club"—saw evolution as connected to religion and philosophy. These metaphysicians strove to reconcile evolution with religion, and empiricism with idealism. They shared a belief in the fact of evolution and the supremacy of science in solving problems over religious dogma. They were not totally in agreement, however. Peirce and James were founders, along with John Dewey, of the philosophy known as pragmatism. They were comfortable with the role of chance and fortuitous elements. Peirce saw chance leading to order while others saw it leading to chaos. Peirce had misgivings that natural selection was not the answer to how evolution actually worked. James saw a place for personal effort and willpower, and wrestled with the knotty problem of determinism versus free will. James, however, did understand that the underlying conflict over evolution was the battle between chance and design.

Theologians and philosophers were not alone in trying to reconcile transmutation with religion. There were evolutionists who thought about the concept of design in the living world. Naturalists including Alfred Russel Wallace, the co-discoverer of natural selection, sought a way around strict Darwinian selection. They were comfortable with allowing that evolution was responsible for life on earth but felt that human consciousness was too sophisticated to have come about by chance. These naturalists argued that while evolution may account for the physical aspect of humans, the human mind was a result of Divine intervention. For example, the early-twentieth-century president of the prestigious American Museum of Natural History in New York, Henry Fairfield Osborn, who had been a student of James McCosh, while a staunch evolutionist, spent most of his career trying to prove that the evolutionary process was the product of God. Osborn believed that God began the evolution process then let it go on its own guided by a kind of internal perfecting mechanism at the genetic level. He was an otherwise respected and influential scientist, but his theistic God-driven evolution ideas were not embraced by the mainstream mostly because he could show no evidence of it that was acceptable to other scientists.

NUMBER CRUNCHING FOR GOD

By the late twentieth century, carrying a pocket calculator and going on about genetics and complexity theory, the argument from design took on a new life as Intelligent Design. The seeds of the modern ID theory were planted in the United States in the late 1970s and early 1980s. It was formed out of the wreckage of the creation science movement combined with the idea that genetic material was a complex coding system. ID proponents wanted—in public at least—to create an argument that was not theological but still questioned the evolutionary paradigm.

The book that became the early movement's Holy Scripture was Michael Denton's *Evolution: A Theory in Crisis* (1985). A biochemist and medical doctor operating out of the chemistry department of Prince of Wales Hospital in Sydney, Australia, Denton began as an evolutionist, but became increasingly bothered by what he considered a lack of evidence supporting evolution. His work has been hailed by supporters in hagiographic terms and he is characterized as a courageous radical like Martin Luther protesting at the gates of the evolution establishment (this is ironic as within the American milieu at least, anti-evolutionist thinking and creationism are embedded well inside the ranks of political and religious conservatism while evolution finds its base of support in the liberal political and religious sphere).

Denton's argument has to do with macro-evolution, which is the large-scale biological changes over long periods of time that result in the appearance of new species as well as larger groups like families and genera. This is in opposition to micro-evolution, that is, the appearance of small-scale changes that by themselves do not produce a new species. (He does not reject the evidence for micro-evolution.) Denton's central critique, which also became the standard critique most other ID supporters would utilize, is that evolution does not explain large-scale evolution or the origins of life. While he does spend time on the biochemical aspects of his argument, most of the book is given over to arguing that evolution is a socially constructed paradigm rather than one built on empirical evidence. He attacks the narrative quality of evolutionary thinking and that its real power comes from being a materialistic and non-theological explanation for life. He accuses evolutionists of being a new priesthood unified in the single-minded pursuit of keeping their idea safe from dissenters.

What makes this work fascinating on one level is that for a movement that wants to take science back to a pre-Enlightenment period of fixed "truth" where science worked in the support of faith and searched for the answers to metaphysical questions, the book is a form of postmodernist deconstruction of an idea, namely the Darwinian "myth." Denton's is a textural analysis arguing that while Darwin himself was an honest broker searching for the truth of transmutation and bothered by not having all the answers, later, scoundrels (presumably T.H. Huxley) came along and turned his work into evolution as a socially constructed dogma based on no evidence and held up by the nefarious forces of conservatism. While Denton's was a weighty scholarly tome, others were working on spreading the word though more popular works and school textbooks.

In 1984 *The Mystery of Life's Origins* was published. It was one of the first books to use Intelligent Design as its central organizing theme. The authors (chemist Charles Thaxton, engineer Walter Bradley, and geochemist Roger Olsen) were prompted to publish by Dallas minister Jon Buell, head of the group Foundations for Thought and Ethics (FTE). Founded in 1980 the FTE had as its primary mission to produce school texts that promoted a Christian and creationist worldview. The manuscript for the book was brought to Dean Kenyon an academically trained biologist from San Francisco State University. Kenyon was already pondering the idea of intelligent causes for biological structures and enthusiastically reviewed it. He liked the idea because it argued for a designer without overtly involving Christian theological baggage and that it sidestepped the issue of saying who the designer was. This allowed believers to have it both ways: You could be a creationist without being a creationist. Thaxton met oil industry geophysicist-turned-historian and philosopher of science Stephen Meyer and the two would eventually join forces to found the Center for the Renewal of Science and Culture of which Meyer became director. In 1983 following *The Mystery of Life's Origins* the FTE began working on putting out another creationist book to be promoted to high school science classes. Buell, who had also been a member of the evangelical Campus Crusades for Christ, enlisted Kenyon and Percival Davis to co-author the work originally titled *Creation Biology*. As work progressed on the book the *Edwards v. Aguillard* case worked its way through the court, eventually ruling against the inclusion of creationist concepts in science classes. The FTE then excised all the overtly creationist references in the manuscript and replaced them with what was thought to be more palatable design language. They also changed the title to *Of Pandas and People*. Published in 1989 it became the book of choice for creationists, ID supporters, and high school science curriculum reconstructionists in state boards of education across America. Charles Thaxton, also a fellow of the American Scientific Affiliation (another creationist organization), is credited with coming up with the term Intelligent Design though he admits he borrowed it from engineers who were using it for other areas and applied it to what he was working on. Thaxton was drawn to ID theory by work being done on DNA sequencing and its similarity to the way words and letters are arranged in a book. As language is something that is consciously designed then DNA must also have been designed.

The FTE had the book published by a Texas firm called Haughton Publishing, a printing house which normally produced packaging materials for the agricultural industry. Funds to subsidize its publication were provided by donations to the FTE from Christian followers. They then began a campaign to get the book out by donating free copies to various high school libraries and by approaching local school board members they thought would be sympathetic to their cause. While praised by ID and creationist supporters *Of Pandas and People* was immediately criticized by the mainstream as pseudo-science. Berkeley paleontologist Kevin Padian called it a "catalog of errors" and cautioned that "it is hard to say what is worst in this book: the misconceptions of its sub-text [of Christian creationism], the intolerance for honest science, or the incompetence with which it is presented." The book took what would become the standard approach of ID publications: Show what is wrong with evolution rather than what is right about Intelligent Design.

The next work of the ID canon came along in 1991: Philip Johnson's *Darwin on Trial*. Johnson was a Berkeley law professor who wanted to deconstruct evolution along legal lines. Johnson attempted, like Matthew Hale so many centuries before, to formulate a purely intellectual and legal argument (though he never really says why Darwin should be on trial. Presumably it is for undermining Johnson's already weak faith). Johnson became a Christian late in life and attacked evolution with all the zeal of a recent convert. He came to believe that modern culture was responsible for a growing secularization of the education system that in turn was fostering a general breakdown in American society. At the heart of this breakdown was the teaching of evolution. He claimed his examination of evolutionary mechanics was on its own terms, not in light of any religious framework.

Johnson starts off with a number of fallacies. In the first paragraph of the first chapter he shows a fundamental lack of knowledge about current evolution studies. He explains that "scientific orthodoxy" teaches that life evolved "from nonliving matter to simple microorganisms, leading eventually to man" (Johnson, p. 3). The idea that evolution showed a linear track from simplest to most complex with humans as the epitome of the process went out of favor with scientists decades before Johnson made his remark. It is the equivalent of someone saying that astronomers believe the sun is at the center of the universe. While that was believed early on, modern astronomers understand that the sun is only the center of our specific solar system not the universe.

Johnson also argues that according to the tenets of science and the words of evolutionists themselves natural selection as a scientific explanation for life is unscientific. Many evolutionists, he argues, do not believe in evolution because of questions they have raised over various parts of it. He fails to appreciate that just because a scientist might express doubts about some of the mechanics of evolution or wonder why a particular explanation does not seem to cover every answer, or how a particular fossil has strange aspects to it does not mean they are questioning the overall idea of evolution. Asking tough questions is how science is advanced and how knowledge is gained. This fundamental aspect of intellectual inquiry—doubting and asking questions and being ready to change position—is viewed with deep suspicion by anti-evolutionists and fundamentalist religionists. They tend to prefer knowledge to remain static and unchanging the way Holy Scripture does. Johnson also suggests that when scientists do take the time to put forward answers to the uncomfortable questions anti-evolutionists ask they are doing it as part of a cover-up: a pernicious establishment flailing away at any and all opponents.

Johnson works hard throughout his text at retaining a judicial objectivity and begins by stating that evolution contradicts creation when expressed as naturalistic evolution which contradicts purpose. Johnson is willing to accept evolution to a point as long as it was acceptable as divine law. He uses all tried-and-true objections stating Darwinism is really a religion and that the fossil record is misleading and actually supports creationism. He also uses Darwin's own example of the peacock which, in *The Descent of Man* (1873), the English naturalist used as an example of sexual selection. Johnson argued the bright colors would have attracted predators as well as mates, which would have eliminated the peacock population. Johnson muses, "It seems to me that the peacock and the peahen are just the kind of creatures a whimsical Creator might favor . . . but natural selection would never permit to develop" (Johnson, p. 31).

Like a good lawyer Johnson sets up the opposition as inherently bad and untrustworthy. Speaking to his Christian jury he includes theistic evolutionists (those religionists who have no problem with evolution) in the category of creationist as a way of separating them from the enemy. This way he can hold up evolution as anti-Christian (which it is not). He has to do this for his theologically based argument to work.

Like so many of his colleagues, Philip Johnson refers to anyone who believes in evolution as "Darwinists." This is a clever turn of phrase designed to help his case. As "Christians" worship Jesus, so must "Darwinists" worship Darwin? Evolutionists certainly admire and respect Darwin for his pioneering work and insights, but they do not worship him as a religious figure. To suggest they do is one more way Johnson sets up anyone who believes in evolution as idolatrous heretics no decent Christian should be associated with. This is part of his misconception that anyone who believes in evolution is a strict Darwinist. Few modern evolutionary biologists call themselves a Darwinist. This is because modern evolutionary science has changed a good bit from the way Darwin originally explained it. Modern breakthroughs in genetics, the understanding of DNA and cell structure, and other new information has altered the original idea. Just as astronomers admire and respect the work of Nicolaus Copernicus they do not call themselves Copernicans anymore. In the mock courtroom Johnson creates—where his job is not to establish the "truth" or even a fact-based reality—he makes his opponents seem suspect by calling them *Darwinists* turning the term into a pejorative (a tactic picked up by most subsequent anti-evolutionists). Overall, one is left with the impression that Johnson is either using the rhetorical skills of a lawyer or he does not really understand the technical material he is criticizing. If it is the former his work can be dismissed as simple courtroom antics designed to win a case regardless of the facts. If it is the latter his work can be dismissed as the incoherent ramblings of one unqualified to make such a critique. For all his lawyerly acumen and metaphorical adroitness, Johnson ends his book by arguing that evolution should not be accepted because evolutionists do not have all the answers. Despite his pleas to objectivity, Johnson is well within the religious fundamentalist anti-evolutionist camp, so much so that he became a revered spokesman and elder statesman for the cause.

In 1993 Johnson called together a group of men who would become the who's who of the Intelligent Design movement for a quiet get-together. Michael Behe, William Dembski, Jonathan Wells, and others met quietly with Johnson at a California seaside resort called Pajaro Dunes to discuss a strategy for putting their ideas forward. They came to the conclusion that they had to be careful to reduce as much of the religious content of their work as possible to make it palatable to a wider audience. The group began arranging a series of public meetings, notably "The Death of Materialism and the Renewal of Culture" conference which became the basis of the Discovery Institute.

While enthusiastic in his denunciation of evolution, Johnson was a lawyer and not a biologist so lacked the deeper knowledge needed to critique evolution beyond simple semantics. The core scientific intellectual of the ID movement is the philosopher-theologian-mathematician William Dembski who wrote extensively on complex biological systems and their relation to evolution and argued that microorganisms and their parts, though superficially simple, are still highly complex creatures, too complex to have been formed by chance. Like Johnson, Dembski became an active Christian later in life as he searched for answers to higher philosophical questions. He employs statistics and other mathematical models to show that the odds of life evolving were so astronomically high as to be impossible. As a result, the only alternative answer to why life is here at all is that it was designed by a higher power.

Dembski argued that there was a "design inference" where the designer occasionally stepped in to do things. He conceded that not everything in nature was designed and that some things did happen by chance. To work out the differences he assembled a kind of algorithmic table to compare natural phenomena to. If something happens on a regular basis in nature it is not designed. He sees the creation of life as a very rare event and so the probability that it was designed by a higher intelligence is great. There are two kinds of design patterns in nature. Specifications are those whose probabilities for existence eliminate the possibility of having come about by chance. Fabrications are those whose probabilities do not eliminate chance. The design inference is a hypothetical filtering device used to determine which one is which. Dembski divided events into three categories: laws, chance, and design. Laws govern events that will always happen under certain circumstances. Chance governs those events that happen as much as they would not happen. Design governs those events that do not fall under the first two categories. Arranged like a funnel the design inference allows for events to be dumped in from the top. Laws are separated out from the stream and shunted off to the side. The few that make it past the first filter are separated out as chance. The minuscule numbers of events that make it through both filters fall out the bottom of the device. By Dembski's definition these events have no other explanation than that they were the product of a conscious design. Unlike most ID publications Dembki's *The Design Inference* (1998) was published by the prestigious Cambridge University Press.

The intellectual inspiration for Dembski's work came from the writings of the Hungarian émigré philosopher Michael Polanyi (1891–1976). Studying medicine and then chemistry at Budapest, Polanyi left home for Germany after World War I and set up a career as an experimental chemist. Being Jewish he left Germany following the rise of the Nazis and landed in England where he worked for some years before going to the United States. Though initially working as a practicing scientist he became fascinated by the philosophy of science and turned to that field permanently. He had begun to question the basic methodological assumptions of science, especially its grounding in positivism and the belief of its practitioners that what they were doing was an objective, value-free enterprise that produced untainted truth. Polanyi's ponderings on the meaning of science resulted in *Personal Knowledge: Towards a Post-Critical Philosophy* (1958). He argued that while they might be trying to be objective scientists were not. Their personal feelings interfered with their objectivity. This meant that science was not a value-free enterprise after all. His suggestion was to accept this personal knowledge aspect of science and move forward. What Polanyi wanted was to create a new epistemology—a way of knowing things—which was more intuitive and took into account unseen aspects and influences.

Polanyi's work was in a similar vein to Thomas Kuhn whose widely influential *Structure of Scientific Revolutions* came out a few years later. Kuhn argued that scientific ideas, or paradigms, are supported by facts and empirical evidence but also by the scientific community's attachment to those ideas and their willingness to hold on to them and protect them. Only when there was enough new information, and the community was ready to do so, would one idea replace another in an event Kuhn called a paradigm shift.

A central tenet of Polanyi's scheme was the concept of "tacit knowledge." This was the intangible, creative aspect of scientific research. Tacit knowledge was something that could not be communicated through books, but only shown or explained by one person to another like the apprentice system of medieval (pre-Enlightenment) Europe. He also saw the concept of God as a structure or system related to the structure of life on earth. Intuition and hunches help scientists make the intellectual leaps necessary for knowledge to really move forward and create new models to explain the universe. Polanyi's suggestion was to join formal thinking and experimentation with tacit knowledge. He was looking for a way to examine the universe without having to rely on the intellectual traditions of Enlightenment objectivity and modernity and its rejection of the metaphysical. He wanted to create an artful blend of science and metaphysics that would give a deeper and more satisfying account of the universe. In *Meaning* (1975) he pondered the place of God in modern science and society and how people search for meaning in a cold society based on objectivity and rationalism devoid of the mysteries of life. William Dembski imbibed Polanyi's work, as well as that of Kuhn, into his own critique of modern science and society.

Dembski was drawn to Polanyi's work because to his mind it undermined the modern notion of what science is all about, something Dembski himself and the other ID theorists wish to do. For Dembski and other ID theorists Enlightenment ideals such as positivism, secularism, and the removal of theology from the scientific enterprise are the things that have brought on society's downward slide. Modern science, evolution in particular, is a social construction with no relationship to reality and facts. It is simply a model to explain the universe without God or the supernatural. He argued that scientists are not rebels looking for new knowledge but conservatives protecting an establishment that rejects anything that might question their position. Dembski wants to put God back into science, but he will do it with mathematics, probability theory, and genetic sequencing. Using these methods Dembski is trying to give concrete form to the tacit knowledge that God designed life on earth.

The notion of Enlightenment ideals is crucial to understanding modern anti-evolutionism. The Enlightenment—roughly the seventeenth and eighteenth centuries—was the period where theology began to lose its control over intellectual inquiry in the Western tradition. Enlightenment philosophers were intent on focusing on studying nature by applying methods of reason and empirical fact gathering over superstition and religious belief in order to separate theology, politics, and science into separate spheres. The resistance to evolution then concerns not just evolution itself but the very nature of science. In its modern, post-Enlightenment definition, science is the pursuit of knowledge about the nature of the universe. Evidence is generated by collecting facts through observation, experimentation, and discovery. Data so collected is then collated and used to generate theories and intellectual models to explain how and why things work the way they do. These models and hypotheses are always subject to revision, change, and even rejection if new evidence appears to warrant it.

This approach is alternately called naturalism and materialism. Broadly this approach says that nature must be explained by natural not supernatural causes. There are two types of naturalism/materialism. Methodological naturalism explains how nature works without any reference to the supernatural because science should only deal with those things it has the tools to explain (the supernatural is referred to as metaphysics because it is beyond the power of science to explain). Methodological naturalism does not necessarily reject metaphysics, only ignores it. Philosophical naturalism does not just avoid reference to the supernatural, but argues there is no such thing as the supernatural. As ID is in large part an attempt to prove that life was created by a supernatural entity neither of these approaches are of much interest to ID theorists in particular and creationists in general. In fact, these two approaches undermine the creationist position. As a result, ID theorists have been attempting to rewrite the definition of science to include a pro-supernatural attitude that allows for metaphysical explanations beyond science. At a gathering of ID researchers in the late 1990s, called the Mere Creation Convention, Dembski told his colleagues that what they needed to do was "redesign" science. Scientists see this as a cheap excuse for getting around the more difficult parts of the universe. Those opposed to ID argue this approach cheats both science and religion as whenever you cannot explain how a biological system works you can simply fall back on saying, "well, God must have done it." Naturalism gives no such easy way out thus forcing investigators to work harder to find possible answers.

Along with William Dembski the most public defender of design theory and a lightning rod of the intelligent design school is Michael Behe and his book *Darwin's Black Box: The Biochemical Challenge to Evolution* (1996). A professor of biochemistry at Lehigh University, Pennsylvania, Behe built upon the notion that the "irreducible complexity" of a cell showed that life was designed. Irreducible complexity describes biological systems that consist of connected parts that must all be in place for the larger system to operate. Behe took the title of his book from Charles Darwin's reference to the cell as a black box which contained processes unknowable at the time. Building upon an old anti-evolutionist argument about what good was half an eye or half a heart while it waited for all the necessary parts to evolve, Behe claimed that since an eye would not work unless all the constituent parts were present that alone would prove a designer had to have put the whole thing together at once. As few people carry pocket watches anymore Behe used the analogy of the mousetrap, which consists of several interlocking parts that must all be present for the device to function; take one out and it does not work. If the mousetrap (if it were a living thing) had evolved one piece at a time, what good would it have been until it could perform its function? But evolutionists argue it doesn't work that way. If the wooden board of the mousetrap were to evolve first, it would have performed some action. As each piece was added the device would do what it did more efficiently or might take on other functions altogether. Once enough parts were present it would then be possible to start catching mice. As more parts were added, the trap might lose its ability to catch mice and would then perform yet some other function. What many anti-evolutionists often relied on was the notion that evolution was trying to do something specific, like catch mice. The trap was not meant to catch mice; it did so only when the circumstances of structure and environment made it possible. It could just as easily have performed some other function.

Numerous critical reviews of Behe's work appeared which in tone ran the gamut from the coolly professional to the vituperative. Most made the same critique. Reviewers noted several problems with Behe's approach and conclusions. For example, he argued that the scientific literature on molecular evolution was scanty. He also argued that if evolution were to be proven at the molecular level, only biochemistry would be able to do it. So far, Behe claimed, it had been unable to. Reviewers countered that the scientific literature of biochemistry and molecular biology was full of papers on evolutionary explanations, that many conferences of biochemists dealt with the origins of life and that evolution had been explained at the molecular level. They saw Behe's use of the mousetrap analogy as misleading. Reviewers were also troubled by Behe's acceptance of the normal give-and-take of scientific inquiry as somehow evidence that scientists did not believe evolution worked. There are genuine questions being asked about the mechanics of evolution within pro-evolution circles and debates over the finer points. Debate and disagreement do not mean scientists secretly believe evolution invalid or that there is a conspiracy of silence.

Part of the underlying thesis of ID is that if you crunch the numbers you will see conscious order in the universe. This accounts for why so many ID proponents are mathematicians, biochemists, and lawyers. Math and the law are built on the notion of taking complex interplays and reducing them to simple expressions. In this way variables can be eliminated and standard questions produce standard and rigorous answers. A formula such as $1 + 1 = 2$ makes for an uncomplicated answer in which interpretation is eliminated, the outcome can be predicted, and the questioner can relax in the warm glow of certainty. If the law says "if you do X then Y will be the result," interpretation and randomness are reduced. Part of why evolution annoys so many is because it is difficult to predict long term and so the questioner is left with endless possibilities and a level of uncertainty about outcomes and meanings. The bulk of ID's "evidence" lies in the realm of mathematical modeling. Indeed, the heart of the Intelligent Design argument is mathematical and philosophical not biological. Followers of ID are engaged in a program to use cryptological and algorithmic methodologies to discern the face of God.

The natural theologians of the nineteenth century had the strength of their convictions to say who they thought the designer was. The reason for the modern two-stepping over the 'G' word is because if Intelligent Design proponents want to have their work accepted for teaching in the public school system in America under the U.S. Constitution they must show it to be secular and not religious. While they attempt to scrub ID of any public whiff of the Divine, behind closed doors the situation seems different. Johnson, Behe, and Dembski have all made remarks at private meetings of various religious groups that ID is indeed a movement closely aligned with Christian fundamentalism and supports the redemptive power of Jesus. Dembski often writes on strictly Christian theological topics. Religionists seem more willing to admit that the ultimate goal of the creationist movement is to eliminate evolutionary thinking and replace it with a biblically based understanding of the universe not only in biology classes but also throughout the fabric of American society and ultimately the world. The religious grounding of ID has been acknowledged by Michael Behe in an interview where he said, "Although I find it congenial to think that it's God, others might prefer to think it as an alien—or who knows? An angel, or some satanic force, some new age power." He eliminated any ambiguity at the Dover trial when he testified that "the designer is in fact God" (Postman, 2006). William Dembski has said Intelligent Design leads believers to Jesus.

THE DISCOVERY INSTITUTE

A prime corporate mover in the Intelligent Design movement is the Seattle, Washington–based Discovery Institute. Begun in 1991 by ex-politician and former Reagan administration advisor Bruce Chapman, the Discovery Institute grew out of a group called the Hudson Institute that worked to improve Seattle's mass transit system and promote local conservative political issues. Chapman had been director of the U.S. Census Bureau and was introduced to the Intelligent Design concept in 1994 by

Steven Meyer. The next year they met with Howard Ahmanson—to whose children Meyer had been a personal tutor—and received money to begin the Center for the Renewal of Science and Culture (CRSC) which appeared in 1996. This offshoot of the Discovery Institute was begun specifically to attack evolutionary biology, to remove its pernicious influence from American culture, and to make science more religious. For them, to renew science is to take it back to a pre-Enlightenment era where science was seen as something supporting Christian theology. CRSC founders felt "modern" science had lost its way and become too secular. As a result American society had become depressingly materialist. A CRSC document stated that they seek the overthrow of materialism itself.

The Discovery Institute and the CRSC are a modern-day temperance movement. They believe all social ills find their root in a belief in evolution and materialism the way the temperance people believed alcoholic beverages did. So to get rid of social ills, get rid of evolution and materialism. In 2002 fearing they might have tipped people off too much about their intentions, the word renewal was dropped from the title and it became simply the Center for Science and Culture (CSC). Along with pushing the Intelligent Design agenda the CSC has been trying to redefine science to be more in line with their Christian worldview. Their strategy is to take terms like "academic freedom" and "fairness" and turn them on their ears. William Demski has admitted the only way to get ID accepted as science is to redefine what science is. In 2000, Discovery Institute vice president Stephen Meyer published an article in the *Utah Law Review* in which he argued that as ID was not a form of creationism it did not fall under the strictures of *Edwards v. Aguillard*. As a result, he said, denying ID in science class was discrimination of free thought and expression of new ideas.

The most ubiquitous catch phrase employed by the Discovery Institute and thus the ID movement is "teach the controversy." This term is ingenious in that it suggests all they want is an honest debate pitting two sides against each other in the arena of ideas. The underlying conceit is that there is no controversy between evolution and ID because ID is not a science that can be compared to or debated by evolution: It is rather a carefully orchestrated promotional campaign designed to achieve a particular end. To further their end the Discovery Institute has been a major supporter of all the leading ID proponents discussed here.

Along with the Discovery Institute another vocal anti-evolution society is the University of Minnesota–based MacLaurin Institute. More openly religious, MacLaurin sees evolution as a threat to American culture and considers the Bible "authoritative and infallible." The MacLaurin Institute and Discovery Institute are linked in that they share not only ideologies but members and spokespersons. Michael Behe was a special guest speaker at their 2003 ID symposium held at the University of Minnesota, and they run think pieces in their newsletter by Bob Osburn, who is the president of the Intelligent Design Network. A Discovery Institute fellow is Nancy Pearcy who has co-authored a book with convicted Watergate criminal Charles Colson, and whose husband is a senior MacLaurin Institute writer. Pearcy argues, "America has undergone a dramatic cultural decline—abortion, family breakdown, and morally decadent entertainment . . . a host of issues" (Pearcy Web site, 2006). She worries that America is "hostile to Christian truth" and she is unhappy with the separation of public culture and private religious worship. She sees a need to "transform" U.S. culture into a kind of Christian fundamentalist utopia. In another MacLaurin Institute article Bob Osburn asked, "What makes a great university?" His answer was that it was one that was ready to "make room for Christian (or, more broadly, theistic) scholarship as a worthy academic competitor to the naturalistic and post modern world views that dominate academia." The political Right has long harbored distaste for college campuses, viewing them as breeding grounds for such ills as independent thinking, liberal politics, and other general unpleasantness.

ACADEMIC ACCEPTABILITY

As part of the Discovery Institute's strategy to gain scientific acceptance for ID, they are trying to get scientific papers describing research done on design published in respected, academically rigorous, peer-reviewed journals. If this were achieved some of the sting of the critique of ID not being real science could be eliminated. In 2004 that dream came true, sort of. Discovery Institute fellow Stephen Meyer managed to have a pro–Intelligent Design paper of his published in the *Proceedings of the Biological Society of Washington*. The Discovery Institute touted this as a major breakthrough because their ideas had been accepted into mainstream science. Critics quickly saw two problems with the article. Titled "The Origin of Biological Information and the Higher Taxonomic Categories" the article argued that in the decade prior to its publication a number of evolutionists had published papers that questioned the source of biological innovation, in other words, where new species come from. Meyer also said that "many biologists" questioned the ability of natural selection to explain evolution. What Meyer was intending to do was put forward just the sort of explanation which would answer these fundamental questions of biological diversity. He argued that because the probability of new genes appearing in DNA is so astronomically high it couldn't be accounted for by natural selection. After supposedly proving that evolution cannot account for biological diversity he failed to explain how ID could. In the end, critics pointed out that the paper was more a denunciation of evolution than it was an explanation of Intelligent Design.

The other element critics pointed to was the way the paper was published. The *Proceedings of the Biological Society of Washington* is a small, generally respected, but not widely distributed publication. The journal's editor, Richard Von Sternberg, was accused of being a creationist and ID supporter with ties to the Discovery Institute. He chose to edit the paper himself instead of assigning it to another as is the usual practice. He sent the paper out for blind review, but critics charged he may have sent it to reviewers sympathetic to creationism. Members of the journal's board were upset such an article was published, according to them, without their knowledge. The board declared that they would never have authorized such a publication. So while Meyer and the Discovery Institute tout this article as a breakthrough in ID history, the mainstream sees its publication as unworthy of serious consideration. The one thing that has always dogged the ID movement is that for all their talk of scientific values and methodology they have had a great deal of trouble convincing other scientists of this point. They have put forward no real evidence of their contentions that other scientists find of value. In mainstream science respect for a point of view or a new idea comes from being able to publish an article in a refereed scientific journal. This grueling process requires articles to be reviewed by a series of blind reviewers for accuracy, scientific merit, and intellectual plausibility. The author has to be able to show specific examples and convincing evidence of their contention. If this is not present in the article it does not get published. This sort of rigorous oversight helps ensure the scientific bona fides of the author and generates respect for their work. Not being able to publish in such high-level venues is a sign that one's work is not taken seriously by other scientists. If ID is really a scientific endeavor the way believers claim, it must prove, like all new ideas, that it has the strength to go the distance in a scientific discourse.

Trouble began for the ID movement in 1999 when a copy-store employee leaked an extensive internal Discovery Institute memo to the public. Known as the "Wedge Document," it laid out the organization's short- and long-term goals. The document caused a stir, as it seemed to be a manifesto for remaking, if not taking over, American society and culture. Their goal was to replace evolution with Intelligent Design theory as the dominant approach to science. A number of people called the document an alarming conspiracy of religious fanatics. The CSC tried to wave off the text as an early fundraising proposal of no real importance. They said that the Wedge articulates a plan for reasoned persuasion, not political control. According to the document the Discovery Institute wants to move in careful steps: First making the broader public aware of

ID, then getting scientists to support ID, then gaining parity with evolution, and finally ousting evolution altogether. Zealous civilian supporters who often on their own have thrown ID into the public school teaching campaign ahead of schedule are undermining this careful stratagem, however. The numerous legal defeats of those trying to insert ID into the American public school system have brought publicity to the cause and made ID a household word, but it has also generated a backlash that could ultimately prove the movement's undoing.

Claims by the Discovery Institute, the CSC, and the Intelligent Design movement in general that they are secular and not religious come across as disingenuous when one looks at the underlining political and religious connections the organization has. A serious contributor of funding to the Discovery Institute has been the California millionaire Howard Ahmanson, who has also given money to a number of conservative political causes. (Ahmanson was eventually appointed to the Discovery Institute's board of directors). Phillip Johnson dedicated his book *Defeating Darwinism* (1997) to Ahmanson and his wife. Ahmanson was a student and close confidant of Rousas J. Rushdoony (1916–2001), a prime mover in the Christian reconstructionist and dominionist movements as well as a pioneer in the American home school movement. Reconstruction/Dominion followers believe in replacing the U.S. Constitution with Mosaic law and in taking dominion over the U.S. government and culture. They see themselves surrounded by Godless heathens and their perverted ways. Rushdoony preached a severe eschatology (end of the world belief) in which the Second Coming of Christ could only occur after biblical law reigns over the earth. A major obstacle to this plan, as they see it, is the teaching of and belief in evolution.

Combining such realms as religious and political influence has not been without problems. As governor of Texas and then presidential-hopeful George Bush said, alternatives to evolution should be taught in public schools so children would get all the sides of the argument. At the height of the Dover case in 2005, President Bush reiterated his position and added that ID was good science. At the same time his science advisor, John Marburger, said ID was not a genuine scientific concept—backpedaling Marburger then said Bush meant it should be taught as part of the "social context" of science. The elected official who was probably the most outspoken supporter of ID in the United States was the former Republican senator from Pennsylvania, Rick Santorum. As a neo-conservative Santorum gravitated to the fundamentalist end of the religious spectrum and involved himself in the creation/evolution debate by associating himself with the Discovery Institute and the Dover, Pennsylvania, court case. In 2000 the Discovery Institute held a special briefing in Washington, D.C., to promote their position and explain ID to a group of sympathetic Congress members including Santorum. The next year Santorum became involved in the writing of George Bush's controversial "No Child Left Behind Act" which was promoted as a major positive overhaul of the U.S. public school system meant to ensure an adequate elementary education for all American children. Santorum wrote a special amendment to the act that included language that promoted ID as an alternative to teaching evolution in the classroom. The "Santorum amendment" was eventually attached marginally to the No Child Left Behind Act—over strenuous objections from a wide range of critics—but only in an altered form that did not have the force of law. The Discovery Institute held this up as a major leap forward for ID supporters despite that fact that just what the Santorum amendment accomplished was in doubt.

Santorum also supported the Dover school board's desire to include ID in the official curriculum saying in newspaper accounts that it was "a legitimate scientific theory that should be taught." Santorum drew upon the language and philosophy of the Discovery Institute. He took the position that students should be made aware of the "gaps and problems" in evolution theory and insisted that schools be allowed to "teach the controversy." He also joined the Thomas More Law Center, the group that backed the Dover school board, as an advisor. When the plaintiffs won the Dover case, Santorum quickly jumped ship by dropping his support for the Dover board and resigning his position on the Thomas More Law Center board. Taking the Discovery Institute's party line he said that while teaching the controversy was still a good idea trying to force school boards and teachers to teach ID was not. He claimed he was bothered by the religious aspects of the testimony of some ID supporters (mostly Michael Behe). At the time, Santorum was preparing to mount his next congressional election campaign. His opponent, Pennsylvania state treasurer Robert Casey, said he thought Santorum was changing allegiances in a case of political expediency. The Thomas More Law Center was also a bit peeved at Santorum's actions saying he was playing politics with their cause. Still not sure where he wanted to stand on the ID issue, in 2006 Santorum wrote a glowing introduction to a book praising the career of ID legend Phillip Johnson. Santorum lost his subsequent bid for reelection.

Another example of the religious underpinnings of the ID movement is Discovery Institute senior fellow Jonathan Wells. With a pair of science doctorates from the University of California at Berkeley he is the author of the other book titled *Icons of Evolution* (2000) that purported to have debunked many of the most famous icons of evolutionary thought including peppered moths, *Archaeopteryx*, the branching tree model of organic development, and other subjects discussed in the book you are now reading. Wells is also a devoted follower and confidant of the Reverend Sun Yung Moon of the Unification Church. Wells speaks openly and proudly of how "Father," that is, Rev. Moon, personally picked him to go to Berkeley to get a doctorate in biology for the express purpose of destroying Darwinism. Wells believes that empirical evidence proves Intelligent Design, but that science can never show who the designer is. Like so many ID proponents he wants to have it both ways by saying that science can prove enough to support his position but not enough to get him in trouble publicly. This trouble was growing increasing public and coming increasingly from a surprising source.

THE OTHER SIDE OF THE COIN

Along with mainstream scientists the less obvious and more ironic opponent to Intelligent Design are other religionists. Superficially the ID movement and traditional Christian fundamentalists would appear to be natural allies, but they are not, certainly not at the deeper theological level. ID is only part of the wider continuum of anti-evolutionary thinking. While it may seem so, the religious response to evolution is not monolithic. Not all religious people are necessarily opposed to evolution. There are several blocks of influence currently active in the creation/evolution debate: the evangelical Christian fundamentalist block, the Intelligent Design block, and the liberal Christian block (who might also be termed liberal creationists).

Traditional Christian fundamentalists are mostly biblical literalists who believe that the story of Genesis is to be taken exactly as it is written to mean exactly what it implies. In other words, they believe God made the world in six twenty-four-hour days—period. As far as they are concerned there is no need for further explanation. These people are known as "young earth" creationists as they hold the earth to be at most 10,000 years old, probably less. Over the years some biblical literalists have tried to incorporate and account for geological evidence that suggests the earth is much older. One way to do this is to use the "gap" theory in which the twenty-four-hour creation days were then followed by a gap of indeterminate length (which could have been hundreds of thousands or even millions of years) before the historical period of the Bible began. Another approach is the "day-age" theory that believers argue supports the idea that each of the Genesis days was of indeterminate length, not twenty-four-hour days. These concepts allow believers to hold a literalist interpretation of the Bible without established scientific facts getting in the way of scripture. Gap and day-age theory are, however, often looked at askance by other fundamentalists as dangerous backsliding and a capitulation to materialism. The confrontations between various schools of fundamentalist creationists are as vociferous as those directed against evolution with some of the more fantastic theorists accusing their colleagues of un-Godliness for even gently questioning their wilder claims.

The liberal creationists differ from their more conservative brethren in that they are mostly *not* biblical literalists. As such they are more open to accepting evolution as part of God's wider plan or do not see it as such a threat to their spirituality. They include a large portion of the Protestant Christian world as well as Catholics, Jews, Muslims, and indeed most of the world's religious communities. A number of nineteenth-century theologians and scientists came under this banner as they accepted the physical evidence of evolution as a working biological process yet wanted to say that ultimately God was responsible for it all.

Increasingly, in the early twenty-first century as the evangelical fundamentalists vigorously advanced their agendas non-fundamentalist religionists began to speak out, not so much because they supported evolution—though some did—but because they were growing increasingly alarmed at the attempt by a narrowly focused section of Christianity to exercise its control over the wider culture and government. They feared that if the fundamentalists ignored or repealed the separation of church and state in America it would leave non-fundamentalist religious groups open to the same kind of attacks as were being made against evolution. Hard-line fundamentalists made it clear they were uncomfortable with any religion not believing exactly as they did. The far religious Right wanted to do away with "false" religions as well as evolution. In response to this a Christian apologetics organization called Reasons to Believe came out against ID. A biochemist spokesman for this creationist group argued that ID was theology not science and introducing a theologically based idea into school science classes created far more problems than it solved. The American Scientific Affiliation—an older collection of scientists who consider themselves staunch Christians—said the growing acrimony over evolution was needlessly setting one Christian group against another. They also argued that ID is neither good science nor good theology.

While pro-evolution forces were quick to point out the glaring scientific faults in ID some thoughtful Christian philosophers have seen the inherent theological problem with it as well. If God is not responsible for all of creation, if God only steps in where needed—as William Dembski and others suggest—then what accounts for the other aspects of life? This "God of the gaps," theologians argue, takes the grandeur and omnipotence out of the Divine and reduces God to a sort of part-time repairman who only comes along to install devices or fix things from time to time.

In an article for *The National Catholic Bioethics Quarterly*, "Intelligent Design: The Scientific Alternative to Evolution" (2003), authors William Harris and John Calvert argue that there is an element of chance with the purposeful design. This suggests that the creator is not omnipotent. They also admit that organisms have changed over time from non-intelligent causes. That ID proponents accept that the universe is billions of years old is in direct contravention to what the literalists believe. Worst still, because the ID proponents refuse to say, at least publicly, who the designer is, their view rejects Jesus as the savior and the only route to redemption and salvation. The designer—as framed by ID theorists—could conceivably be the Judeo-Christian God, it could also be Allah, Shiva, Zeus, space aliens, or even a giant hedgehog named Spiney Norman. If God is only used to explain questions of origins and biological processes that seem to have no rational answer or are not explained by natural selection what happens when rational answers to those questions are found? Will God then disappear or go into retirement? ID leaders want to be able to teach their views in public schools so they are forced to water down the religiosity of their message. This pushes them even further from their natural allies in the Christian fundamentalist camp. What some are only now realizing is that ID poses far more profound problems for creationists than it does for evolutionists.

A leading organization of anti-evolutionist rhetoric, which, unlike the Discovery Institute, wears its Christian fundamentalism openly and proudly, is Answers in Genesis (AIG). Begun in Australia in the 1970s it was originally known as Creation Science Education Media Services. It combined with the Creation Science Association to become the Creation Science Foundation. The central character in the Australian organization was evangelical preacher and ex-school science teacher Ken Ham, who split his time between the CSF and the American Institute for Creation Research. By 1994 he founded AIG. Ham then took up residence in the United States his organization quickly becoming a multimillion-dollar worldwide operation. Besides generally spreading Christianity, AIG's primary reason for being is to take on evolution with scripture. Their view of science is that it takes second place to scripture—as does the U.S. Constitution—and that if science is "properly" interpreted it supports the young earth creation model as well as the historicity of Genesis. This puts them, as it does all young earth fundamentalists, in an awkward relationship to Intelligent Design.

AIG's approach to ID was to keep a tone of outward bonhomie and mutual respect. However, as time goes by, AIG seems to be growing increasingly uncomfortable with the basic tenants of ID. In a number of its publications, as well as its 2005 Mega Creation Conference held at Jerry Falwell's Liberty University, AIG spokespersons smiled through their descriptions of ID and gently led their members away from it. Acceptance of a universe billions of years old, the idea that some life on earth evolved, coupled with the public resistance to accepting Jesus as savior and the Judeo-Christian God as creator undermines everything AIG stands for. Not only has Ham accused mainstream churches of abandoning their duty to Christ by becoming secular, he accused the ID proponents of riding the coattails of the creation movement to popularity.

This inherent conundrum has been picked up on by many traditional biblical literalists and caused them to view ID with as much suspicion as they do evolution.

Playing a Creation Game

At the "Creation Mega Conference" sponsored by the group Answers In Genesis and held at Liberty University, Lynchburg, Virginia, in August 2005 there was a substantial concession area. There was a wide variety of creationist books, DVDs, and other materials showing how the Neandertals were descendants of Adam and Eve, how the extensive stratigraphy of the earth was proof of its young age, and how scientific interpretations of fossils are wrong. Amid this horn of creationist plenty, there were no board games or toys. In 2007, however, that situation was remedied. The Christian fundamentalist publisher Living Waters released *Intelligent Design vs. Evolution*. The game is promoted by popular biblical exegete Ray Comfort and Kirk Cameron, the former teen television star of *Growing Pains*, who turned to religious works and has starred in the *Left Behind* movies, a set of eschatological films based on the novels of the same name.

The game itself is fairly simple; a kind of religious Trivial Pursuit®. Two players or teams take the sides of Intelligent Design and evolution respectively. A series of question cards are consulted with correct answers receiving points. Dice casts get the teams—each represented by a little rubber brain—moving around the board and reading off instructions they "immediately obey." The first to reach the end of the trail, or who has accumulated the most points, win.

The central dynamics of the game swirl around the questions on the "Brain Teaser" cards. On these cards are found the usual theological self-congratulation, statements of scientists taken out of context, proclamations from creationists, and snippets from scripture. Many of the cards make it clear that when scientists change their minds because new information is discovered this is evidence of the inherently faulty nature of science. The cards reinforce believer's positions and hammer home how silly sounding and befuddled evolutionists are. According to the cards, there is no fossil evidence for human evolution and carbon dating does not work. One of the cards proclaims rather tellingly that it was not Christians who perpetrated the Crusades, but Catholics.

The greatest irony of the game is that they use Intelligent Design at all. In the early twenty-first century there is a growing rift in the anti-evolution movement in America between young earth traditionalists, who wear their Christianity on their sleeve—the way this game does—and Intelligent Design proponents who have been trying the best they can to distance themselves, at least publicly, from the kind of evangelical fundamentalism the game rides on. That they use Intelligent Design here is an example of the quirky place it holds in the world of creationism. The game box art states that the Christian God is the Intelligent Designer. The Discovery Institute—the primary promoter of ID—has been disavowing statements like these in an effort to win legal cases. Dubious logic aside, the audience for this game is unclear. It's reliance on Intelligent Design puts off many in the fundamentalist world and its overbearing fundamentalism makes it more difficult for Intelligent Design proponents to argue they are not religious.

An indication of the fragmentation and growing infighting of religious groups over ID is that the Discovery Institute advised the local El Tejon, California, school district in 2006 to drop its plans to offer a course called "Philosophy of Design" because it commingled ID with young earth creationism and the institute did not want its ideas of ID mixed up with traditional fundamentalist views. Cracks were showing in the ID movement in a number of ways. In 1999 William Dembski was invited to start an ID research center on the campus of Baylor University. Baylor is a respected Baptist-centered college outside Waco, Texas. The college's president Robert Sloan, a devout Christian, invited him. The two met because Dembski was a tutor to Sloan's daughter at a Christian summer camp. The Baylor faculty became agitated over the new addition, called the Polanyi Center, because Sloan bypassed the usual faculty consultation normally done before a new college office is opened. The furor grew in part because the science faculty and others disagreed greatly with the idea that this new operation was a "science" center at all. An outside advisory board suggested the center be made part of the college's existing Institute for Faith and Learning and that the name Polanyi be taken off as Polanyi's work, in their estimation, had nothing to do with what the center's proposed mission—looking for evidence of Intelligent Design—was set up to do (the "center" only had two members, Dembski and one administrator). The college president followed the board's advice and the changes were made. Dembski saw this as a great victory and immediately taunted his Baylor colleagues calling them "dogmatic" opponents of design theory, which upset people more. The school then removed Dembski from the position because of these remarks. He later left Baylor and went to the Center for Science and Theology at Southern Methodist University, Kentucky. The work meant to be done by Dembski at the center was quietly dropped at Baylor.

Despite this debacle, Dembski, Baylor, and controversy still had some time to spend together. In March of 2006 Baylor turned down the tenure request of theologian and philosopher Francis J. Beckwith. Supporters said the school was doing this in response to Beckwith's open support of

Intelligent Design, his relationship to the Discovery Institute, and his contention that teaching ID in public schools was not unconstitutional. Discovery Institute fellow Nancy Pearcy ran an opinion piece in her newsletter charging that Beckwith had been denied his tenure because of his pro-life stance. Attacking their own, Pearcy's newsletter then went on to suggest that liberal secularization had infected Baylor and undermined its Christian mission and that it had been taken over by the nefarious forces of darkness. William Dembski too weighed in that Baylor was on the fast track to perdition. He agreed that secularization was killing a once-proud Christian university. Dembski's suggestion that Baylor was turning its back on proper Christianity is odd coming from a man who claims his ID work has nothing to do with religion. Days after the Beckwith story came out Dembski left Southern Methodist and went to Southwestern Baptist Theological Seminary at Fort Worth, Texas. He was replaced at Southern Baptist by Kurt Wise, a traditional young earth creationist, who was not particularly enamored of ID. This has all lead to a curious situation of opposing forces. Evolutionists see ID as disguised creationism while some creationists (young earth adherents especially) see ID as disguised evolutionism, while neither side sees it as both.

CONCLUSION

Taken in the aggregate Intelligent Design is a political movement which masks itself as a pseudo-religion with some of the trappings of science. It does not promote, but ultimately offends both sensibilities. While research into complexity theory could have the potential for generating new insights into biological mechanics, the attempt to link it to a search for evidence of a designer undermines its scientific acceptability and alienates it from theology. It trivializes religion and attempts to undermine science. Critics charge that ID proponents fight so passionately against evolution because they are really fighting against modernity (at least the philosophical implications of modernity) in its broadest sense. There is a fear of the rough-and-tumble and ever-changing nature of modern society where truths and absolutes are hard to come by and where people long for a time when things seemed clearer, less ambiguous, and more stable. People naturally search for meaning and peace of mind in a world that rarely offers it. We are profoundly troubled by the apparent social ills that seem everywhere to inundate us: promiscuity, violence, lack of focus, and rapid change (things which trouble evolutionists as well).

The course Design theorists have chosen is to attack what they think is at the heart of it all, the belief in and teaching of evolution. They are trying to force a paradigm shift, but as Thomas Kuhn pointed out, paradigms only change organically when the conditions are right. For anti-evolutionists evolution is an icon of everything they fear, everything they think is ruining us all. Evolution, they say, teaches us we are animals and so we act that way. Anti-evolutionists believe that evolution leads to lawlessness and the breakdown of social order. The reality, however, is whether a person is a saint or a sinner has nothing to do with whether they believe in evolution or not. The argument that Adolf Hitler and the Nazis used evolution to justify their horrors is a spurious one. While he did make references, early in his régime, to a perverted form of evolution Hitler also perverted the tenets of Christianity and the Bible as well. The many excuses Hitler used were just that, excuses that had nothing to do with the subjects themselves. Also, while the loathsome serial killer and cannibal Jeffrey Dahmer once made a vague reference to evolution in an interview, his father was a noted creationist who often gave public lectures on the topic. On the other hand, the examples are legion where individuals and groups rampaged, raped, and murdered their way across human history and claimed they did it in the name of one religion or another or because they claimed God told them to. So should we ban religion? There are just as many instances of religious belief leading to great acts of kindness, sacrifice, and efforts to improve the human condition.

Creationists like to say that evolutionists "fear" Intelligent Design. They are right, but for the wrong reasons. The fear evolutionists have is not that evolutionary science will be shown to be a hollow sham, or a social construction—reading this book should help show how strongly they feel about their work and how much overwhelming evidence there is for evolution—but that something masquerading as science will be used to fool the unenlightened and sway them to intellectual stagnation, and stultifying and undermining the education system and returning us to a dark age of intellectual backwardness which will help eliminate the human rights free people hold so dear.

In his *Doubts About Darwin: A History of Intelligent Design* (2003) Thomas Woodward said, "Awakening to the power of hidden metaphysical foundations [brought on by ID] is the most inevitable and immediately beneficial sort of "arousing from dogmatic slumber" that Design could advance" (p. 210). In other words belief in ID will make the world a better place: The implication being that evolution does the opposite. Eliminating belief in evolution and substituting it with Intelligent Design will not save the world. As evolution is a natural biological system which works regardless of whether we believe it or not, it will go on. (Even if you do not believe in evolution you will still evolve.) In the end there may very well be a God or gods or an Intelligent Designer, but by its very nature, science lacks the tools to make the determination.

Belief in the divine does not come from fossils, genetic sequences, geologic formations, irreducible complexity theory, or state legislation, it comes from the heart.

In November 2005 the American Museum of Natural History in New York opened a major exhibit on the life and work of Charles Darwin that had been in the making for years. For the first time visitors had the chance to see many of Darwin's original letters and notebooks as well as a meticulous reconstruction of his personal library in which he worked out many of his ideas. As part of the opening festivities a public panel discussion was held by three of the people most responsible for putting the exhibit together. Niles Eldredge (who along with Stephen Jay Gould developed the idea of Punctuated Equilibria), David Kohn (who spent most of his career studying and publishing Darwin's notebooks), and Randall Keynes (author and Charles Darwin's great-great grandson) gave presentations and answered questions from the audience. At the end of the discussion the question was asked about how Charles Darwin would have reacted to the current furor over Intelligent Design. All three men—whose intimate knowledge of the subject allowed them a greater insight into a possible answer—smiled. Answering for them Randall Keynes mused that if Darwin were alive today he would not be surprised by the controversy, but might be a little disappointed in the low quality of the opposing argument.

FURTHER READING

Dembski, William. *The Design Inference: Eliminating Chance Through Small Probabilities* (Cambridge: Cambridge University Press, 1998).

Forrest, Barbara, and Gross, Paul A. *Creationism's Trojan Horse: the Wedge of Intelligent Design* (New York: Oxford University Press, 2004).

Padian, Kevin. (1989). "Review, *Of Pandas and People*," *Bookwatch Reviews* 2: 11.

Postman, David. "Seattle's Discovery Institute Scrambling to Rebound after Intelligent Design Ruling," *Seattle Times* (April 27, 2006).

Regal, Brian. *Human Evolution: a Guide to the Debates* (Oxford: ABC-CLIO, 2004).

Wasley, Paula, Bartlett, Thomas, and Labi, Aisha. "Peer Review," *Chronicle of Higher Education* (April 14, 2006).

Woodward, Thomas. *Doubts about Darwin: A History of Intelligent Design* (Grand Rapids, MI: Baker Books, 2003).

Bibliography

Albarello, Bruno. *L'Affaire de l'Homme de La Chapelle-aux-Saints, 1905–1909* (Treignac: Editions 'Les Monédières,' 1987).

Albritton, Claude C. *The Abyss of Time* (San Francisco: Freeman, 1980).

Alemseged, Z., Spoor, F., Kimbel, W. H., Bobe, R., Geraads, D., Reed, D., and Wynn, J. G. "A Juvenile Early Hominin Skeleton from Dikika, Ethiopia," *Nature* 443 (2006): 296–301.

Alsberg, P. "The Taungs Puzzle: A Biological Essay," *Man* 34 (Oct. 1934): 154–159.

Altholz, Josef. *Anatomy of a Controversy: The Debate over Essays and Reviews 1860–1864* (Brookfield, VT: Ashgate, 1994).

Alvarez, W. *T. Rex and the Crater of Doom* (Princeton: Princeton University Press, 1997).

Anon. "Am I a Man and a Brother?" (cartoon), *Punch* (18 May 1861): 206.

Anon. "Lion of the Season" (cartoon), *Punch* 40 (25 May 1861): 213.

Anon. "The Philosophical Institution and Professor Huxley," *The Witness* (14 January 1862): 2377; (18 January 1862): 2379.

Anon. "Review of Man's Place," *The Athenaeum* (28 February 1863): 287–288.

Anon. "Man's Origin and Nature" Part 1 (Review), *The National Reformer* (28 March 1863): 5.

Anon. "Man's Origin and Nature" Part 2 (Review), *The National Reformer* (4 April 1863): 5.

Anon. "Man's Place in Nature" (Review), *Natural History Review* (April 1863): 381–384.

Anon. "Evidences as to Man's Place in Nature" (Review), *North American Review* (July 1863): 293.

Anon. "Huxley's Origin of Species" (Review), *New Englander and Yale Review* (July 1863): 592.

Anon. "The Neanderthal Man," *Harper's Weekly* 17 (864) (19 July 1873): 617–618.

Anon. "Palaeontologische onderzoekingen oop Java," *Bataviaasch Nieuwsblad* (6 February 1893), no. 57.

Anon. "The Most Important Anthropological Discovery for Fifty Years," *The Illustrated London News* (27 February 1909): 300–301, 312–313.

Anon. "U.N.E.S.C.O. on Race," *Man* 50 (1950): 138–139.
Anon. "U.N.E.S.C.O.'s New Statement on Race," *Man* 51 (1951): 154–155.
Appel, Toby A. *The Cuvier-Geoffroy Debate: French Biology in the Decades Before Darwin* (Oxford: Oxford University Press, 1987).
Asimov, Isaac. "Lastborn (The Ugly Little Boy)," *Galaxy* (September 1958).
Auel, Jean. *The Clan of the Cave Bear* (London: Hodder and Stoughton, 1980).
Bakker, R. T. "Dinosaur Renaissance," *Scientific American* 232 (1975): 58–78.
Bakker, R. T. *The Dinosaur Heresies* (New York: William Morrow, 1986).
Bakker, R. T., and Galton, P. M. "Dinosaur Monophyly and a New Class of Vertebrates," *Nature* 248 (1974): 168–172.
Bannister, Robert C. *Social Darwinism: Science and Myth in Anglo-American Social Thought* (Philadelphia: Temple University Press, 1979).
Barbour, Erwin Hinckley. "Evidence of Loess Man in Nebraska," *Nebraska Geological Survey* 2(6): (1907).
Barther, F. A. "The Word 'Australopithcus' and Others," *Nature* 115(2903) (20 June 1925): 947.
Basalla, George, Coleman, William, and Kargon, Robert H. (eds.). *Victorian Science: A Self-Portrait from the Presidential Addresses of the British Association for the Advancement of Science* (Garden City, NY: Doubleday, 1970).
Bateson, W. *Mendel's Principle of Heredity* (Cambridge: University of Cambridge Press, 1930).
Benen, Steve. "From Genesis to Dominion," *Church & State* (July/August, 2000).
Bennett, D. K. "Stripes Do Not a Zebra Make. Part I: A Cladistic Analysis of *Equus*," *Systematic Zoology* 29 (1980): 272–287.
Berger, L. R., and Clarke, R. J. "Eagle Involvement in Accumulation of the Taung Child Fauna," *Journal of Human Evolution* 29(3) (Sept. 1995): 275–299.
Berry, R. J. "Industrial Melanism and Peppered Moths (*Biston betularia* (L))," *Biological Journal of the Linnean Society* 39 (1990): 301–322.
Berry, W.B.N. *The Growth of a Prehistoric Time Scale* (San Francisco: Freeman, 1968).
Bibby, Cyril. *Scientist Extraordinary: The Life and Scientific Work of Thomas Henry Huxley 1825–1895*. Oxford: Pergamon, 1972.
Bishop, B. E. "Mendel's Opposition to Evolution and to Darwin," *Journal of Heredity* 87 (1996): 205–213.
Blake, C. C. "On the Cranium of the Most Ancient Races of Man," *Geologist* (June 1862).
Blake, Carter. "Professor Huxley on Man's Place in Nature," *Edinburgh Review* 177(1863): 541–569.
Blanckaert, Claude, Cohen, Claudine, Corsi, Pietro, and Fischer, Jean-Louis (eds.). *Le Muséum au Premier Siècle de son Histoire* [The Museum in the First Century of its History] (Paris: Editions du Muséum d'histoire naturelle, 1997).
Blinderman, Charles. *The Piltdown Inquest* (Buffalo, NY: Prometheus Books, 1986).
Blinderman, Charles. *The Huxley File*. Available online at http://aleph0.clarku.edu/huxley.
Boaz, Noel T., and Ciochan, Russell L. *Dragon Bone Hill: An Ice Age Saga of Homo Erectus* (New York: Oxford University Press, 2004).

Boomgaard, P. "Morbidity and Mortality in Java, 1820–1880: The Evidence of the Colonial Reports," in Norman G. Owen (ed.). *Death and Disease in Southeast Asia; Explorations in Social, Medical and Demographic History* (Singapore, Sudeny: Oxford University Press, 1987), 48–69.

Boswell, P.G.H. "Human Remains from Kanam and Kanjera, Kenya Colony," *Nature* 135 (1935): 371.

Boule, Marcellin. "L'Homme fossile de La Chapelle-aux-Saints (Corrèze)," *L'Anthropologie* 19 (1908): 519–525.

Boule, Marcellin. "L'Homme fossile de La Chapelle-aux-Saints," *Annales de Paléontologie* 6, 7, 8 (1911, 1912, 1913): 1–64, 65–208, 209–279.

Boule, Marcellin. "La guerre," *L'Anthropologie* 25 (1914): 575–580.

Boule, Marcellin. "L'Homo Néanderthalensis et sa place dans la nature," in *Comptes rendus, 14e Congrès International d'Anthropologie et d'Archéologie Préhistoriques, Genève* (Geneva: Kündig, 1914).

Boule, Marcellin. "Les hommes fossiles," *Éléments de paléontologie humaine*, 2nd ed. (Paris: Masson, 1923).

Bouyssonie, A., Bouyssonie, J., and Bardon, L. "Découverte d'un squelette humain Moustérien à la bouffia de La Chapelle-aux-Saints (Corrèze)," *L'Anthropologie* 19 (1908): 513–518.

Bowater, W. "Heredity of Melanism in the Lepidoptera," *Journal of Genetics* 3 (1914): 299–315.

Bowden, M. *The Rise of the Evolution Fraud* (San Diego: Creation-Life Publishers, 1982).

Bowler, Peter J. "Darwin's Concepts of Variation," *J. Hist. Medicine* 29 (1974): 196–212.

Bowler, Peter J. *Fossils and Progress. Paleontology and the Idea of Progressive Evolution in the 19th Century* (New York: Science History Publications, 1976).

Bowler, Peter. "Malthus, Darwin and the Concept of Struggle," *J. Hist. Ideas* 37 (1976): 631–650.

Bowler, Peter. "Alfred Russel Wallace's Concepts of Variation," *J. Hist. Medicine* 31 (1976): 17–29.

Bowler, Peter J. *The Eclipse of Darwinism: Anti-Darwin Evolutionary Theories in the Decades Around 1900* (Baltimore: Johns Hopkins, 1983).

Bowler, Peter J. *Evolution: The History of an Idea* (Berkeley: University of California Press, 1984).

Bowler, Peter J. *Theories of Human Evolution: A Century of Debate 1844–1944* (Baltimore: Johns Hopkins, 1986).

Bowler, Peter. *The Non-Darwinian Revolution: Reinterpreting a Historical Myth* (Baltimore: Johns Hopkins, 1988).

Bowler, Peter. *Life's Splendid Drama: Evolutionary Biology and the Reconstruction of Life's Ancestry, 1860–1940* (Chicago: University of Chicago Press, 1996).

Bowler, P. J. *Evolution: History of an Idea*, 3rd ed. (Berkeley: University of California Press, 2003).

Brace, Loring C. "The Fate of the 'Classic' Neanderthals. A Consideration of Hominid Catastrophism," *Current Anthropology* 5 (1964): 3–43.

Brain, C. K. *The Hunters of the Hunted. An Introduction to African Cave Taphonomy* (Chicago: University of Chicago Press, 1981).

Bräuer, G. "Africa's Place in the Evolution of Homo Sapiens," in G. Bräuer, Y. Yokoyama, C. Falguères, and E. Mbua (1997), "Modern Human Origins Backdated," *Nature* 386 (1992): 337–338.

Bräuer, G., and Smith, F. *Continuity or Replacement? Controversies in* Homo Sapiens *Evolution* (Rotterdam: Balkema, 1992), pp. 83–98.

Broberg, Gunnar. *Homo Sapiens L: Studier i Carl von Linnés naturuppfattning och människolära* [Homo Sapiens L.: Studies in Carolus Linnaeus' Anthropology and Conception of Nature] (Stockholm: Almqvist & Wiksell, 1975).

Brook, John Hedley. *Science and Religion: Some Historical Perspectives* (New York: Cambridge University Press, 1993).

Broom, R. "Some Notes on the Taung Skull," *Nature* 115(2894) (18 April 1925): 569–571.

Broom, R. "A New Fossil Anthropoid Skull from South Africa," *Nature* 138(1), (19 September 1936): 486–488.

Broom, R. "Pleistocene Anthropoid Apes of South Africa," *Nature* 142 (1938): 377–379.

Broom, R., and Robinson, J. T. "Swartkrans Ape-Man: *Paranthropus crassidens*," *Transvaal Museum Memoir* 6 (1952): 1–123.

Brown, Lee Rust. "The Emerson Museum," *Representations* 40 (1992): 57–80.

Brown, P., et al. "A New Small-Bodied Hominin from the Late Pleistocene of Flores, Indonesia," *Nature* 431 (2004): 1055–1061.

Browne, Janet. "Darwin's Botanical Arithmetic and the 'Principle of Divergence,' 1854–1858," *J. Hist. Biology* 13 (1980): 53–89.

Browne, Janet. *Charles Darwin: Voyaging* and *Charles Darwin: The Power of Place* (New York: Knopf, 1995 and 2002).

Browne, Janet. *Charles Darwin, Biography: Vol. I, Voyaging* (London: Princeton University Press, 1996).

Browne, Janet. *Charles Darwin, Biography: Vol. II, The Power of Place* (London: Knopf, 2002).

Brunet, M., et al. "A New Hominid from the Upper Miocene of Chad, Central Africa," *Nature* 8 (2002): 145–151.

Brush, A. H. "The Beginnings of Feathers," in J. Gauthier, and L. F. Gall (eds.), *New Perspectives on the Origin and Early Evolution of Birds: Proceedings of the International Symposium in Honor of John H. Ostrom* (New Haven, CT: Peabody Museum of Natural History, 2001), pp. 171–179.

Bryson, Bill. *A Short History of Nearly Everything* (New York: Broadway Books, 2003).

Buckland, William. "Geology and Mineralogy Considered with Reference to Natural Theology," *The Bridgewater Treatises VI, On the Power, Wisdom, and Goodness of God, as Manifested in the Creation* (London: William Pickering, 1836).

Budiansky, S. *The Nature of Horses: Exploring Equine Evolution, Intelligence and Behavior* (New York: The Free Press, 1997).

Burchfield, Joe D. *Lord Kelvin and the Age of the Earth* (Chicago: University of Chicago Press, 1990).

Burkhardt, Frederick, and Duncan Porter (eds.). *The Correspondence of Charles Darwin, vols vii–xi* (Cambridge: Cambridge University Press, 1911–1999).

Burkhardt, F., and Smith, S. (eds.). *A Calendar of the Correspondence of Charles Darwin, 1821–1882* (New York: Garland, 1985).

Burkhardt, F., and Smith, S. *The Correspondence of Charles Darwin*, 14 vols. (Cambridge: Cambridge University Press, 2005).
Burroughs, Edgar Rice. "The Land that Time Forgot," *Blue Book Magazine* (August 1918).
Burroughs, Edgar Rice. "The People that Time Forgot," *Blue Book Magazine* (October 1918).
Burroughs, Edgar Rice. "Out of Time's Abyss," *Blue Book Magazine* (December 1918).
Busk, G. "D. Schaafhausen [*sic*], On the Crania of the most Ancient Races of Man," *Natural History Review* (1 April 1861), no 2. Translation of Schaaffhausen, H. "Zur Kentniss der ältesten Rassenschädel," *Archiv. Verbindung Mehreren Gelehrten* (1858): 453–488.
Cadbury, Deborah. *Terrible Lizard: The First Dinosaur Hunters and the Birth of a New Science* (New York: Holt, 2001).
Cain, A. J. and P. Sheppard. "Selection in the Polymorphic Land Snail *Cepaea nemoralis*," *Heredity* 4 (1950): 275–294.
Cain, J. "Ernst Mayr as Community Architect: Launching the Society for the Study of Evolution and the Journal *Evolution*," *Biology and Philosophy* 9 (1994): 387–427.
Camp, C. L., and Smith, N. "Phylogeny and Functions of the Digital Ligaments of the Horse," *University of California Memoir* 13 (1942): 69–124.
Campbell, Bernard. "The Centenary of Neandertal Man. Part I, II," *Man* 56 (November, December 1956): 156–158, 171–173.
Cann, Rebecca L. "Mothers, Labels, and Misogyny," in Lori D. Hager, (ed.), *Women in Human Evolution* (London: Routledge, 1997).
Cann, Rebecca L., Stoneking, Mark, and Wilson, Allan C. "Mitochondrial DNA and Human Evolution," *Nature* 325 (1987): 32–36.
Cartmill, Matt. "Taxonomic Revolutions and the Animal-Human Boundary," in R. Corbey and W. Roebroeks (eds.), *Studying Human Origins. Disciplinary History and Epistemology* (Amsterdam: Amsterdam University Press, 2001).
Castle, W. "Piebald Rats and Selection, A Correction," *Am. Nat.* 53 (1919): 570–576.
Charig, A. J., Greenaway, F., Milner, A. C., Walker, C. A., and Whybrow, P. J. "*Archaeopteryx* Is Not A Forgery," *Science* 232 (1986): 622–626.
Chatterjee, S., and Templin, R. J. "Feathered Coelurosaurs from China: New Light on the Arboreal Origin of Avian Flight," in P. J. Currie, E. B. Koppelhus, M. A. Shugar, and J. L. Wright (eds.), *Feathered Dragons. Studies on the Transition from Dinosaurs to Birds*. (Bloomington: Indiana University Press, 2004).
Chiappe, L. M., and Witmer, L. M. (eds.). *Mesozoic Birds* (Los Angeles: University of California Press, 2002).
"China Exclusive: New Clues Loom on Missing Peking Man Skulls," *People's Daily*, English language version, available online (September 6, 2005).
Christiansen, P., and Bonde, N. "Body Plumage in *Archaeopteryx*: A Review, and New Evidence from the Berlin Specimen," *Comptes Rendus Palevol* 3 (2004): 99–118.
Clark, Constance Areson. "Evolution for John Doe: Pictures, the Public, and the Scopes Trial Debate," *Journal of American History* (March 2001): 1275–1303.

Clark, W. E. Le Gros. "Anatomical Studies of Fossil Hominoidea from Africa," in L.S.B. Leakeyand S. Cole (eds.), *Proceedings of the Pan-African Congress on Prehistory, 1947* (Oxford: Basil Blackwell, 1952), 111–115.

Clarke, B. C. "The Art of Innuendo," *Heredity* 90 (2003): 279–280.

Clarke, R. J. "First Ever Discovery of a Well-Preserved Skull and Associated Skeleton of *Australopithecus*," *South African Journal of Science* 94 (1998): 460–463.

Cohen, Claudine. "Stratégies et Rhétoriques de la Preuve dans les Recherches sur les ossements Fossiles de Quadrupèdes" [Strategies and Rhetorics of Proof in the Recherches], in Blanckaert, Cohen, Corsi, and Fischer (eds.), *Le Muséum au Premier Siècle de son Histoire* (Paris: Editions du Muséum d'histoire naturelle, 1997).

Cole, Sonia. *Leakey's Luck. The Life of Louis Seymour Bazett Leakey 1903–1972* (London: Collins, 1975).

Coleman, William. *Georges Cuvier, Zoologist* (Cambridge, MA: Cambridge University Press, 1964).

Coleman, William. "William Bateson," in C. C. Gillespie (ed.), *Dictionary of Scientific Biography*, vol I (New York: Charles Scribner's Sons, 1970), pp. 505–506.

Coleman, William. "Morphology Between Type Concept and Descent Theory," *Journal of the History of Medicine and Allied Sciences* 31(2): (1976).

Cook, Harold J. *Tales of the 04 Ranch* (Lincoln: University of Nebraska Press, 1968).

Cook, L. M. "The Rise and Fall of the Carbonaria Form of the Peppered Moth," *The Quarterly Review of Biology* 78 (2003): 399–417.

Cooke, N. "On Melanism in Lepidoptera," *The Entomologist's Monthly Magazine* 10 (1877): 92–96, 151–153.

Coon, Carleton S. *The Origin of Races* (New York: Alfred A. Knopf, 1962).

Copans, Jean, and Jamin, Jean. *Aux origines de l'anthropologie Française* (Paris: J. M. Place, 1994).

Corbey, Raymond. *The Metaphysics of Apes. Negotiating the Animal Human Boundary* (Cambridge: Cambridge University Press, 2005).

Correns, Carl. *Gregor Mendels Briefe an Carl Nägeli 1866–1873, ein Nachtrag zu den Veröffentlichten Bastardierungsversuchen Mendels* (Leipzig: B.G. Teubner, 1905).

Corsi, Pietro. "Le Muséum et l'Europe" [The Museum and Europe], in Blanckaert, Cohen, Corsi, and Fischer (eds.), *Le Muséum au Premier Siècle de son Histoire* (Paris: Editions du Muséum d'histoire naturelle , 1997).

Cox, William. *The Life of William Colenso, D.D., Bishop of Natal*, 2 vol. (London: W. Ridgway, 1888).

Coyne, J. "Not Black and White," *Nature* 396 (1998): 35–36.

Coyne, J. "Evolution Under Pressure: A Look at the Controversy About Industrial Melanism in the Peppered Moth," *Nature* 418 (2002): 19–20.

Cronin, Helena. *The Ant and the Peacock: Altruism and Sexual Selection from Darwin to Today* (Cambridge: Cambridge University Press, 1991).

Crook, Paul. *Darwinism, War and History* (Cambridge: Cambridge University Press, 1996).

Currie, P. J., and Padian, K. (eds.). *Encyclopedia of Dinosaurs* (New York: Academic Press, 1997).

Cutler, Alan. *The Seashell on the Mountaintop: A Story of Science, Sainthood, and the Humble Genius Who Discovered a New History of the Earth* (New York: Dutton, 2003).

Cuvier, Georges. *Le Règne Animal* (Paris, 1817).

Dale, Richard. *Walking with Cavemen* (GB: BBC Warner, 2003).

Dalrymple, G. Brent. *The Age of the Earth* (Stanford: Stanford University Press, 1994).

Dalrymple, G. Brent. *Ancient Earth, Ancient Skies* (Stanford: Stanford University Press, 2004).

Dames, W. "Über Brüstbein, Schulter- und Beckengürtel der *Archaeopteryx*," *Sitzungsberichte der Königlich Preussischen Akademie der Wissenschaften zu Berlin* (1897) (1897): 818–834.

Damuth, J., and Janis, C. M. "Paleoecological Inferences Using Tooth Wear Rates, Hypsodonty and Life History in Ungulates," *Journal of Vertebrate Paleontology* 25 (2005): 48A.

Dana, James Dwight. "On Man's Zoological Position," *New Englander and Yale Review* (April 1863): 283–287.

Dana, James Dwight. "Review of *Evidence as to Man's Place in Nature* by T. H. Huxley, F.R.S.," *American Journal of Science and the Arts* 35 (1863): 451–454.

Dart, R. A. "*Australopithecus africanus*: The Man-Ape of South Africa," *Nature* 115(2884) (7 February 1925): 195–199.

Dart, R. A. "Faunal and Climatic Fluctuations in the Makapansgat Valley," in L.S.B. Leakey and S. Cole (eds.), *Proceedings of the Pan-African Congress on Prehistory, 1947* (Oxford: Basil Blackwell, 1952), pp. 96–106.

Dart, R. A. "The Predatory Transition from Ape to Man," *International Anthropological and Linguistic Review* 1(4) (1953): 201–219.

Dart, R. A. and Craigh, D. *Adventures with the Missing Link* (London: Hamish Hamilton, 1959).

Darwin, Charles. *On the Origin of Species by Means of Natural Selection: Or the Preservation of Favoured Races in the Struggle for Life*, 6th ed. (London, John Murray, 1872); reprinted with an introduction by Ernst Mayr (Cambridge, MA: Harvard University Press, 1964).

Darwin, Charles. *The Variation of Animals and Plants under Domestication*, 2 vols. (London: John Murray, 1868).

Darwin, Charles. *The Descent of Man and Selection in Relation to Sex*, 2 vols. (London: John Murray, 1871).

Darwin, Charles. *The Autobiography of Charles Darwin*. Edited by Nora Barlow. (New York: Norton, 1992).

Darwin, Charles, and Wallace, Alfred Russel. *Evolution by Natural Selection* (Cambridge: Cambridge University Press, 1958).

Darwin, F. (ed.). *Life & Letters of Charles Darwin*, 3 vols. (London: Murray, 1887).

Davis, P. G. "The Taphonomy of *Archaeopteryx*," *Bulletin of the National Science Museum, Tokyo, Series C* (22) (1996): 91–106.

Dawkins, R. *The Selfish Gene* (New York: Oxford University Press, 1976).

Delisle, Richard G. "Human Palaeontology and the Evolutionary Synthesis During the Decade 1950–1960," in R. Corbey and B. Theunissen (eds.), *Ape, Man, Apeman. Changing Views Since 1600: Evaluative Proceedings of the Symposium Ape, Man, Apeman*, Leiden, the Netherlands, 28 June–1 July, 1993 (Leiden: Department of Prehistory, Leiden University, 1995).

Delisle, Richard G. "Adaptationism Versus Cladism in Human Evolution Studies," in R. Corbey and W. Roebroeks (eds.), *Studying Human Origins. Disciplinary History and Epistemology* (Amsterdam: Amsterdam University Press, 2001).

Dembski, William. *The Design Inference: Eliminating Chance through Small Probabilities* (Cambridge: Cambridge University Press, 1998).

de Moor, J. A. "An Extra Ration of Gin for the Troops; the Army Doctor and Colonial Warfare in the Archipelago, 1830–1880," in G. M. van Heteren, A. de Knecht-van Eekelen, and M.J.D. Poulissen (eds.), *Dutch Medicine in the Malay Archipelago 1816–1942: Articles Presented at a Symposium Held in Honor of Professor De Moulin* (Amsterdam: Rodopi, 1989), pp. 133–152.

Desmond, Adrian. *Archetypes and Ancestors* (Chicago: University of Chicago Press, 1982).

Desmond, Adrian. *The Politics of Evolution* (Chicago: Chicago University Press, 1989).

Desmond, Adrian. *Huxley: From Devil's Disciple to Evolution's High Priest* (Reading, MA: Addison-Wesley, 1997).

Desmond, Adrian, and Moore, James. *Darwin: The Life of a Tormented Evolutionist* (New York: Warner Books, 1992).

Desmond, A. J. *The Hot-Blooded Dinosaurs* (New York: The Dial Press/James Wade, 1977).

Di Gregorio, Mario A. *T. H. Huxley's Place in Natural History* (New Haven: Yale, 1984).

Dingus, L., and Rowe, T. *The Mistaken Extinction* (New York: W. H. Freeman, 1998).

Dobzhansky, T. *Genetics and the Origin of Species* (New York: Columbia University Press, 1937).

Dobzhansky, T. "On Species and Races of Living and Fossil Man," *American Journal of Physical Anthropology* 2 (1944): 251–265.

Dobzhansky, T. *Mankind Evolving: The Evolution of the Human Species* (New Haven: Yale University Press, 1962).

Dobzhansky, T. "Nothing in Biology Makes Sense Except in the Light of Evolution," *The American Biology Teache*r 35(3) (1973): 125–129.

Doncaster, L. "Collective Inquiry As To Progressive Melanism in Lepidoptera: Summary of Evidence," *Entomologist's Record and Journal of Variation* 18 (1906): 165–170; 206–208; 222–226; 248–264.

Doyle, Sir Arthur Conan. "The Lost World," *The Strand* (April–October 1912): 43–44.

Driver, Felix. *Geography Militant: Cultures of Exploration and Empire* (Oxford: Blackwell, 2001).

Dubois, M.E.F.T. "Over de wenschelijkheid van een onderzoek naar de diluviale fauna van den Nederlandsch Indië, in het bijzonder van Sumatra," *Natuurkundig Tijdschrift voor Nederlandsch-Indië* 48 (1888):148–165.

Dubois, M.E.F.T. "Palaeontologische onderzoekingen op Java," *Verslag van het Mijnweizen. Extra bijvoegel der Javansche courant.* 3rd Quarterly Report of 1891, 1892a.

Dubois, M.E.F.T. "Palaeontologische onderzoekingen op Java," *Verslag van het Mijnweizen. Extra bijvoegel der Javansche courant.* 4th Quarterly Report of 1891, 1892b.

Dubois, M.E.F.T. *Pithecanthropus erectus, eine menschenaehnliche Uebergangsform aus Java* (Batavia: Landsrukkerij, 1894).

Duckworth, W.L.H. "The Fossil Anthropoid from Taungs," *Nature* 115(2885) (14 February 1925): 236.

Dunn, L. "Preface," in T. Dobzhansky, *Genetics and the Origin of Species* (New York: Columbia University Press, 1937).

Dyster, Frederick. "Evidence as to Man's Place in Nature" (review), *The Reader* 1(7 March 1863): 234.

East, Edward M. "A Mendelian Interpretation of Variation That Is Apparently Continuous," *The American Naturalist* 44 (1910): 65–82.

Edleston, R. S. "88. Amphydasis betularie." *The Entomologist* 2 (1864): 150.

Edwards, A.W.F. "Perspectives: The Genetical Theory of Natural Selection," *Genetics* 154 (2000): 1419–1426.

Edwards, A.W.F. "Darwin and Mendel United: The Contributions of Fisher, Haldane and Wright up to 1932," in E.C.R. Reeve (ed.), *The Encyclopaedia of Genetics* (London: Fitzroy Dearborn, 2001).

Eggerton, Sir Phillip (attributed). "Monkeyana by 'Gorilla Geological Gardens'" (satirical poem), *Punch* (18 May 1861): 206.

Eisenmann, V. "Comparative Osteology of Modern and Fossils Horses, Half-Asses, and Asses," in H. Meadow and H.-P. Uerpman (eds.), *Equids in the Ancient World* (Wiesbaden, Germany: Dr. Ludwig Reichart Verlag, 1980), 67–116.

Eldredge, N., and Cracraft, J. *Phylogenetic Patterns and the Evolutionary Process* (New York: Columbia University Press, 1980).

Ellegård, Alvar. *The Readership of the Periodical Press in Mid-Victorian Britain* (Göteborg: Universitatis Gothoburgensis, 1957).

Ellegård, Alvar. *Darwin and the General Reader* (Göteborg: Universitatis Gothoburgensis, 1958).

Elzanowski, A. "A New Genus and Species For the Largest Specimen of *Archaeopteryx*," *Acta Palaeontologica Polonica* 46 (2001): 519–532.

Elzanowski, A. "Archaeopterygidae (Upper Jurassic of Germany)," in L. M. Chiappeand and L. M. Witmer (eds.), *Mesozoic Birds: Above the Heads of Dinosaurs* (Berkeley: University of California Press, 2002), 129–159.

Endler, J. A. *Natural Selection in the Wild* (Princeton: Princeton University Press, 1986).

Etler, Dennis. "Implications of New Fossil Material Attributed to Plio-Pleistocene Asian Hominidae," available online at www.chineseprehistory.org.

Fairbanks, Daniel J., and Rytting, Bryce. "Mendelian Controversies: A Botanical and Historical Review," *American Journal of Botany* 88(5) (2001): 737–752.

Falk, D., Hildebolt, C., Smith, K., Morwood, M. J., Sutikna, T., Brown, P., Jatmiko, E., Saptomo, W., Brunsden, B., and Prior, F. "The Brain of LB1, *Homo floresiensis*," *Science* 308(5719) (2005): 242–245.

Farlow, J. O., and Brett-Surman, M. K. (eds.). *The Complete Dinosaur* (Bloomington: Indiana University Press, 1997).

Fastovsky, D. E., and Sheehan, P. M. "The Extinction of the Dinosaurs in North America," *GSA Today* 15 (2005): 4–10.

Feduccia, A., and Tordoff, H. B. "Feathers of *Archaeopteryx*: Assymetric Vanes Indicate Aerodynamic Function," *Science* 203 (1979): 1021–1022.

Findlen, Paula. *Possessing Nature: Museums, Collecting, and Scientific Culture in Early Modern Italy* (Berkeley: University of California Press, 1994).

"First Hominid from the Miocene (Lukeino Formation, Kenya)," *Comptes Rendus de l'Academie des Sciences Series IIA Earth and Planetary Science* 332(2)(2001): 137–144.

Fisher, R. A. *The Genetical Theory of Natural Selection* (Oxford: Clarendon Press, 1930).

Fisher, R. A. *The Genetical Theory of Natural Selection*, reprint (New York: Dover, 1958).

Fisher, R. A. "Natural Selection from the Genetical Standpoint," *Anat. J. Sci.* 22 (1959): 444–449.

Fisher, R. A., and Ford, E. B. "The Spread of a Gene in Natural Conditions in a Colony of the Moth *Panaxia Dominula L.*," *Heredity* 1(2) (1947): 143–174.

Fiske, John. *The Destiny of Man Viewed in the Light of His Origin* (Boston: Houghton Mifflin, 1890).

Ford, E. B. "Problems of Heredity in the Lepidoptera," *Biological Reviews* 12 (1937): 461–503.

Ford, E. B. "The Experimental Study of Evolution," *Australian and New Zealand Association for the Advancement of Science* 28 (1953): 143–154.

Ford, E. B. *Ecological Genetics* (4th ed.) (New York: Chapman and Hall, 1975).

Forrest, Barbara. " 'The Vise Strategy' Undone: Kitzmiller et al. v. Dover Area School District," *Creation and Intelligent Design Watch*, available online at www.csicop.org/intelligentdesignwatch.

Forrest, Barbara, and Gross, Paul A. *Creationism's Trojan Horse: the Wedge of Intelligent Design* (New York: Oxford University Press, 2004).

Forsten, A. "The Taxonomic Status of the Miocene Horse *Sinohippus*," *Palaeontology* 25 (1982): 673–679.

Foster, Michael, and Lankester, E. Ray (eds.). *The Scientific Memoirs of T. H. Huxley v. ii* (London: Macmillan, 1899).

Fraipont, Charles. "Les hommes fossiles d'Engis," *Archives de l'Institut de Paléontologie Humaine* 16 (1936): 1–52.

Fraipont, Julien. "The Imaginary Race of Canstadt or Neanderthal," *Science* 22(568) (1893): 346.

Fraipont, J., and Lohest, M. "La race humaine de Néanderthal ou de Cannstadt en Belgique—Recherches Ethnographiques sur des ossements humains découverts dans les depôts quaternaries d'une grotte à Spy et détermination de leur âge géologique," *Archive de Biologie* 7 (1887): 587–757.

Froehlich, D. J. "Quo Vadis Eohippus? The Systematics and Taxonomy of the early Eocene Equids (Perrissodactyla)," *Zoological Journal of the Linnean Society* (2002): 134, 141, 256.

Futuyma, D. J. *Evolutionary Biology*, 3rd ed. (Sunderland, MA: Sinaur Associates, 1997).

Gardiner, Peter, and Oey, Mayling. "Morbidity and Mortality in Java 1880–1849: The Evidence of the Colonial Reports," in Norman G. Owen (ed.), *Death and Disease in Southeast Asia; Explorations in Social, Medical and Demographic History* (Singapore, Sudeny: Oxford University Press, 1987), 70–90.

Garrod, A. *Inborn Errors of Metabolism* (Oxford: Oxford University Press, 1909).

Gasman, Daniel. *The Scientific Origins of National Socialism: Social Darwinism in Ernst Haeckel and the Monist League* (New York: American Elsevier, 1971).

Gatesy, S. M., and Dial, K. P. "From Frond to Fan: *Archaeopteryx* and the Evolution of Short-Tailed Birds," *Evolution* 50(5) (1996): 2037–2048.

Gaudry, A. *Animaux Fossiles et Géologie de l'Attique*, 4th ed. (Paris, 1867).

Gauthier, J. A. "Saurischian Monophyly and the Origin of Birds," *Memoirs of the California Academy of Sciences* 8 (1986): 1–55.

Gauthier, J. A., and Gall, L. F. (eds.). "New Perspectives on the Origin and Early Evolution of Birds," *Special Publication of the Peabody Museum of Natural History* (New Haven: Yale University, 2001).

Gayon, Jean. *Darwinism's Struggle for Survival: Heredity and the Hypothesis of Natural Selection* (Cambridge: Cambridge University Press, 1998).

"Geological and Palaeontological Context of a Pliocene Juvenile Hominin at Dikika, Ethiopia," *Nature* 443 (2006): 332–336.

George, M., and Ryder, O. A. "Mitochondrial DNA Evolution in the Genus *Equus*," *Molecular Biology and Evolution* 3 (1986): 535–546.

Gibbons, A. "A Rare Meeting of the Minds," *Science* 312 (2006a): 1739–1740.

Gibbons, A. *The First Humans* (New York: Doubleday, 2006b).

Gibbons, A. "Lucy's Tour Abroad Sparks Protests," *Science* 314 (2006c): 574–575.

Gilbert, S., Opitz, J. M., and Raff, R. "Resynthesizing Evolutionary and Developmental Biology," *Development* 173 (1996): 357–372.

Gilder, Robert F. "A Primitive Human Type in America," *Putnam's Monthly* (January 1907): 407–409.

Gish, D. T. *The Origin of Mammals*. ICR Impact no. 87 (1980).

Gliboff, Sander. "Gregor Mendel and the Laws of Evolution," *History of Science* 37 (1999): 217–235.

Glick, Thomas F. (ed.). *The Comparative Reception of Darwinism* (Austin, TX: University of Chicago Press, 1972).

Golding, William. *The Inheritors* (London: Faber and Faber, 1955).

Gore, R. "The First Pioneer?," *National Geographic* (August 2002).

"The Gorilla's Dilemma" (satirical poem), *Punch* 43 (October 18 1862): 164.

Gorjanović-Kramberger, Dragutin. "Der Diluviale Mensch von Krapina in Kroatien," *Ein Beitrag zur Palaeoanthropologie* (Wiesbaden: Kreidels, 1906).

Gottlieb, Carl. *Caveman* (USA: Metro-Goldwyn-Mayer, 1981).

Gould, Stephen Jay. "G. G. Simpson, Paleontology, and the Modern Synthesis," in E. Mayr and W. B. Provine (eds.), *The Evolutionary Synthesis. Perspectives on the Unification of Biology* (Cambridge, MA: Harvard University Press, 1980).

Gould, Stephen Jay. *Time's Arrow, Time's Cycle* (Cambridge, MA: Harvard University Press, 1987).

Gould, Stephen Jay. *Wonderful Life* (New York: W. W. Norton, 1989).
Gould, Stephen Jay "Case Two: Life's Little Joke," in *Full House* (New York: Three Rivers Press, 1996), 57–73.
Gradstein, Felix M., Ogg, James G., and Smith, Alan G. *A Geological Time Scale 2004* (Cambridge: Cambridge University Press, 2005).
Grant, B. S. "Fine Tuning the Peppered Moth Paradigm," *Evolution* 53(3) (1999): 980–984.
Grant, B. S. "Sour Grapes of Wrath," *Science* 297 (2002): 940–941.
Grant, B. S., Owen, D. F., and Clarke, C. A. "Parallel Rise and Fall of Melanic Peppered Moths in America and Britain," *Journal of Heredity* 87 (1996): 351–357.
Grant, P. *Evolution on Islands* (Oxford: Oxford University Press, 1998).
Green, R. E., Krause, J., Ptak, S. E., Briggs, A. W., Ronan, M. T., Simons, J. F., Du, L., Egholm, M., Rothberg, J. M., Paunovic, M., and Pääbo, S. "Analysis of One Million Base Pairs of Neanderthal DNA," *Nature* 444 (2006): 330–336.
Greene, John C. *The Death of Adam: Evolution and Its Impact on Western Thought* (Ames: Iowa State University Press, 1959).
Gregory, William King. "The Origin of Man from a Brachiating Anthropoid Stock," *Science* 71(1852) (27 June 1930): 645–650.
Gregory, William King. "Studies in Comparative Myology and Osteology, no. V: On the Anatomy of the Preorbital Fossae of Equidae and Other Ungulate," *Bulletin of the American Museum of Natural History* 42 (1920): 265–284.
Griffiths, P. J. "The Isolated *Archaeopteryx* Feather," *Archaeopteryx* 14 (1996): 1–26.
Gunther, A. E. *The Founders of Science at the British Museum, 1753–1900* (Halesworth: Halesworth Press 1980).
Hackett, Abigail, and Dennell, Robin. "Neanderthals as Fiction in Archaeological Narrative," *Antiquity* 77(298) (2003): 816–827.
Haeckel, Ernst. *The Evolution of Man: A Popular Exposition on the Principle Points of Human Ontogeny and Phylogeny* (New York: Appleton, 1896).
Hagen, J. "H.B.D. Kettlewell & the Peppered Moths," in J. B. Hagen, D. Allchin, and F. Singer (eds.), *Doing Biology* (Glenview, IL: HarperCollins, 1996), 1–20.
Hagen, J. "Retelling Experiments: H.B.D. Kettlewell's Studies of Industrial Melanism in Peppered Moths," *Biology and Philosophy* 14(1) (1999): 39–54.
Haldane, J.B.S. "A Mathematical Theory of Natural and Artificial Selection," *Transactions of the Cambridge Philosophical Society* 23 (1924): 19–41.
Haldane, J.B.S. *The Causes of Evolution* (New York: Longmans, Green, 1932).
Haldane, J.B.S. *Heredity and Politics* (London: Allen and Unwin, 1938).
Haldane, J.B.S. *The Causes of Evolution*, reprint (Ithaca, NY: Cornell University Press, 1966).
Hallam, A. *Great Geological Controversies*, 2nd ed. (Oxford: Oxford University Press, 2005).
Haller, Mark H. *Eugenics: Hereditarian Attitudes in American Thought* (New Brunswick, NJ: Rutgers University Press, 1963).

Hammond, Michael. "Anthropology as a Weapon of Social Combat in Late-Nineteenth-Century France," *Journal of the History of the Behavioral Sciences* 16 (1980): 118–132.

Hammond, Michael. "The Expulsion of the Neanderthals from Human Ancestry. Marcellin Boule and the Social Context of Scientific Research," *Social Studies of Science* 12(1) (1982): 1–36.

Hammond, Michael. "The Shadow Man Paradigm in Paleoanthropology 1911–1945," in G. W. Stocking Jr. (ed.), *Bones, Bodies, Behavior. Essays on Biological Anthropology* (Madison: University of Wisconsin Press, 1988), 117–137.

Harris, H. "Enzyme Polymorphisms in Man," *Proc. Roy. Soc. London B* (164) (1966): 298–310.

Harris, M. *The English Lepidoptera: Or, the Aurelian's Pocket Companion*, reprint (Hampton, England: E. W. Classey, 1969).

Harris, M. *The Aurelian or Natural History of English Insects; Namely, Moths and Butterflies*, reprint (Twickenham, UK: Newnes, 1986).

Harrison, J.W.H. "A Further Induction of Melanism in the Lepidopterist Insect, *Selenia bilunaria* Esp. and Its Inheritance," *Proceedings Royal Society B* 102 (1927–1928): 338–347.

Harrison, J.W.H., and Garrett, F. C. "The Induction of Melanism in the Lepidoptera and its Subsequent Inheritance," *Proceedings Royal Society B 99* (1925–1926): 241–263.

Häusler, M., and Schmid, P. "Comparison of the Pelves of Sts 14 and AL 288-1: Implications for Birth and Sexual Dimorphism in Australopithecines," *Journal of Human Evolution* 29 (1995): 363–383.

Hawkins, Mike. *Social Darwinism in European and American Thought, 1860–1945: Nature as Model, Nature as Threat* (Cambridge: Cambridge University Press, 1997).

Hecht, M. K., Ostrom, J. H., Viohl, G., and Wellnhofer, P. *The Beginnings of Birds. Proceedings of the International Archaeopteryx Conference Eichstätt 1984* (Eichstätt: Freunde des Jura-Museums Eichstätt, 1985).

Heilmann, G. "Vor nuværende Viden om Fuglenes Afstamning I-V," *Dansk Ornitologisk Forenings Tidsskrift* 7, 8, 9, 10 (1912–1916): 1–71; 1–92; 1–160; 73–144.

Heilmann, G. *The Origin of Birds* (London: Witherby, 1926).

Heller, F. "Ein dritter *Archaeopteryx* Fund aus den Solnhofener Plattenkalken von Langenaltheim/Mfr," *Erlanger Geologische Abhandlungen* 31 (1959): 1–25.

Hennig, W. *Phylogenetic Systematics* (Chicago: University of Chicago Press, 1966).

Herbert, Sandra. *Charles Darwin: Geologist* (Ithaca, NY: Cornell University Press, 2005).

Herfkens, J.W.F. *De Atjeh-Oorlog van 1873 tot 1896 omgewerkt en aangevuld door J.C. Pabst* (Breda: De Koninklijke Militaire Academie, 1905).

Hermanson, J. W., and MacFadden, B. J. "Evolutionary and Functional Morphology of the Shoulder Joint and Stay Apparatus in Fossil and Extant Horses (Equidae)," *Journal of Vertebrate Paleontology* 12 (1992): 337–386.

Hiebert, H. *Evolution: Its Collapse in View?* (Norwood, MA: Horizon House Publishing, 1979).

Hilgartner, Stephen. "The Dominant View of Popularization. Conceptual Problems, Political Uses," *Social Studies of Science* 20(3) (1990): 519–539.

Himmelfarb, G. *Darwin and the Darwinian Revolution* (New York: W. W. Norton, 1959).

Hofstadter, Richard. *Social Darwinism in American Thought*, revised (New York: George Braziller, 1959).

Honoré, F. "La crâne du plus vieil ancêtre connu de l'humanité," *L'Illustration* 3443(20 February 1909): 125–129.

Hooper, J. *Of Moths and Men: An Evolution Tale: The Untold Story of Science and the Peppered Moth* (New York: W. W. Norton, 2002).

Hooper, Judith. *Of Moths and Men: Intrigue, Tragedy and the Peppered Moth* (London: Fourth Estate, 2002).

Hopkins, M. (2001-2002). "Is 'Dawn Horse' a Hyrax? Examining a Common Creationist Claim about *Hyracotherium*," available online at The TalkOrigins Archive: www.talkorigins.org/faqs/horses/eohippus_hyrax.html.

Hopp, T. P., and Orsen, M. J. "Dinosaur Brooding Behavior and the Origin of Flight Feathers," in: P. J. Currie, E. B. Koppelhus, M. A. Shugar, and J. L. Wright (eds.), *Feathered Dragons. Studies on the Transition from Dinosaurs to Birds* (Bloomington: Indiana University Press, 2004), 234–250.

Horner, J. R., and Gorman, J. *Digging Dinosaurs* (New York: Workman Press, 1988).

Horner, John R., Schweitzer, Mary H., Toporski, Jan K., and Wittmeyer, Jennifer L. "Soft-Tissue Vessels and Cellular Preservation in *Tyrannosaurus rex*," *Science Magazine* (March 2005).

Howell, Clark F. "The Evolutionary Significance of Variation and Varieties of 'Neanderthal' Man," *The Quarterly Review of Biology* 32(4) (1957): 330–347.

Hrdlička, Aleš. "The Neanderthal Phase of Man," *Journal of the Royal Anthropological Institute of Great Britain and Ireland* 57 (1927): 249–274.

Hughes, A. W. "Induced Melanism in the Lepidoptera," *Proc. R. Soc. B* 110 (1932): 378–402.

Hull, D. L., Tessner, P. D., and Diamond, A. M. "Planck's Principle: Do Younger Scientists Accept New Ideas with Greater Alacrity Than Older Scientists?," *Science* 202 (1978): 717–723.

Hunt, James. "The Negro's Place in Nature" (editorial letter), *The Reader* 3(19 March 1864): 368.

Husemann, Dirk. *Die Neandertaler: Genies der Eiszeit* (Frankfurt am Main: Campus, 2005).

Hussain, S. T. "Evolutionary and Functional Anatomy of the Pelvic Limb in Fossil and Recent Equidae (Perissodactyla, Mammalia)," *Anatomy, Histology, and Embryology* 4 (1975): 179–222.

Huxley, Julian S. *Evolution: The Modern Synthesis* (New York: Harper Brothers, 1943).

Huxley, Leonard. *The Life and Letters of Thomas Henry Huxley*, 2 vols. (New York: Appleton, 1901).

Huxley, Thomas Henry. "Time and Life: Mr. Darwin's 'Origin of Species'," *Macmillan's Magazine* 1(1859): 142–148.

Huxley, Thomas Henry. "The Darwinian Hypothesis," *The Times* (26 December 1859).

Huxley, Thomas Henry. "The Origin of the Species," *Westminster Review* 17 (n.s.) (1860): 541–570.
Huxley, Thomas Henry. "On the Brain of Ateles Pansicus," *Proceedings of the Zoologoical Society* (1861).
Huxley, Thomas Henry. "Letter to Henrietta Huxley." (10 April 1861).
Huxley, Thomas Henry. "On the Zoological Relations of Man with the Lower Animals," *Natural History Review* (1861): 67–68.
Huxley, Thomas Henry. "The Brain of Man and Apes," *Medical Times and gazette* (14 October 1862).
Huxley, Thomas Henry. *Man's Place in Nature* (New York: D. Appleton, 1863).
Huxley, Thomas Henry. "On the Animals Which Are Most Nearly Intermediate Between the Birds and Reptiles," *Annual Magazine, Natural History* 4 (1868): 66–75.
Huxley, Thomas Henry. "Remarks upon *Archaeopteryx lithographica*," *Proceedings, Royal Society London* 16 (1868): 243–248.
Huxley, Thomas Henry "Further Evidence of the Affinity Between the Dinosaurian Reptiles and Birds," *Quarterly Journal of the Geological Society, London* 26 (1870): 12–31.
Huxley, Thomas Henry "On the Methods and Results of Ethnology," in T. H. Huxley, *Man's Place in Nature and Other Anthropological Essays* (New York: D. Appleton, 1890).
Huxley, Thomas Henry. *Man's Place in Nature and Other Anthropological Essays*. From *The Selected Works of Thomas H. Huxley*, Westminster ed. (New York: Appleton, 1893–1894).
Huxley, Thomas Henry. "Evidence as to Man's Place in Nature," in T.H. Huxley, *Man's Place in Nature and Other Anthropological Essays* (New York: D. Appleton, 1900).
Huxley, Thomas Henry. *Autobiography and Essays*. Ed. by Ada F. Snell (Boston: Houghton-Mifflin, 1909).
Iltis, Hugo. *Life of Mendel* (New York: W. W. Norton, 1932).
Ingman, M., Kaessmann, H., Pääbo, S., and Gyllensten, U. "Mitochondrial Genome Variation and the Origin of Modern Humans," *Nature* 408 (2000): 708–713.
Jablonski, N. *Skin: A Natural History* (Los Angeles: University of California Press, 2006).
Jackson, Patrick Wyse. *The Chronologer's Quest: The Search for the Age of the Earth* (Cambridge: Cambridge University Press, 2006).
Jaeger, J., Thein, Tin, Benammi, M., Chaimanee, Y., Soe, Aung Naing, Lwin, Thit, Wai, San, and Ducrocq, S. "A New Primate from the Middle Eocene of Myanmar and the Asian Early Origins of Anthropoids," *Science* 286 (15 October 1999): 528–530.
Jahme, Carole. *Beauty and the Beasts. Woman, Ape and Evolution* (London: Virago Press, 2000).
James, William. "Huxley's Comparative Anatomy," *North American Review* (January 1865): 290.
Janis, C. M. *Aspects of the Evolution of Herbivory in Ungulate Mammals*. Unpublished Ph.D. thesis, Harvard University (1979).
Janis, C. M. "The Correlation Between Diet and Dental Wear in Herbivorous Mammals, and Its Relationship to the Determination of Diets of Extinct

Species," in A. J. Boucot, (ed.), *Evolutionary Paleobiology of Behavior and Coevolution* (New York: Elsevier, 1990), 241–259.

Janis, C. M. "Correlation Between Craniodental Morphology and Feeding Behaviour in Ungulates: Reciprocal Illumination Between Living and Fossil Taxa," in J. J. Thomason (ed.), *Functional Morphology in Vertebrate Paleontology* (Cambridge: Cambridge University Press, 1995), 76–98.

Janis, C. M. "Paleoecology and Evolutionary Trends," in D. R. Prothero and S. E. Foss (ed.), *Artiodactyla* (Baltimore: John Hopkins University Press, 2007).

Janis, C. M., Gordon, I., and Illius, A. "Modelling Equid/Ruminant Competition in the Fossil Record," *Historical Biology* 8 (1994): 15–29.

Janis, C. M., Hulbert, R., and Milhbachler, M. "Addendum to Volume 1," in C. M. Janis, G. F. Gunnell, and M. D. Uhen (eds.), *Evolution of Tertiary Mammals of North America*, vol. 2 (Cambridge: Cambridge University Press, in press).

Janis, C. M., and Wilhelm, P. B. "Were There Mammalian Pursuit Predators in the Tertiary? Dances with Wolf Avatars," *Journal of Mammalian Evolution* 1 (1993): 103–125.

Janus, Christopher G., and Brashler, William. *The Search for Peking Man* (New York: Macmillan, 1975).

Jardine, N., Secord, J. A., and Spary, E. C. (eds.). *Cultures of Natural History* (Cambridge, MA: Cambridge University Press, 1996).

Johanson, D., Johanson, L., and Edgar, B. *Ancestors: In Search of Human Origins* (New York: Villard Books, 1994).

Johanson, D. C. "Lucy, Thirty Years Later: An Expanded View of *Australopithecus afarensis*," *Journal of Anthropological Research* 60(4) (2004): 465–486.

Johanson, D. C., and Edey, M. *Lucy: The Beginnings of Humankind* (New York: Simon and Schuster, 1981).

Johanson, D. C., and Shreeve, J. *Lucy's Child: The Discovery of a Human Ancestor* (New York: William Morrow, 1989).

Johanson, D. C., and Taieb, M. "Plio-Pleistocene Hominid Discoveries in Hadar, Ethiopia," *Nature* 260 (1976): 293–297.

Johanson, D. C., Taieb, M., and Coppens, Y. (eds.). "Pliocene Hominids from Hadar, Ethiopia," *American Journal of Physical Anthropology* 57 (1982): 373–719.

Johanson, D. C., and White, T. D. "A Systematic Assessment of Early African Hominids," *Science* 203 (1979): 321–330.

Johanson, D. C., White, T. D., and Coppens, Y. "A New Species of the Genus *Australopithecus* (Primates: Hominidae) from the Pliocene of Eastern Africa," *Kirtlandia* 28 (1978): 1–14.

Jones, Steve, Robert Martin, David Pilbeam, and Sarah Bunney (eds.). *The Cambridge Encyclopedia of Human Evolution* (New York: Cambridge University Press, 1992).

Kalb, J. *Adventures in the Bone Trade* (New York: Copernicus, 2001).

Kardong, K. V. *Vertebrates* (New York: McGraw Hill, 1998).

Keith, A. "The New Missing Link," *The British Medical Journal* (14 February 1925): 325–326.

Keith, A. "Australopithecine or Dartians," *Nature* 159(4037 (15 March 1947): 377.

Keith, Arthur. *The Antiquity of Man* (London: Williams and Norgate, 1915).
Keith, Arthur. "History from Caves. A New Theory of the Origin of the Modern Races of Mankind. Presidential Address to the First Speleological Conference," British Speleological Association 25 July 1936.
Keith, Arthur. *A New Theory of Human Evolution* (London: Watts, 1948).
Kelly, Alfred. *The Descent of Darwin: The Popularization of Darwinism in Germany, 1860–1914* (Chapel Hill, NC: University of North Carolina Press, 1981).
Kerkhoff, A.H.M. "The Organization of the Military and Civil Medical Service in the 19th Century," in G. M. van Heteren, A. de Knecht-van Eekelen, and M.J.D. Poulissen (eds.), *Dutch Medicine in the Malay Archipelago 1816–1942: Articles Presented at a Symposium Held in Honor of Professor De Moulin* (Amsterdam: Rodopi, 1989), 9–24.
Kettlewell, H.B.D. "Selection Experiments on Industrial Melanism in the Lepidoptera," *Heredity* 9 (1955): 323–342.
Kettlewell, H.B.D. "Further Selection Experiments on Industrial Melanism in the Lepidoptera," *Heredity* 10 (1956): 287–301.
Kettlewell, H.B.D. "A Survey of the Frequencies of *Biston betularia* (L.) (Lep.) and its Melanic Forms in Great Britain." *Heredity* 12 (1958): 51–72.
Kettlewell, H.B.D. "Darwin's Missing Evidence," *Scientific American* (March 2003) (1959): 48–53.
Kettlewell, H.B.D. *The Evolution of Melanism: The Study of a Recurring Necessity* (Oxford: Clarendon Press, 1973).
Kevles, Daniel. *In the Name of Eugenics: Genetics and the Uses of Human Heredity* (New York: Knopf, 1985).
Keynes, Randal. *Darwin, His Daughter, and Human Evolution* (New York: Riverside, 2002).
Keynes, Richard Darwin. *Fossils, Finches and Fuegians: Darwin's Adventures and Discoveries on the Beagle* (Oxford: Oxford University Press, 2006).
Kimbel, W. H., et al. "Late Pliocene *Homo* and Oldowan tools from the Hadar Formation (Kada Hadar Member), Ethiopia," *Journal of Human Evolution* 31 (1996): 549–561.
Kimbel, W. H., Johanson, D. C., and Rak, Y. The First Skull and Other New Discoveries of *Australopithecus afarensis* at Hadar, Ethiopia," *Nature* 368 (1994): 449–451.
King, William. "The Reputed Fossil Man of the Neanderthal," *Quarterly Journal of Science* 1(January 1864): 88–97. "Lectures to Working Men" (Advertisement), *The Athenaeum* (1738) (16 February 1861).
Kings, Matthias, Stone, Anne, Schmitz, Ralf W., Krainitzki, Heike, Stoneking, Mark, and Pääbo, Svante. "Neandertal DNA Sequences and the Origin of Modern Humans," *Cell* 90 (1997): 19–30.
Kingsley, Charles. "Speech of Lord Dundreary in Section D. on Friday Last, On the Great Hippocampus Question," *Letters and Memories* (1861).
Kitts, D. B. "American *Hyracotherium*," *Bulletin of the American Museum of Natural History* 110 (1956): 1–60.
Knecht-van Eekelen, A. "The Debate about Acclimatization in the Dutch East Indies (1840-1860)," in "Public Health Service in the Dutch East Indies," *Medical History*: Supplement v. 20 (2000): 70–85.

Knight, Chris. "Review of *Ancestral Passions: The Leakey Family and the Quest for Humankind's Beginnings*," *The Journal of the Royal Anthropological Institute* 3(2) (June 1997): 387–388.

Kofahl, R. E. *The Handy Dandy Evolution Refuter* (San Diego, CA: Beta Books, 1997).

Kohn, David, (ed.). *The Darwinian Heritage: A Centennial Retrospect* (Princeton: Princeton University Press, 1985).

Koselleck, Reinhart. *Futures Past: On the Semantics of Historical Time* (Cambridge, MA: MIT Press, 1985) [Vergangene Zukunft: Zur Semantik geschichtlicher Zeiten, Frankfurt-am-Main, 1979].

Kottler, Malcolm Jay. "Charles Darwin and Alfred Russel Wallace: Two Decades of Debate over Natural Selection." in D. Kohn (ed.), *The Darwinian Heritage* (Princeton: Princeton University Press, 1985), 367–432.

Krings, M., Capelli, C., Tschentscher, F., Geisert, H., Meyer, S., von Haeseler, A., Grossschmidt, K., Possnert, G., Paunovic, M., and Pääbo, S. "A View of Neandertal Genetic Diversity," *Nature Genetics* 26 (2000): 144–146.

Kritsky, G. "Darwin's *Archaeopteryx* Prophecy," *Archives of Natural History* 19(3) (1992): 407–410.

Kuhn, Thomas. *The Structure of Scientific Revolutions* (Chicago: University of Chicago Press, 1962).

Lahr, M. *The Evolution of Modern Human Diversity: A Study of Cranial Variation* (Cambridge: Cambridge University Press, 1996).

Lahr, M., and Foley, R. "Towards a Theory of Modern Human Origins: Geography, Demography, and Diversity in Recent Human Evolution," *Yearbook of Physical Anthropology* 41 (1998): 137–176.

Landau, M. *Narratives of Human Evolution* (New Haven: Yale University Press, 1991).

Lan-Po, Chia. *The Cave Home of Peking Man* (Beijing, China: Foreign Language Press, 1975).

Lanpo, Jia. *Early Man in China* (Beijing, China: Foreign Language Press, 1980).

Lanpo, Jia, and Weiwen, Huang. *The Story of Peking Man* (Beijing, China: Foreign Language Press, 1990).

Lartet, Édouard, and Christy, Henry. *Reliquiae Aquitanicae; Being Contributions to the Archaeology and Palaeontology of Périgord and the Adjoining Provinces of Southern France* (London: Williams and Norgate, 1875).

Laurent, Goulvan. *Paléontologie et évolution en France 1800–1860: De Cuvier-Lamarck à Darwin* [Paleontology and Evolution in France 1800–1860: From Cuvier-Lamarck to Darwin] (Paris: Editions du CTHS, 1987).

Laurent, Goulvan, (ed.). *Jean-Baptiste Lamarck, 1744–1829* (Paris: Editions du CTHS, 1997).

Lawrence, William. *Lectures on Physiology, Zoology, and the Natural History of Man* (Salem: Foote and Brown, 1828).

Leakey, L.S.B. "The Oldoway Human Skeleton" *Nature* 129 (1932): 721–722.

Leakey, L.S.B. *Adam's Ancestors. An Up-To-Date Outline of What Is Known About the Origin of Man* (London: Methuen & Co., 1934).

Leakey, L.S.B. *White African* (London: Hodder & Stoughton, 1937).

Leakey, L.S.B. "A New Fossil Skull From Olduvai," *Nature* 184 (1959): 491–493.

Leakey, L.S.B. "Finding the World's Earliest Man," *National Geographic* (September 1960): 421–435.
Leakey, L.S.B. "Exploring 1,750,000 Years into Man's Past," *National Geographic* (October 1961): 564–589.
Leakey, L.S.B. "An Early Miocene Member of Hominidae," *Nature* 213 (1967): 156–163.
Leakey, L.S.B. "Bone Smashing by Late Miocene Hominidae" *Nature* 218 (1968): 28–30.
Leakey, L.S.B. "Fort Ternan Hominid" *Nature* 222 (1969): 1202.
Leakey, L.S.B. "'Newly' Recognized Mandible af *Ramapithecus*," *Nature* 225 (1970): 199–200.
Leakey, L.S.B. *By the Evidence. Memoirs 1932–1951* (New York: Harcourt Brace Jovanovich, 1974).
Leakey, L.S.B. *The Southern Kikuyu before 1903*, vol. I. (London: Academic Press, 1977).
Leakey, L.S.B., et al. "New Yields from the Oldoway Bone Beds, Tanganyika Territory" *Nature* 128 (1931): 1075.
Leakey, L.S.B., et al. "A New Species of the Genus *Homo* from Olduvai Gorge," *Nature* 202 (1964): 7–9.
Leakey, M. D., and Hay, R. L. "Pliocene Footprints in the Laetoli Beds at Laetoli, Northern Tanzania," *Nature* 278 (1979): 317–323.
"Lectures to Workingmen" (advertisement), *The Athenaeum* 1738 (16 February 1861): np.
Lesley, Margaret Mann. "Mendel's Letters to Carl Nageli," *The American Naturalist* 61(675) (1927): 370–378.
Lewenstein, Bruce V. "Science and the Media" in Sheila Jasanoff, Gerald E. Markle, James C. Petersen, and Trevor Pinch, *Handbook of Science and Technology Studies* (Thousand Oaks: Sage, 1995).
Lewin, R. "Surprise Findings in the Taung Child's Face," *Science* 228(4695) (5 April 1985): 42–44.
Lewin, R. *Bones of Contention. Controversies in the Search for Human Origins* (Chicago: University of Chicago Press, 1997).
Lewin, R., and Foley, R. *Principles of Human Evolution* (Oxford: Blackwell, 2003).
Lewontin, R. "The Organism As the Subject and As the Object of Evolution," *Scientia* 118 (1983): 65–80.
Liedman, Sven-Eric. *Motsatsernas spel: Friedrich Engels' filosofi och 1800-talets vetenskap* [The Interplay of Oppositions: The Philosophy of Friedrich Engel's and the Science of the 19th century], vol. I–II (Lund, 1977).
Lightman, Bernard, (ed.). *Victorian Science in Context* (Chicago: University of Chicago Press, 1997).
Lightman, Bernard. *The Origins of Agnosticism: Victorian Unbelief and the Limits of Knowledge* (Baltimore: Johns Hopkins, 1987).
London, Jack. "Before Adam," *Everybody's Magazine* (October 1906–February 1907).
Lorenz, K. *On Aggression* (New York: Harcourt, Brace and World, 1966).
Loring Brace, C. "Review of *Leakey's Luck: The Life of Louis Seymour Bazett Leakey, 1903–1972*," *American Anthropologist* 79(1) (1977): 171–172.

Lovejoy, Owen. *The Great Chain of Being: A Study of the History of an Idea* (Cambridge, MA: Harvard University Press, 1936).

Lydekker, R. "Review of Dubois' *Pithecanthropus erectus*, eine menschenähnliche Uebergangsform aus Java," *Nature* 51 (1895): 291.

Lyell, Charles. *The Geological Evidence of the Antiquity of Man*. Edited by R. H. Rastall (London: D. M. Dent, 1914).

Macbeth, Norman. *Darwin Retried: An Appeal to Reason* (Boston: Gambit, 1971).

MacFadden, B. J. "Fossil horses from 'Eohippus' (*Hyracotherium*) to *Equus*: Scaling, Cope's Law, and the Evolution of Body Size," *Paleobiology* 12 (1987): 355–369.

MacFadden, B. J. *Fossil Horses: Systematics, Paleobiology, and Evolution of the Family Equidae* (Cambridge: Cambridge University Press, 1992).

MacFadden, B. J. "Equidae," in C. M. Janis, K. M. Scott, and L. L. Jacobs (eds.), *Evolution of Tertiary Mammals of North America*, vol. 1 (New York: Cambridge University Press, 1998), 537–559.

MacFadden, B. J. "Fossil Horses—Evidence for Evolution," *Science* 307 (2005): 1728–1730.

Majerus, M.E.N. *Melanism: Evolution in Action* (Oxford: Oxford University Press, 1998).

Majerus, M.E.N. "The Peppered Moth: Decline of a Darwinian Disciple," in M.D.E. Fellowes, G. J. Holloway, and J. Rolff (eds.), *Insect Evolutionary Ecology (Proceedings of the Royal Society's 22nd Symposium)* (Cambridge, MA: CABI Publishing, 2005), 371–396.

Marchant, James. *Alfred Russel Wallace: Letters and Reminiscences*, reprint (New York: Arno Press, 1916).

Marsh, O. C. "Notice of New Tertiary Mammals," *V. American Journal of Science* 3(12) (1876): 401–404.

Marsh. O. C. "Polydactyle Horses, Recent and Extinct," *American Journal of Science* 17 (1879): 499–505.

Martin, R. A. *Missing Links* (Sudbury, MA: Jones and Bartlett Publishers, 2004).

Martin, R. D., MacLarnon, A. M., Phillips, J. L., and Dobyns, W. B. "Flores Hominid: New Species or Microcephalic Dwarf?," *The Anatomical Record Part A: Discoveries in Molecular, Cellular, and Evolutionary Biology* 288A(11) (2006): 1123–1145.

Matschie, P. "*Anthropopithecus erectus* E. Dubois," *Naturwissenschaftliche-Wochenschrift* (9) (1894): 122–123.

Matthew, W. D. "The Evolution of the Horse," *American Museum of Natural History, Supplement to the American Museum Journal. Guide Leaflet*, 9(3) (1903): 1–30.

Matthew, W. D. "The Evolution of the Horse: A Record and Its Interpretation," *Quarterly Review of Biology* 1 (1926): 139–185.

Matthews, M. *Science Teaching: The Role of History and Philosophy of Science* (New York: Routledge, 1994).

Mäuser, M. "Der achte *Archaeopteryx*," *Fossilien* 14(3) (1997): 156–157.

Mayr, E. *Systematics and the Origin of Species from the Viewpoint of a Zoologist* (New York: Columbia University Press, 1942).

Mayr, E. "Taxonomic Categories in Fossil Hominids," *Cold Spring Harbor Symposia on Quantitative Biology* 15 (1950): 109–118.

Mayr, E. *Animal Species and Evolution* (Cambridge, MA: Harvard University Press, 1963).

Mayr, E. "Agassiz, Darwin and Evolution," in P. Appleman (ed.), *Darwin* (New York: Norton, 1970), 299–307.

Mayr, E. "G. G. Simpson," in E. Mayr and W. B. Provine (eds.), *The Evolutionary Synthesis: Perspectives on the Unification of Biology* (Cambridge, MA: Harvard University Press, 1980), 452–463.

Mayr, E. "Prologue: Some Thoughts on the History of the Evolutionary Synthesis," in *The Evolutionary Synthesis: Perspectives on the Unification of Biology* (Cambridge, MA: Harvard University Press, 1980), 1–50.

Mayr, E. *The Growth of Biological Thought* (Cambridge, MA: Harvard University Press, 1982).

Mayr, E. *Systematics and the Origin of Species.* Edited by Niles Eldredge and Stephen Jay Gould, Columbia Classics in Evolution (New York: Columbia University Press, 1982).

Mayr, E. *Toward a New Philosophy of Biology* (Cambridge, MA: Harvard University Press, 1988).

Mayr, E. *One Long Argument: Charles Darwin and the Genesis of Modern Evolutionary Thought* (Cambridge, MA: Harvard University Press, 1991).

Mayr, E. "The Establishment of Evolutionary Biology as a Discrete Biological Discipline," *BioEssays* 19 (1997): 263–266.

Mayr, E. "Reminiscences From the First Curator of the Whitney-Rothschild Collection," *BioEssays* 19 (1997): 175–179.

Mayr, E. "Introduction, 1999," *Systematics and the Origin of Species from the Viewpoint of a Zoologist* (Cambridge, MA: Harvard University Press, 1999).

Mayr, E. "Interview," *BioEssays* 24 (2002): 960–973.

Mayr, E. *What Makes Biology Unique? Considerations on the Autonomy of a Scientific Discipline* (Cambridge: Cambridge University Press, 2004).

Mayr, E., and Provine, W. (eds.). *The Evolutionary Synthesis: Perspectives on the Unification of Biology* (Cambridge, MA: Harvard University Press, 1980).

Mayr, G., Pohl, B., Hartman, S., and Peters, D. S. "The Tenth Skeletal Specimen of *Archaeopteryx*," *Zoological Journal of the Linnean Society* 149 (2007): 97–116.

Mayr, G., Pohl, B., and Peters, D. S. "A Well-Preserved *Archaeopteryx* Specimen with Theropod Features" *Science* 310 (2005): 1483–1486.

McBrearty, S., and Brooks, A. "The Revolution That Wasn't: A New Interpretation of the Origin of Modern Human Behavior," *Journal of Human Evolution* 39 (2000): 453–563.

McCosh, James. *The Religious Aspects of Evolution* (New York: Charles Scribner's Sons, 1890).

McLeod, Roy M. "The Support of Victorian Science: the Endowment of a Research Movement in Great Britain 1868–1900," *Minerva* 9 (1971): 197–223.

McPhee, John. *Basin and Range* (New York: Farrar, Straus, and Giroux, 1980).

Mellars, Paul, and Chris Stringer (eds.). *The Human Revolution: Behavioral and Biological Perspectives on the Origins of Modern Humans* (Princeton: Princeton University Press, 1989).

Mendel, Gregor. *Experiments in Plant Hybridisation*, trans. by Sir Ronald Fisher, edited by J. H. Bennett (Edinburgh: Oliver & Boyd, 1965).

Mendoza, M., Janis, C. M., and Palmqvist, P. "Ecological Patterns in the Trophic-Size Structure of Large Mammal Communities: A Taxon-Free Characterization," *Evolutionary Ecology Research* 7 (2005): 1–26.

Mercader, J., Barton, H., Gillespie, J., Harris, J., Kuhn, S., Tyler, R., and Boesch, C. "4,300-Year-Old Chimpanzee Sites and the Origins of Percussive Stone Technology," *Proceedings of the National Academy of Science USA* 104(9) (2007): 3043–3048.

Meyer, Stephen. "The Origin of Biological Information and the Higher Taxonomic Categories," *Proceedings of the Biological Society of Washington* 117(2) (2004): 213–239.

Mitchell, W.T.J. *The Last Dinosaur Book* (Chicago: University of Chicago Press, 1998).

Monroe, J. S. "Basic Created Kinds and the Fossil Record of Perissodactyls," *National Center for Science Education Journal* 5(2) (1988). Available online at www.natcenscied.org/resources/articles/4661_issue_16.

Moore, J. N., and Slusher, H. S. (eds.). *Biology: A Search for Order in Complexity* (Grand Rapids, MI: Zondervan, 1970).

Morell, Virginia. *Ancestral Passions. The Leakey Family and the Quest for Humankind's Beginnings* (New York: Simon and Schuster, 1995).

Morgan, C. Lloyd. *Animal Life and Intelligence* (London: Edward Arnold, 1890).

Morris, D. *The Naked Ape: A Zoologist's Study of the Human Animal* (London: Cape, 1967).

Mortillet, Gabriel de. "Le préhistorique; antiquité de l'homme," *Bibliothèque des sciences contemporaines* VIII (Paris: C. Reinwald, 1883).

Moser, Stephanie. "The Visual Language of Archaeology. A Case Study of the Neanderthals," *Antiquity* 66 (1992): 831–844.

Moser, Stephanie, and Gamble, Clive. "Revolutionary Images. The Iconic Vocabulary for Representing Human Antiquity," in Brian Leigh Molyneaux (ed.), *The Cultural Life of Images. Visual Representation in Archaeology* (London: Routledge, 1997).

Musgrave, Ian "Moonshine: Why the Peppered Moth Remains an Icon of Evolution," *Skeptical Inquirer* 29(2) (2005): 23–28.

Nevins, S. "Origins of Mammals." *Creation Research Quarterly* 10 (1974): 196.

"New Look at Human Evolution," *Scientific American* special issue (June 2003).

Nicholson, A. J. "The Role of Population Dynamics in Natural Selection," in Sol Tax (ed.), *Evolution after Darwin,* vol. I (Chicago: University of Chicago Press, 1960), 477–522.

Noonan, J. P., Coop, G., Kudaravalli, S., Smith, D., Krause, J., Alessi, J., Chen, F., Platt, D., Pääbo, S., Pritchard, J. K., and Rubin, E. M. "Sequencing and Analysis of Neanderthal Genomic DNA," *Science* 314 (2006): 1113–1118.

Notebook of John Melick Van Dyke, 1870, Lecture Notes Collection, James McCosh, AC#52, box 32, Mudd Library, Princeton University.
Nyhart, Lynn K. *Biology Takes Form: Animal Morphology and the German Universities, 1800–1900* (Chicago: University of Chicago Press, 1995).
O'Hara, Robert J. "Telling the Tree: Narrative Representation and the Study of Evolutionary History," *Biology and Philosophy,* 7(2) (1992): 135–160.
Olby, Robert C. *The Origins of Mendelism*, 2nd ed. (Chicago: University of Chicago Press, 1985).
Oldroyd, David R. *Thinking About The Earth: A History of Ideas in Geology* (Cambridge, MA: Harvard University Press, 1996).
Olson, T. R. "Taxonomic Affinities of the Immature Hominid Crania from Hadar and Taung," *Nature* 1316 (8 August 1985): 539–540.
Orel, Vítězslav. *Mendel* (Oxford: Oxford University Press, 1984).
Orel, Vítězslav. *Gregor Mendel: The First Geneticist* (Oxford: Oxford Univeristy Press, 1996).
Osborn, Henry Fairfield. "Hesperopithecus, the First Anthropoid in America," *American Museum Novitates* 37 (25 April 1922): 1–5.
Osborn, Henry Fairfield. *Man Rises to Parnassus* (Princeton: Princeton University Press, 1927).
Osborn, Henry Fairfield, Gregory, W. K., and Pinkley, George. *The Hall of the Age of Man, Guide Leaflet Series* (New York: American Museum of Natural History, 1925–1938).
Ospovat, Dov. *The Development of Darwin's Theory: Natural History, Natural Theology, and Natural Selection, 1838–59* (Cambridge: Cambridge University Press, 1981).
Ostrom, J. H. "The Origin of Birds," *Annual Review of Earth and Planetary Science* 3 (1975): 55–77.
Ostrom, J. H. "*Archaeopteryx* and the Origin of Birds," *Biological Journal of the Linnean Society* 8(2) (1976): 91–182.
Outram, Dorinda. *Georges Cuvier: Vocation, Science and Authority in Post-Revolutionary France* (Manchester: Manchester University Press, 1984).
Outram, Dorinda. "New Spaces in Natural History," in Jardine, Secord, and Spary (eds.), *Cultures of Natural History*, (Cambridge, MA: Cambridge University Press, 1996).
Owen, R. "Description of the Fossil Remains of a Mammal, a Bird, and a Serpent, from the London Clay," *Proceedings Geological Society London* 3 (1840): 162–166.
Padian, Kevin. "Review, *Of Pandas and People*," *Bookwatch Reviews* 2 (1989): 11.
Palmer, Douglas. *Neanderthal* (London: Channel 4 Books, 2000).
Paradis, James, and Williams, George C. (eds.). *Evolution and Ethics* (Princeton: Princeton University Press, 1989).
Pearson, Karl. *The Grammar of Science,* 2nd ed. (London: Walter Scott, 1900).
"The Philosophical Institution and Professor Huxley," *The Witness* (14 January and 18 January 1862).
Popper, K. R. *The Logic of Scientific Discovery* (New York: Harper Torchbooks, 1968).

Postman, David. "Seattle's Discovery Institute Scrambling to Rebound after Intelligent Design Ruling," *Seattle Times* (27 April 2006).

Powell, J. L. *Night Comes to the Cretaceous* (San Francisco: W. H. Freeman, 1998).

Pratt, Mary Louise. *Imperial Eyes: Travel Writing and Transculturation* (London: Routledge, 1992).

Proctor, Robert. "Three Roots of Human Recency. Molecular Anthropology, the Refigured Acheulean, and the UNESCO Response to Auschwitz," *Current Anthropology* 44(2) (2003): 213–239.

Prothero, D. R., and Shubin, N. "The Evolution of Mid-Oligocene Horses," in D. R. Prothero and R. M. Schoch (eds.), *The Evolution of Perissodactyls* (Oxford: Clarendon Press, 1989), 142–175.

Provine, W. B. *The Origins of Theoretical Population Genetics* (Chicago: Chicago University Press, 1971).

Pruetz, J. D., and Bertolani, P. "Savanna Chimpanzees, *Pan troglodytes verus*, Hunt with Tools," *Current Biology* 17 (2007): 412–417.

Prum, R. O. "Development and Evolutionary Origin of Feathers," *Journal of Experimental Zoology (Mol. Dev. Evol.)* 285 (1999): 291–306.

Quatrefages, Jean Louis Armand de, Hamy, Jules Érnest Théodore, and Formant, Henri C. *Crania ethnica. Les crânes des races humaines décris et figurés après les collections du Muséum d'Histoire Naturelle de Paris, de la Société d'Anthropologie de Paris, et les principales collections de la France et de l'étranger*, 2 vols. (Paris: Baillière, 1882).

Radinsky, L. R. "Oldest Horse Brains: More Advanced Than Previously Realized," *Science* 194 (1976): 636–637.

Radinsky, L. R. "Ontogeny and Phylogeny in Horse Skull Evolution," *Evolution* 38 (1984): 1–15.

Rainger, Ronald. *An Agenda for Antiquity. Henry Fairfield Osborn & Vertebrate Paleontology at the American Museum of Natural History, 1890–1935* (Tuscaloosa: University of Alabama Press, 1991).

Reader, J. *Missing Links: The Hunt for Earliest Man* (Boston: Little, Brown and Company, 1981).

Regal, B. *Henry Fairfield Osborn: Race and the Search for the Origins of Man* (Aldershot, UK: Ashgate, 2002).

Regal, B. *Human Evolution: A Guide to the Debates* (Oxford: ABC-CLIO, 2004).

Renders, E. "The Gait of *Hipparion* sp. from Fossil Footprints in Laetoli, Tanzania," *Nature* 308 (1984): 179–181.

Rensch, B. "Neo-Darwinism in Germany," in E. Mayr and W. B. Provine (eds.), *The Evolutionary Synthesis: Perspectives on the Unification of Biology* (Cambridge, MA: Harvard University Press, 1980), 284–303.

Robertson, D. S., McKenna, M., Toon, O. B., Hope, S., and Lillegraven, J. A. "Survival in the First Hours of the Cenozoic," *Geological Society of America Bulletin* 116 (2004): 760–768.

Roebroeks, Wil, and Corbey, Raymond. "Periodisations and Double Standards in the Study of the Palaeolithic," *in Hunters of the Golden Age. The Mid Upper Palaeolithic of Eurasia 30,000–20,000 BP*, edited by W. Roebroeks,

M. Mussi, J. Svoboda, and K. Fennema (Leiden: The University of Leiden Press, 2000).

Romanes, G. J. *Darwin and After Darwin: An Exposition of the Darwinian Theory of a Discussion of Post-Darwinian Questions* (Chicago: Open Court Publishing, 1894).

Romer, Alfred S. "Australopithecus Not a Chimpanzee," *Science* 71(1845) (9 May 1930): 482–483.

Rose, K. D. *The Beginning of the Age of Mammals* (Baltimore: Johns Hopkins, 2006).

Rosny-Aîné, J. H. *The Quest for Fire*. Translated by H. Talbott (Harmondsworth: Penguin, 1982).

Royle, Edward. *Victorian Infidels: The Origins of the British Secularist Movement 1791–1866*. (Manchester: Manchester University Press, 1974).

Rudge, D. W. "Taking the Peppered Moth with a Grain of Salt," *Biology and Philosophy* 14 (1999): 9–37.

Rudge, D. W. "Does Being Wrong Make Kettlewell Wrong for Science Teaching?," *Journal of Biological Education* 35(1) (2000): 5–11.

Rudge, D. W. "Cryptic Designs on the Peppered Moth," *International Journal of Tropical Biology and Conservation* (*Revista de Biología Tropical*) 50(1) (2002): 1–7.

Rudge, D. W. "The Role of Photographs and Films in Kettlewell's Popularizations of the Phenomenon of Industrial Melanism," *Science & Education* 12 (2003): 261–287.

Rudge, D. W. "Using the History of Research on Industrial Melanism To Help Students Better Appreciate the Nature of Science" and "The Mystery Phenomenon: Lesson Plans," in D. Metz (ed.), *Proceedings of the Seventh International History, Philosophy and Science Teaching Group Meeting* (Winnipeg, Canada: 2004), 761–811.

Rudge, D. W. "The Beauty of Kettlewell's Classic Experimental Demonstration of Natural Selection," *BioScience* 55(4) (2005): 369–375.

Rudge, D. W. "Did Kettlewell Commit Fraud? Re-examining the Evidence," *Public Understanding of Science* 14(3) (2005): 249–268.

Rudge, D. W. "H.B.D. Kettlewell's Research 1937–1953: The Influence of E. B. Ford, E. A. Cockayne and P. M. Sheppard," submitted to *History and Philosophy of the Life Sciences*.

Rudolph, J. L. *Scientists in the Classroom: The Cold War Reconstruction of American Science Education* (New York: Palgrave, 2002).

Rudwick, Martin J. S. *The Great Devonian Controversy: The Shaping of Scientific Knowledge Among Gentlemanly Specialists* (Chicago: University of Chicago Press, 1985).

Rudwick, Martin J. S. *Scenes from Deep Time: Early Pictorial Representations of the Prehistoric World* (Chicago: Chicago University Press, 1992).

Rudwick, Martin J. S. *Bursting the Limits of Time* (Chicago: University of Chicago Press, 2005).

Russell, Miles. *Piltdown Man: The Secret Life of Charles Dawson and the World's Greatest Archaeological Hoax* (Glouchestershire, UK: Tempus, 2003).

Sagan, Carl. *The Dragons of Eden* (New York: Ballantine, 1986).

Schaaffhausen, Hermann. "On the Crania of the Most Ancient Races of Man (translated by George Busk) With Remarks, and Original Figures, Taken From a Cast of the Neanderthal Cranium," *The Natural History Review*, New Series 8 (1861): 155–172.

Schmerling, Philippe-Charles. *Recherches sur les ossemens fossiles découverts dans les cavernes de la province de Liège* (Liège: Collardin, 1833–1834).

Schulter, C. M. "Tactics of the Dutch Colonial Army in the Netherlands East Indies," *Revue Internationale d'Histoire Militaire* 7 (1988): 59–67.

Schwab, J. J. *The Teaching of Science as Enquiry* (Cambridge, MA: Harvard University Press, 1962).

Schwalbe, Gustav. *Studien zur Vorgeschichte des Menschen* (Stuttgart: Schweizerbartsche, 1906).

Schwartz, J. H. *The Red Ape: Orangutans and Human Origins* (Boulder, CO: Westview, 2005).

Schwartz, Jeffrey H. "Race and the Odd History of Human Paleontology," *The Anatomical Record* (Part B: New Anat.) 289B (2006): 225–240.

Secord, James A. *Controversy in Victorian Geology* (Princeton: Princeton University Press, 1986).

Secord, James. *Victorian Sensation: The Extraordinary Publication, Reception, and Secret Authorship of Vestiges of the Natural History of Creation* (Chicago: University of Chicago Press, 2000).

Segestralle, U. "Neo-Darwinism," in M. Pagel (ed.), *The Encyclopaedia of Evolution*, vol. 2 (Oxford: Oxford University Press, 2002), 807–810.

Shapiro, Harry. *Peking Man* (New York: Simon & Schuster, 1974).

Shipman, P. *Taking Wing:* Archaeopteryx *and the Evolution of Bird Flight* (New York: Simon & Schuster, 1998).

Shipman, P. *The Man Who Found the Missing Link* (New York: Simon & Schuster, 2001).

Shipman, P., and Storm, P. "Missing Links; Eugène Dubois and the Origins of Palaeoanthropology," *Evolutionary Anthropology* 11 (2002): 106–116.

Shipman, P., and Walker, A. "The Costs of Becoming a Predator," *Journal of Human Evolution* 18 (1989): 373–392.

Shotwell, J. A. "Late Tertiary Biogeography of Horses in the Northern Great Basin," *Journal of Paleontology* 35 (1961): 203–217.

Simpson, G. G. *Tempo and Mode in Evolution* (New York: Columbia University Press, 1944).

Simpson, G. G. *Horses: The Story of the Horse Family in the Modern World and through Sixty Million Years of History* (Oxford: Oxford University Press, 1951).

Simpson, George Gaylord. *Fossils and the History of Life* (New York: Scientific American Library, 1983).

Sloan, C. P. "Meet the Dikika Baby," *National Geographic* (November 2006): 148–159.

Smith, F. "The Role of Continuity in Modern Human Origins," in G. Bräuer & F. Smith (eds.), *Continuity or Replacement? Controversies in* Homo Sapiens *Evolution* (Balkema: Rotterdam, 1992), 145–156.

Smith, G. Elliot. "The Fossil Anthropoid from Taungs," *Nature* 115(2885) (14 February 1925): 235.

Smocovitis, V. B. *Unifying Science: The Evolutionary Synthesis and Evolutionary Biology* (Princeton: Princeton University Press, 1996).

Solecki, Ralph Stefan. *Shanidar: The First Flower People* (New York: Knopf, 1971).

Sollas, W. J. *Ancient Hunters and their Modern Representatives* (London: Macmillan, 1911).

Solounias, N., and Semprebon, G. "Advances in the Reconstruction of Ungulate Ecomorphology with Application to Early Fossil Equids," *American Museum Novitates* 3366 (2002): 1–49.

Sommer, Marianne. "How Cultural Is Heritage? Humanity's Black Sheep from Charles Darwin to Jack London," in *A Cultural History of Heredity III. Nineteenth and Early Twentieth Centuries*, edited by S. Müller-Wille and H.-J. Rheinberger (Berlin: Max Planck Insitute, 2005), 294.

Sommer, Marianne. "Mirror, Mirror on the Wall. Neanderthal as Image and 'Distortion' in Early 20th-Century French Science and Press," *Social Studies of Science* 36(2) (2006): 207–240.

Sommer, Marianne. "Tarzan at the Earth's Core. The Re-Creation of Lost Worlds in Early Twentieth-Century America." Paper read at 'I Tarzan'—Man-apes and Ape-men between Science and Fiction, at ETH Zurich (23–25 June 2006).

Sommer, Marianne. *Bones and Ochre. The Curious Afterlife of the Red Lady of Paviland* (Cambridge: Harvard University Press, 2007).

Sondaar, P. Y. "The Osteology of the Manus of Fossil and Recent Equidae," *Verhand. Koninkijke Nederlandse Akad. Weten., afd. Naturkunde* 25 (1968): 1–76.

Spencer, Frank. "The Neandertals and Their Evolutionary Significance. A Brief Historical Survey," in Frank Spencer and Fred H. Smith (eds.), *The Origins of Modern Humans. A World Survey of the Fossil Evidence* (New York: Alan R. Liss, 1984).

Spencer, Herbert. *Principles of Biology*, 2 vols. (London: Williams and Norgate, 1864).

Springer, M. S., Cleven, G. C., Madsen, O., de Jong, W. W., Waddell, V. G., Amrine, H. M., and Stanhope, M. J. "Endemic African Mammals Shake the Phylogenetic Tree," *Nature* 388 (1997): 61–64.

Stebbins, G. L. *Variation and Evolution in Plants* (New York: Columbia University Press, 1950).

Stenseth, Nils Chr. "The Evolutionary Synthesis" *Science*, New Series 286(5444) (1999): 1490.

Stern, J. T., and Susman, R. L. "The Locomotor Anatomy of *Australopithecus afarensis*," *American Journal of Physical Anthropology* 60 (1983): 279–317.

Stirton, R. A. "Phylogeny of North American Equidae," *University of California Publications, Bulletin of the Department of Geological Sciences* 25 (1940): 165–198.

Stocking, George W. *Victorian Anthropology* (London: Free Press, 1987).

Stoneking, Mark, and Cann, Rebecca. "African Origin of Human Mitochondrial DNA," in P. Mellars and C. Stringer (eds.), *The Human Revolution. Behavioral and Biological Perspectives on the Origins of Modern Humans* (Edinburgh: Edinburgh University Press, 1989).

Stott, Rebecca. *Darwin and the Barnacle: The Story of One Tiny Creature and History's Greatest Scientific Breakthrough* (New York: Norton, 2003).

Straus, W. L., Jr. "Hunters or Hunted?," *Science* 126(3283) (29 November 1957): 1108.

Straus, William L., and Cave, A.J.E. "Pathology and Posture of Neanderthal Man," *Quarterly Review of Biology* 32(4) (1957): 348–363.

Stringer, C. "Modern Human Origins: Progress and Prospects," *Philosophical Transactions of the Royal Society, London (B)* 357 (2002): 563–579.

Stringer, C., and Andrews, P. *The Complete World of Human Evolution* (London: Thames & Hudson, 2005).

Stringer, Christopher. "Population Relationships of Later Pleistocene Hominids: A Multivariate Study of Available Crania," *Journal of Achaeological Science* 1 (1974): 317–342.

Stringer, Christopher, and Gamble, Clive. *In Search of the Neanderthals. Solving the Puzzle of Human Origins* (London: Thames and Hudson, 1993).

Stringer, Christopher, and Robin McKie. *African Exodus: The Origins of Modern Humanity* (New York: Henry Holt, 1997).

Strömberg, C.A.E. "Evolution of Hypsodonty in Equids: Testing a Hypothesis of Adaptation," *Paleobiology* 32 (2006): 236–258.

Swisher III, Carl C., Garnis, Curtis, and Lewin, Roger. *Java Man* (Chicago: University of Chicago Press, 2000).

Tague, R. C., and Lovejoy, C. O. "A.L. 288-1—Lucy or Lucifer: Gender Confusion in the Pliocene," *Journal of Human Evolution* 35 (1998): 75–94.

Taschdjian, Claire. *The Peking Man is Missing* (New York: Ballantine Books, 1977).

Tattersall, I., and Schwartz, J. H. "Is Paleoanthropology Science? Naming New Fossils and Control of Access To Them," *The Anatomical Record* 269 (2002): 239–241.

Tattersall, Ian. *The Fossil Trail. How We Know What We Think We Know about Human Evolution* (New York: Oxford University Press, 1995).

Tattersall, Ian. *Becoming Human: Evolution and Human Uniqueness* (New York: Harcourt Brace, 1998).

Tattersall, Ian. *The Last Neanderthal. The Rise, Success, and Mysterious Extinction of Our Closest Human Relatives*, 2nd ed. (Boulder: Westview, 1999).

Tattersall, Ian, and Eldredge, Niles. "Fact, Theory, and Fantasy in Human Paleontology," *American Scientist* 65 (March–April) (1977): 204–211.

Templeton, A. "Out of Africa Again and Again," *Nature* 416 (2002): 45–51.

Theunissen, B. *Eugène Dubois and the Ape-Man from Java* (Dordrecht: Kluwer, 1989).

Thewissen, J.G.M., and Bajpai, S. "Whale Origins as a Poster Child for Macroevolution," *BioScience* 51 (2001): 1037–1059.

Thomas, R.D.K., and Olson, E. D. (eds.) "A Cold Look at the Warm-Blooded Dinosaurs," *AAAS Selected Symposium* no. 28, 1980.

Thomason, J. J. "The Functional Morphology of the Manus in the Tridactyl Equids *Mesohippus* and *Merychippus*: Paleontological Inferences from Neontological Models," *Journal of Vertebrate Paleontology* 6 (1986): 143–161.

Thomsen, M., and Lemeche, H. "Experimente zur erzielung eines erblichen melanismus bei dem spanner *Selenia bilunaria* Esp," *Biologisches Zentralblatt* 53 (1933): 541–560.

Tobias, P. V. "Introduction," in P. V. Tobias (ed.), "Hominid Evolution—Past, Present and Future," *Proceedings of the Taung Diamond Jubilee International Symposium* (New York: Alan R. Liss, 1985), xix–xxix.

Tocheri, M. W., et al. "Morphological Affinities of the Wrist of *Homo floresiensis*," *PaleoAnthropology* (2007): A1–A35.

Trinkaus, Erik, and Shipman, Pat. *The Neanderthals: Changing the Image of Mankind* (New York: Alfred A. Knopf, 1993).

Tutt, J. W. "Melanism and Melanochroism in British Lepidoptera." *The Entomologist's Record and Journal of Variation 1* (1890): 5–7, 49–56, 84–90, 121–125, 169–172, 228–234, 293–300, 317–325.

Uddenberg, Nils. *Idéer om livet: En biologihistoria* [Ideas About Life: A History of Biology], vol. I–II (Stockholm: Natur och Kultur, 2003).

Uglow, Jenny. *The Lunar Men: Five Men Whose Friendship Changed the World* (New York: Farrar, Straus, and Giroux, 2003).

Underhill P., Passarino, G., Lin, A., Shen P., Mirazon Lahr, M., Foley, R., Oefner, P., and Cavalli-Sforza, L. "The Phylogeography of Y Chromosome Binary Haplotypes and the Origins of Modern Human Populations," *Annals of Human Genetics* 65 (2001): 43–62.

Vallois, Henri V. "Neandertals and Praesapiens," *Journal of the Royal Anthropological Institute of Great Britain and Ireland* 84(1/2) (1954): 111–130.

Van Marle, A. "De Groep der Europeanen in Nederlands-Indië, iets over ontstaan en groei; III," *Indonesië* 5 (1951–1952): 97–121.

Van Oosterzee, Penny. *Dragon Bones: The Story of Peking Man* (Cambridge: Perseus Press, 2000).

Van Riper, A. Bowdoin. *Men Among the Mammoths: Victorian Science and the Discovery of Human Prehistory* (Chicago: University of Chicago Press, 1993).

Vekua, A., Lordkipanidze, D., Rightmire, G. P., Agusti, J., Ferring, R., Maisuradze, G., Mouskhelishvili, A., Nioradze, M., Ponce de Leon, M., Tappan, M., Tvalchrelidze, M., and Zollikofer, C. "A New Skull of Early *Homo* from Dmanisi, Georgia," *Science* 297 (2002): 85–89.

Von Meyer, H. "*Archaeopteryx lithographica* (Vogel-Feder) und *Pterodactylus* von Solnhofen," *Neues Jahrbuch für Mineralogie, Geologie und Palaeontologie* (1861): 678–679.

Von Meyer, H. "Vogel-Federn und *Palpipes priscus* von Solnhofen" *Neues Jahrbuch für Mineralogie, Geologie und Palaeontologie* (1861): 561.

Voorhies, M. R. "Dwarfing the St. Helens Eruption: Ancient Ashfall Creates a Pompeii of Prehistoric Animals," *National Geographic* 159 (1981): 66–75.

Voorhies, M. R. "Vertebrate Biostratigraphy of the Ogallala Group in Nebraska," in T. C. Gustavson (ed.), *Geologic Framework and Regional Hydrology: Upper Cenozoic Blackwater Draw and Ogallala Formations, Great Plains* (Austin: Bureau of Economic Geology, University of Texas, 1990).

Vorzimmer, Peter J. "Darwin and Mendel: The Historical Connection," *Isis* 59(1) (Spring 1968): 77–82.

Wagner, J. A. "Über ein neues, angeblich mit Vogelfedern versehenes Reptil aus dem Solnhofener lithographischen Schiefer," *Sitzungsbericht Bayerische Akademie für Wissenschaften* 2 (1861): 146–154.

Walker, A., and Leakey, R.E.F. *The Nariokotome* Homo erectus *Skeleton* (Cambridge: Harvard University Press, 1993).

Walker, A., and Shipman, P. *The Wisdom of the Bones: In Search of Human Origins* (New York: Knopf, 1996).

Wallace, Alfred Russel. *The Geographical Distribution of Animals*, 2 vols. (London: Macmillan, 1876).

Wallace, Alfred Russel. *Darwinism: An Exposition of the Theory of Natural Selection* (London: Macmillan, 1889).

Walsh, John Evangelist. *Unraveling Piltdown: The Science Fraud of the Century and Its Solution* (New York: Random House, 1996).

Wasley, Paula, Bartlett, Thomas, and Labi, Aisha. "Peer Review," *Chronicle of Higher Education* (April 14, 2006).

Webb, S. D. "Ten Million Years of Mammal Extinctions in North America," in P. S. Martin and R. G. Klein (eds.), *Quaternary Extinctions: A Prehistoric Revolution* (Tuscon: University of Arizona Press, 1984), 189–210.

Weidenreich, Franz. *Apes, Giants, and Man* (Chicago: University of Chicago Press, 1946).

Weiner, J. S. *The Piltdown Forgery* (London: Oxford University Press, 1955).

Weiner, Jonathan. *The Beak of the Finch: A Story of Evolution in Our Time* (New York: Knopf, 1994).

Weishampel, D. B., Dodson, P., and Osmólska, H. *The Dinosauria,* 2nd ed. (Berkeley: University of California Press, 1990).

Weitzel, Robert. "Creationism's Holy Grail: The Intelligent Design of a Peer-Reviewed Paper," *Skeptic* 11(4) (2005): 66–69.

Wellnhofer, P. "Der bayerische Urvogel, *Archaeopteryx bavarica. Alexander von Humboldt Stiftung*," *Mitteilungen* 75 (2000): 3–10.

Wellnhofer, P. "The Plumage of *Archaeopteryx*: Feathers of a Dinosaur?," in P. J. Currie, E. B. Koppelhus, M. A. Shugar, and J. L. Wright (eds.), *Feathered Dragons. Studies on the Transition from Dinosaurs to Birds* (Bloomington: Indiana University Press, 2004), 282–300.

Wells, H. G. "Stories of the Stone Age," *The Idler* (May–September 1897).

Wells, H. G. "The Grisly Folk," *Storyteller Magazine* (April 1921).

Wells, J. *Icons of Evolution: Science or Myth? Why Much of What We Teach About Evolution Is Wrong* (Washington, D.C.: Regnery Press, 2000).

Wells, J. "Second Thoughts about Peppered Moths: This Classical Story of Evolution by Natural Selection Needs Revising," *The True Origin Archive* (2002), available online at www.trueorigin.org/pepmoth1.asp.

White, F. B. "On Melanochroism and Leuochroism," *The Entomologist's Monthly Magazine* 13 (Nov) (1876–1877): 145–149.

White, Paul. *Thomas Huxley: Making the "Man of Science"* (Cambridge, UK: Cambridge University Press, 2003).

White, T. D. "Earliest hominids," in Walter C. Hartwig (ed.), *The Primate Fossil Record* (Cambridge: Cambridge University Press, 2002), 407–417.

White, T. D., WoldeGabriel, G., Asfaw, B., Ambrose, S., Beyene, Y., Bernor, R. L., Boisserie, J-R., Currie, B., Gilbert, H., Haile-Selassie, Y., Hart, W. K., Hlusko, L. J., Howell, F. C., Kono, R. T., Lehmann, T., Louchart, A., Lovejoy, C. O., Renne, P. R., Saegusa, H., Vrba, E. S., Wesselman, H., Suwa, G. "Asa Issie, Aramis and the Origin of *Australopithecus*," *Nature* 440 (2006): 883–889.

Wiley, E. O. *Phylogenetics: The Theory and Practice of Phylogenetic Systematics* (New York: Wiley, 1981).

Wiley, E. O., Siegel-Causey, D., Brooks, D. R., and Funk, V. A. *The Compleat Cladist*. University of Kansas Museum of Natural History Special Publication no. 19 (1991).

Winans, M. C. "A Quantitative Study of the North American Fossil Species of the Genus *Equus*," in D. R. Prothero and R. M. Schoch (eds.), *The Evolution of Perissodactyls* (Oxford: Clarendon Press, 1989), 262–297.

Wolpoff, M., and Caspari, R. *Race and Human Evolution: A Fatal Attraction* (New York: Simon and Schuster, 1997).

Wolpoff, Milford H., and Thorne, Alan G. "The Multiregional Evolution of Humans," *Scientific American* 13(2) (2003): 46–53.

Wong, K. "Lucy's Baby," *Scientific American* 295(6) (2006): 78–85.

Woodward, A. Smith. "The Fossil Anthropoid From Taungs," *Nature* 115(2885) (14 February 1925): 235–236.

Woodward, A. Smith. "Recent Progress in the Study of Early Man," *Science* 82(2131) (1 November 1935): 399–407.

Woodward, Arthur Smith. *The Earliest Englishman* (London: Watts, 1948).

Woodward, Thomas. *Doubts about Darwin: A History of Intelligent Design* (Grand Rapids, MI: Baker Books, 2003).

Wright, S. "Evolution in Mendelian Populations," *Genetics* 16 (1931): 97–159.

Wysong, R. L. *The Creation-Evolution Controversy* (Midland, MI: Inquiry Press, 1981).

Xu, X., Zhou, Z., Wang, X., Kuang, X., Zhang, F., and Du, X. "Four Winged Dinosaurs From China," *Nature* 421 (2003): 335–340.

Young, David. *The Discovery of Evolution* (London: Cambridge University Press, 1992).

Young, M., and Bennett, D. K. "Stripes Do Not a Zebra Make. Part I: A Cladistic Analysis of *Equus*," *Systematic Zoology* 29 (1980): 272–287.

Young, M., and Musgrave, I. "Moonshine: Why the Peppered Moth Remains an Icon of Evolution," *Skeptical Inquirer* 29(2) (2005): 23–28.

About the Editor and Contributors

EDITOR

Brian Regal is assistant professor for the history of science at Kean University, New Jersey, USA. He is the author of *Henry Fairfield Osborn: Race and the Search for the Origins of Man* (2002) and *Human Evolution: A Guide to the Debates* (2005). His work centers on fringe notions, anomalous beliefs, dubious ideas, and the people who propagate and study them. He has written on eugenics, racial anthropology, and creationism and is currently undertaking research into the history of the various theories surrounding the peopling of and discovery of the Americas and their relationship to evolution studies.

CONTRIBUTORS

Peter Bowler is on the faculty of the department of Social Anthropology at Queen's University, Belfast, Ireland. He has written extensively on the history of evolutionary thought and its social implications, and is considered one of the great authorities on the subject. He is the author of *The Eclipse of Darwinism* (1983), *Evolution: The History of an Idea* (1984), and *Charles Darwin: The Man and His Influence* (1990) and others. Many of his writings have become standard reference works in the field.

Dawn Mooney Digrius recently completed her Ph.D. and is currently serving as assistant professor for History of Science at Drew University in Madison, New Jersey. Her current research focuses on the influence of evolutionary theory on the plant sciences.

Holly Dunsworth. When she is not digging at fossil sites in Kenya and the Republic of Georgia, Holly Dunsworth uses clues from anatomy and bone structure to reconstruct the biology, behavior, and evolution of extinct monkeys, apes, and humans. Holly is the author of *Human Origins 101* (Greenwood) and is a postdoctoral fellow in the Department of Anthropology at the Pennsylvania State University in Happy Valley.

David Fastovsky is professor of geosciences at the University of Rhode Island. He studies the ancient environments of Mesozoic vertebrates, mass extinctions, and the paleobiology of dinosaurs. He has studied dinosaurs in the western United States, Argentina, Mexico, and Mongolia. In addition to over thirty peer-reviewed publications he is the author of *The Evolution and Extinction of the Dinosaurs* (1996, 2005) which has become a standard text on the subject.

Anne Katrine Gjerløff received her Ph.D. in history from the University of Copenhagen. She has done research on the history of prehistoric archaeology, paleoanthropology, and the life sciences, including the relationship between science and society and the popularizations of science and has published several articles on these issues. Her 2004 doctoral dissertation *Abens Ansigter* (Faces of the Ape) explored the history and popularization of paleoanthropology in Denmark in the twentieth century. She has since then been engaged in supplementary education of college teachers, and is currently teaching graduate students at the University of Copenhagen and is creating a permanent exhibition on archaeology and identity in the Museum in Sønderborg castle in southern Denmark. In May 2007 she began research funded by the Danish Research Council on the subject of animal roles and animal rights in the late nineteenth and early twentieth century. She is chair of the Historical Society, and member of the editor's board of two Danish history journals. When she has time she is the proud mother of two beautiful children.

Christine Janis is currently a professor of Ecology and Evolutionary Biology at Brown University, where she teaches vertebrate comparative anatomy and evolution. She is a graduate of the University of Cambridge (B.A. Cantab.) and Harvard University (Ph.D.). She has published numerous articles on the evolution of ungulates (hoofed mammals), as well as on other topics in vertebrate paleontology, and is a major contributor to the textbook *Vertebrate Life*.

Bent E. K. Lindow is a vertebrate paleontologist and holds an M.S. in geology from the University of Copenhagen. He studied the evolutionary relationships of extinct and living whales for his Master's thesis. Currently, he is finishing a Ph.D. dissertation at University College Dublin in Ireland, where he studies 54-million-year-old fossil birds from Denmark, and how prehistoric climate change impacted on the early evolution of modern birds. His latest scientific publication, co-authored with Dr. Gareth Dyke, is titled "Bird Evolution in the Eocene: Climate Change in Europe and a Danish Fossil Fauna" and appeared in *Biological Reviews of the Cambridge Philosophical Society* (2006). Bent is also vice president of the Paleontologic Club of the Danish Geological Society, where he arranges seminars and conferences with both local and international researchers. He has previously worked as a nature guide and science interpreter, has given several professional and popular talks, and appeared on national radio and television to comment on a wide range of subjects within paleontology and evolution. In his spare time, when not reading books on nature or historical subjects, he can usually be found beating up the sandbags at the local boxing club.

Olof Ljungström has a Ph.D. in the history of science and ideas from Uppsala University, Sweden. He is currently employed by the Karolinska Institutet, Stockholm, as a researcher. His work is part of the university's upcoming 2010 bicentenary celebration, and is intended to produce a critical appraisal of the university's medical research history in the period 1960–2010. He is also the coordinator of a project to establish an archive for the history of medicine there. Among his research interests can be named the histories of anthropology, natural history, and medicine.

Lisa Nocks, Ph.D., is an historian of science, technology, and media. She has authored a number of articles on the history of media technologies for college students. She has published essays on the relationship between science fiction and science, and frequently presents papers on the diffusion of technical knowledge through science fiction. Her book, *Robot: The Life Story of a Technology* (2007) is part of the Greenwood *Technographies* series. She was a teaching fellow in the department of Communication and Media Studies at Fordham University, and is now on the faculty of the history department of the New Jersey Institute of Technology, Newark, New Jersey.

David W. Rudge is an associate professor in the Department of Biological Sciences with a joint appointment in The Mallinson Institute for Science Education at Western Michigan University. He received his B.S. in zoology from Duke University, an M.S. in biology, and an M.A. and Ph.D. in history and philosophy of science from the University of Pittsburgh. He has written extensively on H.B.D. Kettlewell's classic investigations on the phenomenon of industrial melanism from historical, philosophical, and science education perspectives. He is currently writing a book with particular attention to how the phenomenon of industrial melanism has been portrayed in science textbooks and the popular media.

Jeffrey H. Schwartz is the president-elect of the World Academy of Art and Science and a fellow of the American Association for the Advancement of Science. He is a professor of Physical Anthropology and of History and Philosophy of Science at the University of Pittsburgh and a research associate at the American Museum of Natural History. In addition to over 150 articles and numerous books, including a history of evolutionary thought (*Sudden Origins*, John Wiley, 1999), he recently published the first study of virtually the entire human fossil record (*The Human Fossil Record* with I. Tattersall), a revised edition of *The Red Ape*, which explores the assumptions underlying molecular and morphological approaches to phylogenetic reconstruction, and a revised edition of his textbook on human osteology, *Skeleton Keys*. He also led a project to forensically reconstruct George Washington at three different ages for a new education center at Mount Vernon.

Pat Shipman is a paleoanthropologist and writer who received a B.A. from Smith College in religion and an M.A. and Ph.D. in anthropology from New York University. Her anthropological research has focused on two main themes: 1) the ancient environments and communities in which our ancestors lived and evolved and the ways in which our ancestors exploited their environment; and 2) the history of science with an emphasis on the influence of gender, class, and social standing on the acceptance or rejection of scientific information. This has involved work on fossil sites, collections, or archives from Australia, Bulgaria, Ethiopia, England, France, Holland, Hungary, Indonesia, Italy, Kenya, North America, Romania, Spain, Yugoslavia, Tanzania, and Turkey. In 2001 she published the only full-length biography of Eugene Dubois, a book made possible by the exceptional generosity of the Curator of the Dubois Collection in the Netherlands, John de Vos. She has written seven books for general audiences on a wide range of scientific subjects—two of which have won major literary prizes—and three biographies: *The Man Who Found the Missing Link* about Eugene Dubois; *To the Heart of the Nile* about the Victorian lady explorer Florence Baker; and the forthcoming *Femme Fatale*, about the Oriental dancer and purported spy, Mata Hari.

Marianne Sommer studied biology and English literature and linguistics at the Universities of Zurich (Switzerland) and Coventry (GB), followed by postgraduate studies at the Collegium Helveticum of the ETH Zurich in the history and social studies of science. She wrote her Ph.D. on anthropomorphism in texts on monkeys and apes between science and the media, analyzing among other sources *National Geographic* articles of from 1888 to 1997 (*Foremost in Creation*, Peter Lang, 2000). Between 2000 and 2004 she held postdoctoral research fellowships at the Max Planck Institute for the History of Science in Berlin, and in the Science, Medicine, and Technology in Culture program at Pennsylvania State University. She is now part of the junior faculty at the chair for science studies of the ETH Zurich. During spring 2006 she was a visiting fellow at Stanford University. Her new book on the pre-history of paleoanthropology, which she approaches through the biography of one particularly fickle skeleton, is currently in editing with Harvard University Press (*Bones and Ochre: The Curious Afterlife of the Red Lady of Paviland*).

Christopher Stringer's early research concentrated on the relationship of Neandertals and early modern humans in Europe, but his current research interests are the evolutionary origins of *Homo sapiens,* and the early human occupation of Britain. He has been closely involved in the development of the Out of Africa theory of modern human origins and collaborates with archaeologists, dating specialists, and geneticists in attempting to reconstruct the evolution of modern humans. He has directed or co-directed excavations at Pleistocene sites in England, Wales, and Gibraltar, and is currently directing the Ancient Human Occupation of Britain Project (AHOB), funded by the Leverhulme Trust. AHOB is investigating the pattern of the earliest human colonization of England and Wales. He recently authored the books *The Complete World of Human Evolution* (2005, with Peter Andrews) and *Homo britannicus* (2006).

Bowdoin Van Riper is an historian of science and technology whose scholarly interests include the history of geology, archaeology, and paleoanthropology. He received a bachelor's degree with a dual major in geology and history from Brown University, and a master's and doctorate in the history of science from the University of Wisconsin—Madison. He has taught at Northwestern University, Franklin and Marshall College, and Kennesaw State University and is currently an adjunct professor of Science, Technology, and Society at Southern Polytechnic State University. He is the author of four books, including *Men Among the Mammoths: Victorian Science and the Discovery of Human Prehistory* (1993), and co-editor of the geology entries in the *Dictionary of Nineteenth Century British Scientists* (2004).

Adam S. Wilkins was born in Columbus, Ohio, in 1945, but grew up in New York City. He attended Reed College, receiving his B.A. in 1965 and then did his postgraduate work in the Department of Genetics, the University of Washington, receiving his Ph.D. in 1969. He did experimental work in bacterial physiology, bacteriophage genetics, and slime mould cell biology and development, before turning to writing and editing. He has been editor of *BioEssays* since 1990. His books include *Genetic Analysis of Animal Development* (1986, 1993) and *The Evolution of Developmental Pathways* (2002). His major interests are in developmental biology and animal evolution.

Index